Desktop Grid Computing

Edited by
Christophe Cérin
Gilles Fedak

CRC Press
Taylor & Francis Group
Boca Raton London New York

CRC Press is an imprint of the
Taylor & Francis Group, an **informa** business

A CHAPMAN & HALL BOOK

CHAPMAN & HALL/CRC
Numerical Analysis and Scientific Computing

Aims and scope:
Scientific computing and numerical analysis provide invaluable tools for the sciences and engineering. This series aims to capture new developments and summarize state-of-the-art methods over the whole spectrum of these fields. It will include a broad range of textbooks, monographs, and handbooks. Volumes in theory, including discretisation techniques, numerical algorithms, multiscale techniques, parallel and distributed algorithms, as well as applications of these methods in multi-disciplinary fields, are welcome. The inclusion of concrete real-world examples is highly encouraged. This series is meant to appeal to students and researchers in mathematics, engineering, and computational science.

Proposals for the series should be submitted to one of the series editors above or directly to:
CRC Press, Taylor & Francis Group
4th, Floor, Albert House
1-4 Singer Street
London EC2A 4BQ
UK

Published Titles

Classical and Modern Numerical Analysis: Theory, Methods and Practice
Azmy S. Ackleh, Edward James Allen, Ralph Baker Kearfott, and Padmanabhan Seshaiyer

Computational Fluid Dynamics
Frédéric Magoulès

A Concise Introduction to Image Processing using C++
Meiqing Wang and Choi-Hong Lai

Decomposition Methods for Differential Equations: Theory and Applications
Juergen Geiser

Desktop Grid Computing
Christophe Cérin and Gilles Fedak

Discrete Variational Derivative Method: A Structure-Preserving Numerical Method for Partial Differential Equations
Daisuke Furihata and Takayasu Matsuo

Grid Resource Management: Toward Virtual and Services Compliant Grid Computing
Frédéric Magoulès, Thi-Mai-Huong Nguyen, and Lei Yu

Fundamentals of Grid Computing: Theory, Algorithms and Technologies
Frédéric Magoulès

Handbook of Sinc Numerical Methods
Frank Stenger

Introduction to Grid Computing
Frédéric Magoulès, Jie Pan, Kiat-An Tan, and Abhinit Kumar

Iterative Splitting Methods for Differential Equations
Juergen Geiser

Mathematical Objects in C++: Computational Tools in a Unified Object-Oriented Approach
Yair Shapira

Numerical Linear Approximation in C
Nabih N. Abdelmalek and William A. Malek

Numerical Techniques for Direct and Large-Eddy Simulations
Xi Jiang and Choi-Hong Lai

Parallel Algorithms
Henri Casanova, Arnaud Legrand, and Yves Robert

Parallel Iterative Algorithms: From Sequential to Grid Computing
Jacques M. Bahi, Sylvain Contassot-Vivier, and Raphael Couturier

Particle Swarm Optimisation: Classical and Quantum Perspectives
Jun Sun, Choi-Hong Lai, and Xiao-Jun Wu

CRC Press
Taylor & Francis Group
6000 Broken Sound Parkway NW, Suite 300
Boca Raton, FL 33487-2742

First issued in paperback 2019

ISBN-13: 978-1-4398-6214-8 (hbk)
ISBN-13: 978-0-367-38118-9 (pbk)

Visit the Taylor & Francis Web site at
http://www.taylorandfrancis.com

and the CRC Press Web site at
http://www.crcpress.com

Contents

List of Figures

List of Tables

Preface

Desktop Grids and Volunteer Computing are well-known terms related to the notion of computing on loosely connected resources, usually personal computers, but also sometimes clusters, that are controlled by their owners. These terms started popping up in the Internet and research community at the end of the 1990s. At that time this idea was not easy to defend.

Several projects had explored similar concepts before, but SETI@home was really the game changer. David Anderson is, in my opinion, the real inventor of this concept because he was the first to understand its exceptional potential and to transform the concept into a worldwide phenomenon. But how did SETI@home open the door to an entire new research field in computer science? Well, its application, the research for alien intelligent lives from the analysis of extraterrestrial signals was, of course, breathtaking. It was not only fun, it resonated immediately in every computer science researcher's (and every user's) mind because of the existential questions it raised. But more than the application, the overall system allowing thousands instances of the same code to run concurrently on a vast number of volunteer PCs was extremely intriguing.

At LRI the whole story started precisely at the end of the 1990s. Gilles Fedak was a Ph.D. student and I was his adviser. He started his Ph.D. on a topic that I don't remember, probably something related to HPC, PC clusters, MPI, OpenMP, etc. One day, I showed up in his office to ask Gilles about his progress. The discussion rapidly turned to the thing that was running on his screen. If I remember correctly, I was caught by FFT graphs. That's somehow logical since my education was in electrical engineering. I asked what this application was and Gilles explained SETI@home. I asked several questions and concluded quickly that it would never work for more than one application. I returned to my office, forgetting to give some advice to Gilles on his original Ph.D. topic. A few days later (or maybe one day, I can't remember), I understood that trying to contradict this would never work for more than one application and was a fantastic challenge, opening a whole new research field in computer science. I don't know how this came to the minds of other computer science researchers at the same time, but I hope that they will be able to tell their story in some future venues.

In a dozen years, this domain has developed considerably in academia and industries. What is really surprising for such a mature research field and technology is the lack of a book. I think that this book is a necessity. A necessity for researchers, teachers, students, and for people interested in this topic in the industry. The covered topics represent the questions raised since we started the research on this subject including some very recent ones. I am happy that the past decade has produced enough results on all the key domains to effectively contradict my initial statement: many applications are now running on volunteer computing systems.

This somehow demonstrates that my second thought was cleverer, although research progress has not really followed a straight path. I remember some of the wrong directions that we have taken. A good example was the extension of the desktop grid concept to smartphones. The analysis behind the extension was essentially based on the fast increase of smartphone performance. On paper, smartphone performance increases following Moore's law and the current effect of this phenomenon is the integration of multicore processors in smartphones. So it was appealing to target the millions of smartphones as potential computing devices for volunteer computing.

I still have the photos of the team using the prototype that we developed from an HP Ipaq device in 2001. However, there are several issues with this concept that make it difficult to apply in practice: First, smartphones have a short battery life and using the CPU for volunteer computing may reduce it further dramatically. Second, many smartphones owners are paying for data transfers. Every task executed on a volunteer smartphone has a clear cost (reduction of battery life) and money cost (each parameter and result transfer) for its owner. I think these two important factors transformed what sounded like a nice idea into a wrong direction, at least until now. This worked as a lesson, at least for me: research direction should start not only from technical analysis but also from a study of its applicability. This may apply to the whole domain of desktop grid and volunteer computing, too.

There are some clear trends that may challenge the concept in its fundamental roots. One issue is related to the effort the whole computer science community is developing to reduce power consumption. Typically, in the near future, only the resources needed to run the user (owner) applications and the operating system will be powered on, the rest being dormant. Volunteer computing applications will more clearly than now increase power consumption for resource owners.

Until what point will the resource owner tolerate this increase of power consumption? If the resource owner can configure a maximum power consumption, until what point will the contribution of this resource still be relevant? By extension, what is the minimum level of power overconsumption that all volunteers should accept (in mean) to keep volunteer computing worthwhile?

A second issue is the general trend toward cloud computing, smart terminals (tablets) and the externalization (from the PC to the cloud) of key applications. Large data centers have also demonstrated that they are powerful enough to remove the need to run high demanding applications on your personal computer.

This may push the user to prefer smart, light, and mobile terminals over classic PCs. I believe that this trend is very strong, and we are the observers of a paradigm shift like the transition from centralized computers to PCs that occurred in the 20th century. To what extent can volunteer computing adapt itself to this paradigm shift? One possibility is to reverse the model. I think that what Derrick Kondo is exploring in the last section of this book, making a cloud infrastructure out of a desktop grid, is a very exciting research direction.

The technology seems to be there. Of course, it needs to be adapted. However, there are still significant open questions. I think that the challenge is on the economics: How do you port or adapt the Cloud trust and economic models to this environment? Contrary to volunteer computing, the cloud model suggests that resource owners will be actually rewarded. What will be the reward: Cloud resource cycles? Money? How can an application user trust the security of this infrastructure? How can a Cloud built on top of a desktop grid provide a trustable service level agreement? Convincing answers need to be given before we will see users moving in mass to this new environment.

I think the research challenges are really exciting. This research domain surprises me in its capacity to push its frontiers and redefine itself. I hope that the reader of this book will feel the extraordinary freedom that researchers in desktop grids or volunteer computing enjoy. I hope that students will engage themselves in this research domain and continue to reinvent it.

Franck Cappello
Co-Director of the INRIA-Illinois Joint Laboratory on PetaScale Computing

Presentation of the Book

The goal of this book is to offer an analysis of various approaches in Desktop Grid Computing, a decade after the pioneering work. Desktop Grids use computing, network, and storage resources of idle desktop PCs distributed over multiple LANs or the Internet. Today, this type of computing platform forms one of the largest distributed computing systems, and currently provides scientists with tens of TeraFLOPS from hundreds of thousands of hosts. Several topics will be discussed in detail, which may appeal to a much larger audience than the community of researchers in the field. This book is certainly the first one to make a deep comprehensive and up-to-date presentation about the topic. It also draws perspectives, recent trends in the domain that underline the evolution of the initial concept.

This book is intended primarily for researchers, advanced undergraduate and graduate students in distributed computing, grid computing, scientific computing, high performance computing and related fields, as well as to engineers and other professionals from academia and industry. The book can be used as a textbook for a graduate course, for an advanced undergraduate course, or for a summer school course. It will serve as a self-contained book and a detailed overview of the relevant Desktop Grid models. The general area of Desktop Grid computing and its state of-the-art methods and technologies are essential to many other computational science fields, including health, bio-informatics, medical imaging, physics, astronomy, climate prediction, mathematics, cryptography, and more generally to simulation.

In summary, the book is devoted to a survey of common techniques used in numerous models, algorithms, tools developed during the last decade to implemented the concept of Desktop Grid computing. Such techniques allow us to implement and to solve many important sub-problems for middleware design. Such sub-problems are scheduling, data management, security, volunteer computing, load balancing, model of parallelism, programming models, result certification, fault tolerance. The book has approximately 350 pages of methods and applications, with accompanying tables and illustrations. It seeks to balance the theory of designing Desktop Grid middleware and architecture with practice, applications and real world deployment on large scale platforms. The book is organized in two parts. The first part is related to the birth (initial ideas, basic concepts...) of Desktop Grid computing and includes two chapters that can serve as practical work for a classroom; the second part is related to what we call 'the maturity' of Desktop Grids that is to say, to a more prospective vision of challenging problems for present and the future.

Each chapter includes the presentation of the sub-problems addressed in the chapter, discussions on the theoretical and practical issues, details about chosen implementation and experiments and finally references to further reading and notes.

There are a few successful analyses, including textbooks, about Desktop Grid Computing. This is not proof of the lack of a strong and continuing interest in this topic and related areas because there exists well-established conferences (CCGRID, EUROPAR, HPDC, SC, IPDPS), journals (*FGCS, PARCO, JOGC, JIT, PPL, JNCA, IEEE Transactions on Service Computing*) and workshops (PCGRID, GP2PC, HETEROPAR) dedicated to the field, entirely or even partially. The reader should also note that the community of Desktop Grid users is large; for instance, the community of BOINC users is about 1.8 million users and the field is full of success stories in science that have received a lot of attention in the mass media.

In particular, we would like to mention the following special issues dedicated to Desktop Grid computing:

- *Distributed and Parallel Systems: In Focus: Desktop Grid Computing* (Hardcover) Peter Kacsuk (Editor), Robert Lovas (Editor), Zsolt Nmeth (Editor) selected papers of DAPSYS'2008 conference, Springer

- *Journal of Grid Computing*, Special issue on Desktop Grids and Volunteer Computing, D. Kondo (Editor), Ad Emmen (Editor), Peter Kacsuk (Editor), Springer, 2009, selected papers of PCGRID 2008

- *Future Generation Computer Systems*, Elsevier, Special issue on Desktop Grids, D. Kondo (Editor), Bahman Javadi (Editor), Peter Kacsuk (Editor), 2010, selected paper of PCGRID 2009

Christophe Cérin and Gilles Fedak
University of Paris and INRIA Rhône-Alpes

Part I

The Birth

Chapter 1

Volunteer Computing and BOINC

David Anderson

University of Berkeley, Berkeley, California, USA

1.1 Background

The consumer digital infrastructure (CDI) consists of a) mass-market digital devices (such as desktop, laptop, and tablet computers, game consoles, media players, smartphones, and VTRs) and b) the communication networks that connect them, most notably the Internet. Driven largely by the requirements of Internet streaming video and 3-D graphical games, consumer products have become very powerful. A typical current PC has 4 GB RAM, 1 TB disk, and a 1 TeraFLOPS GPU. Home Internet connections are on the order of 10 Mbps, increasing to 100 Mbps and 1 Gbps as optical fiber is deployed to homes over the next few years.

The CDI currently includes over 1 billion privately owned PCs and 100 million GPUs capable of general-purpose computing. These have a total computing capability of roughly 100 ExaFLOPS, and on the order of 10 Exabytes of free disk space, accessible via 1 Petabit/second of network bandwidth. In order to handle peak loads, the CDI is over-provisioned and hence underutilized. The average CPU and home Internet connection are used only a few percent of the time they are available, which, even with power-saving modes, is a significant fraction of the time.

A modern PC is capable of running most scientific applications, and thus the CDI is a viable platform for high-performance scientific computing. The CDI has several advantages over platforms, such as clouds, supercomputers, clusters, and grids, that are based on organizational resources:

- It has much larger processing and storage capacity, and hence enables otherwise infeasible science.

- Its hardware, electrical power, and HVAC are paid for by consumers, and the hardware is continuously upgraded to state-of-the-art components. The value of the CDIs PCs alone is roughly $1 trillion.

- It is self-maintaining: consumers fix their own software and hardware problems.

1.1.1   Volunteer Computing

To use the excess capacity of the CDI for scientific computing, we need the consent of the resource owners. This requires incentives that reward consumers for the use of their computing resources, and publicity that makes the public aware of this possibility. Experience has shown that consumers can be motivated by their support for the goals of the research, by the chance to participate in online communities, and by competition based on computational contribution [17]. This is called volunteer computing.

Volunteer computing was first proposed by Luis Sarmenta [19]. The first major projects, GIMPS and distributed.net, launched in 1997, and the first scientific computing projects, SETI@home and Folding@home, were launched in 1999 [1, 9, 12, 16]. Today there are about fifty volunteer computing projects in many scientific areas. The volunteer population consists of 500,000 people and 1 million computers, providing an average throughput of 12 PetaFLOPS. This has resulted in a large amount of research and publications, including papers in top journals like *Nature, Science, Physics Review*, and *Cell.* Equivalent computing power on the Amazon EC2 would cost over $4 billion/year.

1.1.2 BOINC: Middleware for Volunteer Computing

Middleware for volunteer computing dispatches jobs from scientists' servers to consumers' computers, and executes jobs on those computers. The early projects developed their own application-specific middleware. The BOINC (Berkeley Open Infrastructure for Network Computing) project was created in 2002 to develop open-source, general-purpose volunteer computing middleware [2]. BOINC provides server software that lets scientists create volunteer computing projects, and client software, available for all major platforms, that lets volunteers participate in any combination of these projects. BOINC software is distributed under the Lesser General Public License (LGPL), allowing it to be used with open-or closed-source applications. It may be used for any purpose, including non-scientific and for-profit applications.

BOINC must handle a variety of factors that are unique to volunteer computing:

- Scale: The system must handle millions of nodes and tens of millions of jobs per day.

- Heterogeneity: Volunteer hosts are highly diverse in terms of software, hardware, and network connectivity.

- Trust and reliability: Incorrect computational results may be intentionally returned by malfunctioning or malicious participants.

- Communication access: Many systems are behind firewalls that allow only outgoing HTTP traffic.

- Availability: Volunteer hosts have sporadic presence and a high churn rate.

- Ease of use: The client software must be extremely simple to install and must work with no configuration or user intervention.

The BOINC software is not designed simply to process jobs; it is also designed to catalyze and shape several human communities:

The community of computing resource donors: BOINC provides numerous mechanisms for attracting volunteers and for providing rewards, social interaction, competition, and recognition that motivate them to continue volunteering.

The community of helpers: Volunteers contribute in many ways other than supplying computing resources: they provide language translations of user interface text, test pre-release versions of the client software, optimize and port project applications, document, debug, and develop BOINC itself, and provide customer support to other volunteers.

The community of software developers: BOINC exposes software interfaces wherever possible, allowing many types of third-party software to be developed. Programmers have written specialized GUIs for the BOINC client; they have developed account manager websites providing a streamlined interface to multiple projects (see Section 2); and they have developed websites that display the statistics of computation contribution, broken down along axes such as country, and computer type (see Section 2).

The community of science: BOINC is designed to foster a computational ecosystem in which a large and dynamic set of science projects compete for volunteered resources. Scientists attract volunteers by educating the public about their research. Computer owners learn about many research projects and areas of science in the course of deciding which projects to support. The dual goals are that a) the best and most important research gets the most computing resources, and b) public awareness of and interest in science are increased.

Thus, BOINC is designed to facilitate several large-scale human systems. Similarly, BOINC's computational universe, the set of 500K or so volunteer hosts and 50 or so project servers, must be treated as a system. For example, if 100K hosts try to contact a server simultaneously, nonlinear effects such as connection dropping and paging can cause the system to grind to a halt. To prevent this, BOINC uses exponential backoff throughout its communication architecture. This ensures that when a server recovers from an outage, its workload remains fairly constant. Although each BOINC project is completely autonomous — this is a key property of BOINC — they interact in the sense that if one project is down, the others experience an increase in load.

BOINC's computational universe is designed for sustainability: all of BOINCs software interfaces provide backward and forward compatibility. Every component of BOINC — client, GUI, server — has changed radically during BOINC's nine years of development. Yet the oldest client version can successfully interact with the latest server version, and the latest client can interact with the oldest server. The same is true for client/application interaction. This is critical because there is no way to force either volunteers or projects to upgrade software. In particular, the client software has no auto-update mechanism; such a mechanism would introduce a perceived security risk.

1.2   Volunteer Features

1.2.1   Projects, Clients, and Attachment

Computer owners participate in volunteer computing by downloading the BOINC client software from the BOINC website (http://boinc.berkeley.edu). The first time the client runs, it prompts the user to select a project from a

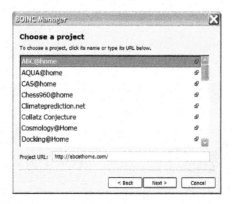

FIGURE 1.1: When the BOINC client first runs, it prompts the user to select a project.

list (see Figure 1.1), and to enter account credentials (an e-mail address and password, or OpenID credentials).

The client then communicates with the projects server to create the account and get jobs. It will continue to get and process jobs indefinitely, with no user intervention. By default, it runs whenever the computer is powered on.

By attaching to a project, the volunteer allows that project to run arbitrary programs on the computer. The client can be configured to run these programs in unprivileged user accounts, but this provides only minimal protection from buggy or malicious applications. Hence the volunteer must place a high degree of trust in the project. The BOINC website emphasizes this to volunteers, and encourages them to investigate the authenticity, credentials, and scientific merit of projects in which they are considering participation.

The list of projects displayed by the BOINC client is not inclusive; it consists of projects whose authenticity is known to the BOINC development team. It is periodically updated (from the BOINC project servers) by the client. Volunteers can attach to projects not in this list by explicitly entering the project's URL.

Users may attach computers to multiple projects by repeating the above process (see Figure 1.2). Each attachment has a user-adjustable resource share. The fraction of a bottleneck resource allocated to a given project is proportional to its resource share. Users may also detach computers from projects.

The BOINC manager provides a graphical interface that allows volunteers to view the status of BOINC on their computer, and to control it in various ways. Users can select two interface variants: the basic view (default) is small and graphical, while the advanced view provides a spreadsheet-style interface showing detailed information; see Figure 1.3.

The BOINC client is designed to be autonomic — to work invisibly, and

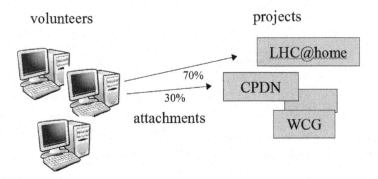

FIGURE 1.2: Volunteers can participate in any of a number of BOINC projects, and can control the resource share of each project.

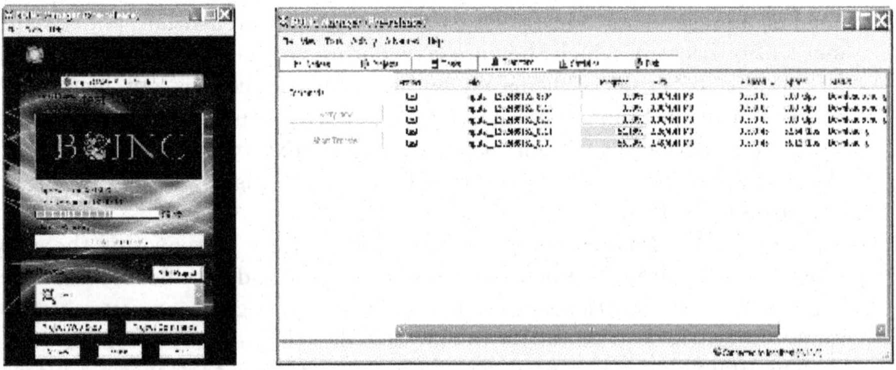

FIGURE 1.3: The BOINC Manager provides basic (left) and advanced (right) views.

to recover from problems without user intervention. However, there are cases where it tries to get the user's attention:

- Computation is not being done because of a situation that the user can remedy, for example by installing a new GPU driver.

- An attached project has important news, such as a scientific discovery.

- There is a social-network event (for example, a new private message) from an attached project.

The Manager has a mechanism called notices to handle these cases. Notices are delivered as RSS feeds, and can contain arbitrary HTML including images. The user is unobtrusively alerted (e.g., by a system tray balloon on Windows) when there is a new notice.

BOINC provides a screensaver. Project applications can optionally include a graphics app that shows a visualization of the computation being done. When the computer is idle and the BOINC screensaver runs, it selects a running job for which a graphics app is available, and runs the graphics app. If no such job exists, it runs a default screensaver, which shows summaries of attached projects and in-progress jobs.

1.2.2 Account Managers

The model in which volunteers attach computers to projects does not scale well with the number of projects, as volunteers must visit the website of each prospective project, and must perform several manual steps to attach to a project. It also does not scale well with the number of computers owned by a volunteer because, to add or change project attachments, manual action is required on each computer.

To address these issues, the BOINC architecture supports account managers: Web services that provide a level of indirection between volunteer and projects. A computer may be attached to an account manager. The account manager then periodically tells the computer which projects to do work for, and the computer gets jobs directly from these projects.

The account manager architecture supports a variety of purposes:

One-stop shopping for projects: Such a site displays a list of BOINC projects and allows the user to explicitly select projects, for example by checking boxes. Examples include BAM! and Gridrepublic [6, 13].

Centralized resource management: A set of resources (organizational and/or consumer) are attached to the account manager; they are then directed to projects according to a centralized policy. Jarifa [14] is an example.

Farm manager: A special case of centralized resource management in which the resources are within an administrative domain. Among other things, this provides a way to track resources and detect hardware problems.

Cause-oriented computing: Such an account manager might let volunteers donate to cancer research, and dynamically determine the set of supported projects and their shares. More generally, it might provide this service for a choice of causes (medicine, environment, physics, and so on).

1.2.3   Project Website Features

Each project offers a public website. Part of this site's content is project-specific: for example, a description of the research being done and of recent results. The remainder is supplied by BOINC, and allows volunteers to log in using their account's credentials. They can then access various features:

View computing information: Descriptions of the computers they have attached to the project, details of in-progress and recently-completed jobs.

Community features: Message boards, teams, friends lists, private messages, profiles.

Leader boards: Lists of the volunteers, hosts, and teams with the most total or recent credit.

1.2.4   Preferences

Volunteers can control when and how BOINC uses their computing resources. These controls are called preferences. There are two groups of preferences: global (which affect all attached projects) and per-project (which affect only one project). Global preferences include

- Limits on when computing can be done. This can be based on whether the computer is in use, on time of day or day of week, and non-BOINC CPU load.

- Similar limits on when network communication can be done.

- Limits on the number of processor cores used, and on the duty cycle of processor usage. The latter control, called CPU throttling, can be used to limit the increase in CPU temperature.

- Limits on network bandwidth used, and on the total transfer over a given period.

- Limits on disk space and RAM usage.

Global preferences can be edited via a Web interface on the website of any attached project. The preferences are propagated to all computers attached to that account, and from there to all projects attached to any of those computers using the same e-mail address. Global preferences can also be edited using a local GUI interface, in which case they apply only to that computer and override the Web-specified preferences.

Each project determines its per-project preferences. Possibilities include:

- Which of the projects applications to run

- Whether to use GPU and/or CPUs

- Controls affecting the appearance of the project's graphics apps

1.2.5 Teams

Project websites allow volunteers to form teams. Each team has a founder and a private message board. The founder can moderate this message board, manage team membership, and delegate these powers to administrators. The project website displays a leader board of teams ordered by credit.

Teams have an important role in volunteer recruitment and retention. Team members recruit their friends and relatives to volunteer and join the team. Top teams compete for creadit leadership. Most teams have a name and identity corresponding to a country or organization (company, university, and so on).

Teams are per-project, and large teams typically want to have an instance on all BOINC projects. To facilitate this, BOINC provides a mechanism called BOINC-wide teams. This is a pseudo-project (http://boinc.berkeley.edu/teams/) whose set of teams (and corresponding founder credentials) is exported as an XML file. Projects, by default, run a periodic task that imports this file and creates the corresponding accounts and teams. Thus, instances of each BOINC-wide team are created on all projects.

1.2.6 Credit

BOINC grants credit for completed jobs. This is a number that reflects the amount of computation performed. Currently it is defined in terms of floating-point operations: an always-on 1 GFLOPS computer gets 200 units of credit per day.

Credit is a critical mechanism for volunteer retention. Increasing credit gives individual volunteers confirmation that their continuing contribution and credit provide a basis for competition among individuals and teams. Credit also provides a measure of the computational throughput of individual projects and of volunteer computing as a whole.

By itself, the credit mechanism has the undesirable effect of discouraging volunteer migration between projects: if a volunteer has accumulated a lot of credit on a project, they are reluctant to start a new project where their credit will initially be zero. To combat this effect, we introduced the concept of cross-project credit, which is the credit for an entity (computer, volunteer, or team) summed over all BOINC projects. This is implemented as follows:

- Each computer and volunteer has a unique cross-project ID (a random character string). This string is determined by a distributed algorithm piggybacked on client/scheduler communication. The identifier of a team is its name.

- Each project exports its credit statistics (lists of computers, volunteers, and teams, together with their cross-project IDs).

- Third-party websites (associated neither with projects nor with BOINC) periodically import these statistics, collate them, and export the resulting cross-project credit in various ways. Most of these sites offer leaderboards of teams, volunteers, countries, and so on, ordered by cross-project credit. Others offer features such as images showing a volunteer's current credit on all projects, used as a signature for their message-board posts.

1.2.7   Volunteer Support

Many volunteers have questions or problems using BOINC. For example, the client's attempts to do network communication may trigger alerts from personal firewall software. Because the BOINC project itself lacks staff to handle these problems, we have devised systems in which experienced volunteers provide support. We set up messages boards on the BOINC website for this purpose. We observed, however, that this medium did not work well for volunteers who did not know the technical vocabulary used by the experienced helpers.

To address this problem, we created a Web-based system based on Skype telephony. Helpers can register, providing their Skype ID, the language(s) they speak, and the operating systems they use. Volunteers seeking help can then go to a Web page (http://boinc.berkeley.edu/help.php), select their spoken language, and see a list of available helpers. They can then contact a helper by voice, chat, or email. Afterward, they can rate the helper; these ratings are publicly visible and are used to order the list of helpers.

1.3   Data Model

The BOINC storage model is based on files. Files may be inputs or outputs of jobs, or the components of an application; they may also exist independently of jobs. Each file has a physical name, unique within its project. Files are immutable; they cannot be modified, and all replicas of a given file are identical. If a file changes, it must be given a new name. Files that are to be downloaded include a list of download URLs. Output files have an associated maximum size; if this size is exceeded, the job is aborted.

Normally, a job's input files are deleted when the job is finished, and its output files are deleted when they have been uploaded to a server. If a file

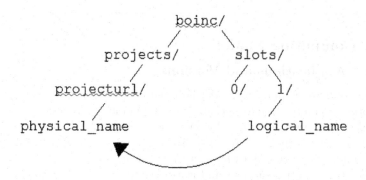

FIGURE 1.4: The BOINC client directory structure.

is marked as sticky, however, it remains resident on the client and is deleted only when requested by the scheduler. Files that are part of an application version are inherently sticky, and are deleted only when the version has been superceded.

The client maintains the directory structure shown in Figure 1.4.

Each project has its own directory (named by the URL of the project) and each job executes in a slot directory.

A file reference is an association between a file and a job or application version. Each file reference includes a logical name by which the application will refer to the file. By default, the file itself exists only in the project directory, and the client creates a link from the slot directory to the project directory prior to starting the job. Applications use a BOINC API function to map the logical name to the physical file. This indirection provides two advantages:

- Multiple concurrent jobs can access a given file without having separate copies of it

- Applications containing hardwired (logical) file names may be run against different files

File references can optionally specify a copy file attribute; if set, the file is copied from project to slot directory before the job is started (input files) or from slot to project directory after the job completes (output files).

Files that are part of an application version must be digitally signed using the private key of a key pair whose public key is issued to the client on the first scheduler RPC. Projects are encouraged to keep the private key only on a physically disconnected and secure computer. This ensures that even if a project server is broken into, the intruders will not be able to distribute and execute malware.

1.4 Computing Model

1.4.1 Applications and Versions

The BOINC computing model is based on the following abstractions: A platform is a compilation target, for example Windows/Intel32. An application is an algorithm that is realized by one or more app versions. Each app version consists of a set of files (always including a main program executable; possibly also including libraries, data files, virtual machine images). Each app version has an integer version number and is associated with a platform. When a job is submitted, it is associated with an application, not an app version. BOINC decides which app versions to use.

The BOINC client recognizes three processor types: CPU, NVIDIA GPU, and ATI GPU. It assumes that all the processors of a given type on a given computer are identical. This is typically true of CPUs. For hosts that have GPUs from the same vendor and different characteristics, BOINC identifies the most powerful one and ignores those that are not equivalent to it.

A project may have multiple app versions for a given platform and application. For example, it might have versions that run on parallel or multiple CPUs, or that use GPUs. Such versions are distinguished by a textual label called a plan class. Versions with an empty plan class are assumed to use one CPU.

Each nonempty plan class has an associated planning function that takes as input a computer description, and decides whether the computer is capable of running an application of that plan class. This decision may involve the GPU version, the size of video RAM, the version of the GPU driver software, and so on. If the computer is capable, the function also returns a description of the processing resources that will be used (number of CPU and GPUs, possibly fractional, and the amount of video RAM) and an estimate of the FLOPS that will be achieved by that app version on that computer.

A project can defines its own plan classes and their planning functions. BOINC supplies a set of predefined classes, including classes for multicore apps, GPU apps for various hardware levels, apps that require particular CPU instruction set extensions, and apps that run in Virtualbox VMs.

When the BOINC scheduler is considering sending a job to a host, it uses a version selection algorithm to decide which app version to use. This involves scanning all the app versions for platforms supported by the host, calling their planning functions if needed, and discarding the versions that use processor types for which no work is being requested. It then estimates the speed of each of the resulting versions, based either on the planning-function estimate or, if the host has already completed sufficient jobs using that app version, the statistics of their elapsed time. It then picks the app version with the greatest predicted speed. It adds a small normally distributed random factor to the speed of each job; this assures that BOINC will eventually discover the fastest version, even if the first few jobs processed using that version happen to be unusually long.

1.4.2 Jobs

A BOINC job includes:

- Lists of input and output files, and corresponding file reference information (see the previous section)

- Estimates and upper bounds on the amount of FLOPs, RAM, and disk space to be used by the job. If one of the upper bounds is exceeded, the job is aborted

- A latency bound, which determines a soft deadline when the job is dispatched (this deadline may be adjusted by the scheduler)

BOINC provides `C++` and command-line APIs for submitting jobs. Systems for submitting batches of jobs, or submitting jobs remotely, may be built on top of these. BOINC provides command-line tools for submitting and awaiting the completion of single jobs. More commonly, projects use programs to submit and asynchronously handle the completion of large numbers of jobs.

By default, jobs are dispatched in a nondeterministic order. Projects can optionally specify job priorities that determine the dispatch order, or they may specify the relative job dispatch rate per application.

1.4.3 Result Validation

Volunteer computers are untrusted: the result of running a job cannot generally be assumed to be correct. Many volunteers overclock their CPUs and GPUs, and this can result in floating-point errors. Malicious users may intentionally return wrong results, either to subvert the research or to get undeserved credit.

Some applications have the property that results can be verified quickly on the server. Others allow plausibility checks, for example that energy has been conserved in a physical simulation. In other cases, projects must use replication to validate results. This means that two instances of each job are run, on computers belonging to different users. If the results agree, they can be assumed to be correct with high probability. If not, BOINC issues a third instance of the job; this continues until a majority of agreeing results is obtained, or until a limit on instances is reached.

Applications that do floating-point arithmetic typically get different results when run on different types of CPUs, or using different numerical libraries. This may make it difficult to decide when two results agree. For algorithms that are numerically stable, the results can be compared with a fuzzy comparison function supplied by the project (see Section 1.7). For unstable algorithms (e.g., most physical simulations), BOINC provides a mechanism called homogeneous redundancy (HR). When HR is used, BOINC ensures that all instances of a given job are issued to hosts that are numerically equivalent for that application. This can be extended to GPU and multicore applications using an analogous homogeneous app version mechanism, which ensures that, in addition to the HR restriction, all instances are run using the same app version.

The disadvantage of replication is that it reduces the effective computational throughput of a project by a factor of at least 2. To address this, BOINC provides a mechanism called adaptive replication. The BOINC server maintains the number N of consecutive validated results returned by each (host, app version) pair (a given host may have a high error rate for GPU jobs but not CPU jobs). If this number exceeds a threshold, BOINC will begin randomly issuing unreplicated jobs to the host. As N increases, the probably of replication goes to zero.

1.4.4   Assigned Jobs

By default, jobs are run on any available host. BOINC also provides a mechanism called assigned jobs that allows the submitter to specify that the job is to be run on a particular host, or on a host belonging to a particular user. One can optionally specify that the job is to be run on all hosts belonging to a user, or an all hosts; one can also specify that the job must be completed successfully before any other jobs are dispatched to the host.

1.4.5   Long-Running Jobs

Some projects have jobs that run for weeks or months on a typical computer. BOINC provides two mechanisms to support such jobs:

- Trickle messages are asynchronous messages between a running application and the project server. A trickle-up message, for example, might contain a summary of the computational state. If the server decides that the job is no longer needed, it can send a trickle-down message telling it to quit.

- Intermediate file upload allows a running application to indicate that an output file is complete and should be uploaded. This could be used, for example, to upload checkpoint files, allowing jobs to be restarted on different hosts.

1.4.6   Non-Compute-Intensive Applications

Applications can be flagged as non-compute-intensive. Jobs for such applications are run whenever the BOINC client is allowed to compute, and they are expected to use little CPU time. This mechanism, together with trickle messages, was used to implement the Quake Catcher Network project, a distributed seismological sensor using USB accelerometers [18].

1.5 Applications

1.5.1 BOINC-Aware Applications

Almost any program can be run as a BOINC application, using the wrapper approach described below. However, there are advantages in making applications BOINC-aware. To explain this idea, we must outline the runtime environment of BOINC applications. This has several aspects:

Directory structure: The BOINC client runs a job by creating an empty slot directory, initializing it with input files or links and executing the app version's main program in the slot directory.

Job control: The client may need to suspend, resume, and kill the application. Not all operating systems provide mechanisms that are adequate for this purpose. So instead, BOINC uses a design in which the client sends job-control commands as messages delivered to the application via shared memory. These messages are handled by the BOINC runtime system, typically running in a separate thread within the application.

Status reporting: For scheduling purposes, the client can benefit from knowing a) when the application has checkpointed, b) the applications current estimate of its completion fraction, and c) the CPU time of the application, including all threads and subprocesses. This communication is also implemented using shared-memory message-passing.

A BOINC-native application is one that makes the following API calls and is linked with the BOINC library. Such applications can be run directly, with no wrapper needed.

- The application calls initialize/finalize functions `boinc_init()` and `boinc_finish()`.

- Before opening a file, the application converts the filename from physical to logical using `boinc_resolve_filename()`.

- The application checkpoints only when `boinc_time_to_checkpoint()` returns true, and calls `boinc_checkpoint_finished()` when the checkpoint is complete.

- The application periodically `calls boinc\index{boinc}_fraction_done()` to report its completion fraction.

The BOINC API is implemented in C++, and bindings are available for C, FORTRAN, Python, and Java.

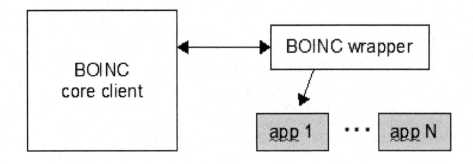

FIGURE 1.5: The BOINC wrapper allows any executable to be run under BOINC.

1.5.2   The BOINC Wrapper

For applications that cannot easily be made BOINC-aware (e.g., because the source code is not available), the BOINC wrapper can be used. The wrapper takes as input an XML file describing a sequence of sub-jobs. It runs these in sequence and handles communication with the BOINC client (see Figure 1.5).

The wrapper does its own checkpoint/restart at the level of sub-job. If any of the sub-jobs does checkpointing, the wrapper can be instruct to check for updates to the restart file, and to report this to the client.

1.5.3   The BOINC Virtualbox Wrapper

BOINC supports applications that run in virtual machines, using the Virtualbox system from Oracle. This is implemented using a BOINC-aware program called the Virtualbox wrapper. In this scheme, an app version consists of

- The Virtualbox wrapper, compiled for a particular platform

- A VM image (the same VM image can be used across multiple apps and versions, in which case it is downloaded only once to a given host)

- An main-program executable to be run within the VM

When the wrapper runs, it creates a subdirectory of the slot directory, and arranges for the subdirectory to be shared with the guest OS. The main program and the jobs input files are copied to this directory. The wrapper then requests the Virtualbox executive to boot the VM image in a new VM. The VM image must be configured so that it runs the main program after booting, and shuts down the VM when the program is done.

VM apps provide two benefits:

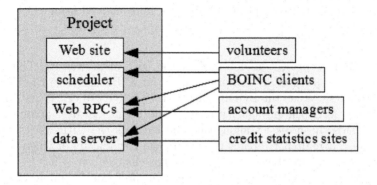

FIGURE 1.6: A BOINC project exports several interfaces using HTTP.

- Projects can develop applications in their target environment of choice (e.g., a particular version of Linux, with a particular set of libraries and packages installed). There is no need to build versions for multiple platforms.

- VMs provide a strong sandbox, so it is feasible to use untrusted applications. For example, a BOINC project serving a science portal could allow portal users to run their own applications.

The BOINC client detects the presence of Virtualbox and reports its version number to the scheduler. Hence the plan class mechanism can be used to decide whether to send VM app versions to a given host.

Currently volunteers must manually install Virtualbox. In the future we will include Virtualbox as part of the standard BOINC client installer.

1.6 Communication Structure

About 90% of BOINC clients are behind firewalls. Many of these do not allow incoming connections, but essentially all of them allow outgoing HTTP and HTTPS connections. Therefore, all communication between the BOINC client and the BOINC server is initiated by the client, and uses HTTP or HTTPS.

A BOINC project exports several HTTP-based interfaces (see Figure 1.6):

- Its website (human-readable pages)

- A set of Web services, implemented in the REST style (arguments encoded in the URL, results returned as XML documents)

- A scheduler RPC interface, implemented using HTTP POST (XML arguments and results)

- A protected file upload mechanism, implemented using HTTP POST

- File download

A BOINC project is identified by the URL of its website. This page has a secondary purpose: It contains (in special XML elements) the URLs of its scheduler(s). When the BOINC client attaches to a project, it first issues Web RPCs to fetch the project parameters and create the user account. Then it downloads the project's main page and extracts the URLs of its scheduling servers. Thereafter it rereads the main page periodically or when a scheduler RPC fails, in case the set of schedulers has changed.

The scheduler RPC is BOINC's central protocol. To minimize the number of RPCs, each transaction can do multiple things. The request message includes:

- A description of the computer's hardware and operating system

- A list of completed jobs

- A description of current workload: running and nonrunning jobs, their processor resource usage, and their estimated remaining runtime

- For each processor type, a work request consisting of a) the number of instance-seconds being requested and b) the number of currently idle instances (the scheduler attempts to send enough jobs to use this number of instances, even if that exceeds the number of seconds being requested)

- Trickle-up messages

The reply message may include

- A list of new jobs, including descriptions of the app versions to be used for each job

- Commands to delete sticky files

- Commands to abort existing jobs

- Trickle-down messages

1.7   Server Software

The BOINC server software consists of a set of Web services, a set of daemon programs that share (and communicate through) a relational database, and a set of utility programs. A server computer can host several projects; each one has its own database, and a directory hierarchy where its software, data

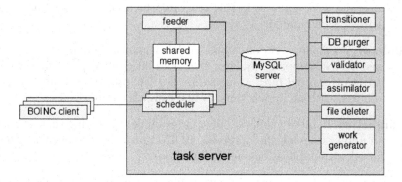

FIGURE 1.7: The components of a BOINC server

files, and configuration information are stored. A project's database stores descriptions of its volunteers, hosts, apps, app versions, jobs and job instances, and so on. BOINC provides utilities for creating projects (database and directory structure), for starting and stopping a project, and for upgrading a project's server software.

A BOINC server consists of several interacting components; see Figure 1.7.

The scheduler handles scheduler RPCs and dispatches jobs to clients. It runs as a CGI program, and many instances of it (serving different clients) may exist at one time. To reduce database access and contention, the scheduler uses a cache of ready-to-send jobs in a shared-memory structure that is replenished by a single feeder program. An instance of the scheduler scans the job cache to find the best jobs to send to its client. The scheduler can be configured to use a policy called locality scheduling that reduces server load and network traffic by preferentially sending jobs to hosts on which the input files are already present.

The scheduler implements a number of other policies: a) the replication and homogeneous replication policies described in Section 1.5; b) project-configurable settings for limiting the number of in-progress jobs for a given application and host, per-processor or in total; c) a mechanism called host punishment that limits the number of jobs per day, for a given app version, issued to hosts that have recently experienced errors with that app version. This limits the system impact of malfunctioning hosts.

Projects use some or all of the following daemons:

Work generator: For applications with unbounded numbers of jobs, a program that submits jobs. The creation process may be flow-controlled based on available disk space (for data files), by the number of unsent jobs, or by other factors.

Transitioner: Handles job state transitions, including timeout of job instances, error handling, and creation of new instances. By centralizing most database updates in this program, the need for locking and transactions is eliminated.

Validator: Compares sets of completed job instances, looking for a consensus. BOINC supplies validator that does bitwise comparison. For applications that need fuzzy comparison, BOINC provides a framework that is linked with an application-specific comparison function.

Assimilator: Handles validated jobs, for example, by storing their results in an application-specific database. BOINC provides a framework that is linked with application-specific functions.

File deleter: Deletes input and output files that are no longer needed.

Database purger: Deletes job and job instance records that are no longer needed. Many projects process hundreds of thousands of jobs per day, and without purging the job table would soon become so large as to impact system performance.

Trickle-up message handler: An application-specific program that handles trickle-up messages from applications.

The BOINC server software is designed for scalability. The daemons, Web servers, and database server can all be moved to different machines. In addition, all the daemons can be replicated: if there are N instances, instance i handles jobs where mod(ID, N) = i. A single server-class computer can handle thousands of clients, and a complex of three or four server computers can typically handle a million clients.

1.8   Client Software

Although the BOINC client software appears as a single entity to volunteers, it consists of several interacting components (see Figure 1.8):

- The core client, which manages computation and communication

- The Manager, which provides the GUI

- A screensaver coordinator, which if enabled by the user runs when the computer is idle. It communicates with the core client to see if a graphics-enabled job is running, and if so it selects one of these jobs and runs its graphics module

- A default screensaver, which provides screensaver graphics if no graphics-enabled job is running

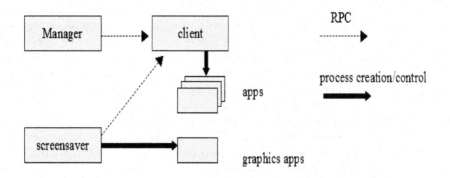

FIGURE 1.8: The structure of the BOINC client software.

The Manager and screensaver coordinator communicate with the core client via a set of RPC interfaces exported by the core client. These RPCs use XML messages sent on a TCP connection; the Manager can control remote as well as local core clients. The interface is documented, and open-source developers have created a number of alternative BOINC GUIs. For example, BoincTasks is a GUI providing an interface to core clients running on a number of hosts.

The BOINC installer allows the user to select two aspects of BOINC's behavior:

- Whether to use account-based sandboxing: if enabled, the installer creates an unprivileged user account, and the core client runs jobs under that account. This provides a measure of protection against malicious or buggy applications, as they can access only world-readable files. (On Windows, GPU applications do not work with account-based sandboxing, so this option is disabled by default).

- Whether the BOINC client can be controlled only by the installing user, or by all users on the computer.

The core client is implemented as a sequential program. It uses select() and other non-blocking interfaces to multiplex its various activities: job execution, communication with running jobs, file transfers, RPC handling, scheduler RPC calls, RSS feed fetching, and so on.

The core client embodies two related scheduling policies:

Job scheduling: This is the decision, given a set of queued jobs and a set of processing resources, of which jobs to run using which resources. The default policy is round-robin time-slicing, weighted by project resource shares. To avoid missing job deadlines, the scheduler periodically runs a simulation of the current workload under the default policy; any jobs that miss their deadline in this simulation are then run with high priority.

Work fetch: This is the decision of when to ask a project for jobs, which project to ask, and how much work of each processor type to request. The current policy is as follows. There are user-configurable upper and lower bounds U and L on the amount of work queued for each processor type. Using the scheduling simulation described above, the client estimates the occupancy of each processor instance. The client maintains the share deficit of each project: roughly speaking, the difference between how much processing the project should have received recently and the amount it actually got. If, for any instance, the occupancy falls below L, the client chooses the project P with the greatest share deficit that may have jobs for that resource type, and asks P for enough work to bring the occupancy up to U. This hysteresis reduces the number of scheduler RPCs, hence server load.

1.9   Research Tools and Directions

BOINC has provided tools and data for various research related to volunteer computing:

- Host availability: The BOINC client records detailed availability information in a log file. Projects can arrange to upload these log files. We have collected availability traces from SETI@home in this way and have used them to study and model availability [15]

- We have developed a client emulator that can be used to study client scheduling policies under a wide range of scenarios [5]

- Michela Taufer's group has developed EmBOINC, a server emulator that can be used to study server scheduling policies [11]

There are a number of area of ongoing development and research involving BOINC:

Peer-to-peer data distribution: Currently all files are downloaded from project data servers. This may impose a bottleneck, especially if large files such as virtual machine images are being distributed. Researchers have studied the use of Bittorrent for file distribution in BOINC [8]. The Attic project has developed a peer-to-peer mechanism designed for volunteer computing[10].

Multi-user projects: Some BOINC projects serve multiple users, who may contend for the project's computer resource. We are designing a mechanism that fairly allocates resources among users.

Predictable and optimized batch completion: We are designing a mechanism that allows users to submit batches of jobs, that provides realistic estimates of batch completion time, and that optimizes batch completion (addressing, in particular, the straggler effect in which the last few jobs of a batch must be retried). This problem has also been studied by Silberstein [20].

Distributed storage: BOINC is designed to support volunteer storage as well as volunteer computing. It offers primitives for transferring files between server and client, for allowing the server to monitor the set of files on each client, and for allowing the server to delete files on a client. These primitives can be used to implement storage systems that use replication (and possibly striping or coding) to provide target levels of reliability, latency, and/or throughput.

Computing with mobile devices: The processing and storage capacity of mobile devices such as smartphones is increasing rapidly, and the boundary between mobile and stationary devices is becoming blurred. Mobile devices typically are more energy efficient, so they may offer an attractive platform for distributed computing. We are developing a version of BOINC that will run on mobile devices using the Android operating system.

Bibliography

[]   Distributednet. http://distributed.net.

[ACA06]   David P. Anderson, Carl Christensen, and Bruce Allen. Grid resource management - designing a runtime system for volunteer computing. In *SC* [con06], page 126.

[ACK⁺02] David P. Anderson, Jeff Cobb, Eric Korpela, Matt Lebofsky, and Dan Werthimer. Seti@. *Commun. ACM*, 45(11):56–61, 2002.

[AET11]   Ian Kelley Abdelhamid Elwaer, Andrew Harrison and Ian Taylor. Attic: A case study for distributing data in BOINC projects. 2011.

[AKW05]   David P. Anderson, Eric Korpela, and Rom Walton. High-performance task distribution for volunteer computing. In *e-Science* [eSc05], pages 196–203.

[And04]   David P. Anderson. BOINC: A system for public-resource computing and storage. In Buyya [Buy04], pages 4–10.

[And11]   David P. Anderson. Emulating volunteer computing scheduling policies. 2011.

[bam]   Bam. http://bam.boincstats.com/.

[Buy04]   Rajkumar Buyya, editor. *5th International Workshop on Grid Computing (GRID 2004), 8 November 2004*, Pittsburgh, PA, USA, Proceedings. IEEE Computer Society, 2004.

[con06]   *Proceedings of the ACM/IEEE SC2006 Conference on High Performance Networking and Computing*, November 11-17, 2006, Tampa, FL, USA. ACM Press, 2006.

[con08a]   *22nd IEEE International Symposium on Parallel and Distributed Processing, IPDPS 2008*, Miami, Florida USA, April 14-18, 2008. IEEE, 2008.

[con08b]   *9th IEEE/ACM International Conference on Grid Computing (Grid 2008)*, Tsukuba, Japan, September 29 - October 1, 2008. IEEE, 2008.

[con09]   *23rd IEEE International Symposium on Parallel and Distributed Processing, IPDPS 2009*, Rome, Italy, May 23-29, 2009. IEEE, 2009.

[con11]   *11th IEEE/ACM International Symposium on Cluster, Cloud and Grid Computing, CCGrid 2011*, Newport Beach, CA, USA, May 23-26, 2011. IEEE, 2011.

[CSKF08]   Fernando Costa, Luís Moura Silva, Ian Kelley, and Gilles Fedak. Optimizing the data distribution layer of BOINC with bittorrent. In *IPDPS* [con08a], pages 1–8.

[eSc05]   *First International Conference on e-Science and Grid Technologies (e-Science 2005)*, 5-8 December 2005, Melbourne, Australia. IEEE Computer Society, 2005.

[ETRA09]   Trilce Estrada, Michela Taufer, Kevin Reed, and David P. Anderson. Emboinc: An emulator for performance analysis of BOINC projects. In *IPDPS* [con09], pages 1–8.

[gim]   Gimps. http://www.mersenne.org/prime.htm.

[gri]   Gridrepublic. http://www.gridrepublic.org/.

[jar]   Jarifa. https://github.com/teleyinex/jarifa/wiki.

[KAA08]   Derrick Kondo, Artur Andrzejak, and David P. Anderson. On correlated availability in internet-distributed systems. In *GRID* [con08b], pages 276–283.

[MKT98] Yoshifumi Masunaga, Takuya Katayama, and Michiharu Tsukamoto, editors. *Worldwide Computing and Its Applications, International Conference, WWCA '98, Second International Conference, Tsukuba, Japan, March 4-5, 1998, Proceedings*, volume 1368 of *Lecture Notes in Computer Science*. Springer, 1998.

[NAA10] Oded Nov, David Anderson, and Ofer Arazy. Volunteer computing: a model of the factors determining contribution to community-based scientific research. In Rappa et al. [RJFC10], pages 741–750.

[Qua] Quake catcher network. http://qcn.stanford.edu/.

[RJFC10] Michael Rappa, Paul Jones, Juliana Freire, and Soumen Chakrabarti, editors. *Proceedings of the 19th International Conference on World Wide Web, WWW 2010*, Raleigh, NC, USA, April 26-30, 2010. ACM, 2010.

[Sar98] Luis F. G. Sarmenta. Bayanihan: Web-based volunteer computing using java. In Masunaga et al. [MKT98], pages 444–461.

[Sil11] Mark Silberstein. Building an online domain-specific computing service over non-dedicated grid and cloud resources: The superlink-online experience. In *CCGRID* [con11], pages 174–183.

[SLP02] M. Shirts S.M. Larson, C.D. Snow and V.S. Pande. Folding@home and genome@home: Using distributed computing to tackle previously intractible problems in computational biology. *Computational Genomics*, Horizon Press, 2002.

[tf] Boinctasks. http://www.efmer.eu/boinc/boinc_tasks/.

[TMN11] David Toth, Russell Mayer, and Wendy Nichols. Increasing participation in volunteer computing. In *Proceedings of the 5th Workshop on Desktop Grids and Volunteer Computing Systems (PCGrid 2011)*, Anchorage, AK. IEEE, 2011.

[vir] Virtualbox. http://www.virtualbox.org/.

Chapter 2

Open, Scalable and Self-Regulated Federations of Desktop Grids with OurGrid

Francisco Brasileiro

Universidade Federal de Campina Grande, Campina Grande, Brazil

Nazareno Andrade

Universidade Federal de Campina Grande, Campina Grande, Brazil

2.1 Case for a Desktop Grid for and by Small and Medium-Sized Laboratories

New developments in communication and computing technologies have substantially impacted the way scientific research is conducted nowadays. The use of computers for data analysis, simulations, and visualization of results are currently playing a fundamental role in the working methodology adopted by many research groups. As a consequence, having access to high-performance computing facilities has become a must for many research areas.

Due to this demand, research in computing science has, for some time now, sought ways to expand the reach of high-performance computing. In particular, many solutions have been proposed to use the widespread availability of reasonably powerful desktop computers to address the computational requirements of a simple, yet very common, class of applications, named *embarrassingly parallel*, or *bag-of-tasks* (BoT) applications. These are parallel applications that can be decoupled in a large number of independent tasks that do not need to communicate with each other. Thus, they can be simultaneously executed in a large number of desktops, geographically dispersed, greatly reducing their completion time. Despite being simple, BoT applications are used in a variety of settings, including data mining, massive searches, parameter sweeps, simulations, fractal calculations, bioinformatics, and computer imaging, among others.

One of the first initiatives in this sense provided a way to aggregate unused computing power in the desktops connected by a local area network [LLM88]. The next step was to increase the scale on the number of resources aggregated by harvesting the idle computing power in the Internet [ACK+02], what has been dubbed *public resource computing* or *voluntary computing*.

The number of resources that can be assembled in these *desktop grids* varies widely. It can span from a handful of desktops connected by a local area network in a small research lab, all the way to voluntary computing systems connecting thousands or even millions of home users' desktops in the edges of the Internet [And04]. Clearly, the larger the grid, the more complex are the challenges that need to be faced to build and manage it, and the higher is the cost of assembling it.

For instance, large voluntary computing systems require a lot of investment in publicity to attract and maintain volunteers supporting them. Typically, these systems must maintain an active website where not only statistics about the resource donors are kept, but, most importantly, feedback is provided about the impact of the computation that is performed with their resources. This requires that they support applications with high social or scientific impact, such as the cure of diseases[1,2], anticipation of natural disasters[3], the modeling of the universe[4], etc. Some effort is also required to keep the back-end servers running, port applications to the voluntary computing platform, validate the security of the applications ported, and provide adequate support for the users to submit their jobs and access the results of their executions. Finally, although not directly related to the cost of setting up a voluntary computing system, another point that is important to attract volunteers is the reputation of the institution that is behind the project. This is because volunteers must trust that the software that they are installing has been suffi-

---

[1]http://folding.stanford.edu/.
[2]http://boinc.bakerlab.org/rosetta/.
[3]http://climateprediction.net/.
[4]http://www.cosmologyathome.org/

ciently tested to guarantee that it will not cause any harm to their computers, and that the security policies put in place by the software provider are strong enough to prevent any attacker from tampering with the software. On the other hand, smaller desktop grids based on machines that are owned by a single institution and connected via a local area or campus-wide network can be set up with relative ease, and very low cost. The drawback in this case is the small scale that can be achieved.

Therefore, if for one side the massive use of computers by researchers fosters, at an ever-faster pace, amazing developments for the society at large, for the other side it contributes to increase even more the gap between the research that can be conducted at the few large research labs that can afford the costs of deploying large desktop grids, and that conducted by the majority of small and medium-sized labs that cannot. Yet, the fact that a lab is not large does not need to have any relation to the importance of the research it develops. As a matter of fact, in most countries, a substantial portion of all research developed is conducted by research groups organized around small and medium-sized labs.

Peer-to-peer systems have been proposed as a low-cost alternative to assembling large communitarian systems. Following this approach, peer-to-peer desktop grids have been proposed to allow small and medium-sized labs (and naturally large-sized labs as well) to join, with minimal additional cost, their local desktop grids in a very large system that can be cooperatively exploited by them. The rationale behind a peer-to-peer grid is that each peer representing a research lab, donates its idle computational resources in exchange for accessing other peers' idle resources when needed [CBA+06]. This chapter is devoted to discuss the design and implementation of this kind of system.

The discussion is focused on a particular middleware, named Our-Grid[5] [CBA+06]. OurGrid can be used to create, in a simple way, a very large peer-to-peer computational grid for the execution of BoT applications. OurGrid leverages from the fact that most computers are not fully used on a continuous basis. Even when actively using computers as research tools, researchers alternate between application execution (when they demand a lot of computational power) and result analysis (when their computational resources go mostly idle). In fact, in most small and medium-sized labs, researchers run relatively small applications, for which they seek fast turn-around times, so as to speed up the run/analyze iterative cycle.

OurGrid was designed to be simple, scalable, secure, and fast. Assembling a grid based on OurGrid is simple and scalable because it requires no negotiation. OurGrid allows peers to join the grid at their will, making their idle computational resources available to other peers at the same time that they gain access to the idle cycles of the resources that have been added to the grid by other peers. OurGrid uses a reciprocation mechanism to prevent free-riding [ABCM07] and a sandboxing mechanism to provide security against

---

[5]http://www.ourgrid.org/.

malicious applications [CAG⁺06]. Moreover, it uses replication schedulers to
achieve very good application turn-around time even in the absence of information about the application and the resources [PCB03, SNCBL04]. These
and other important mechanisms are thoroughly discussed in the following
sections.

2.2   Bird's Eye View of OurGrid

An overview of OurGrid's architecture is necessary for the presentation
of the subsequent sections. This architecture is presented in Figure 2.1 and
is composed of four main components: worker, peer, broker, and discovery
service.

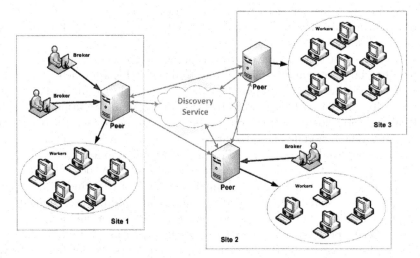

FIGURE 2.1: OurGrid's architecture.

An OurGrid instance is a network of *OurGrid Peers*, each of them managing the resources of a site. Every OurGrid instance operates a *Discovery
Service* that acts as the rendesvouz for the peers in that instance. There exists a public instance named the OurGrid Community[6], which anyone can
join, and which does not rely on strong identities, such as credentials issued
by certification authorities. Groups of sites who prefer not to participate in the
OurGrid Community can deploy alternative instances with controlled membership. ShareGrid[7] is an example of an OurGrid system operated in this
manner.

Each Peer manages a number of *OurGrid Workers*, the basic unit of resource provision in the system. A worker controls the resources of a desktop

---

[6]http://status.ourgrid.org/.
[7]http://dcs.di.unipmn.it/sharegrid/.

in the grid. The worker monitors the usage of the desktop, so as to comply with the sharing policies defined by each resource owner. When running grid applications, the worker instantiates a virtual machine, so as to isolate the physical machine from the application. If the sharing policy indicates that a resource that is available to the grid should no longer be, then any virtual machine that is running in the resource is immediately aborted.

Users rely on *OurGrid Brokers* to execute their applications on the grid. A user is typically authorized to request resources to one peer, and the broker is responsible for communicating with the peer to obtain enough resources matching the application requirements. A peer provides to the broker both the resources available in its sites and resources obtained from other peers in the grid. Upon receiving such resources, the broker uses them to schedule the tasks that comprise the application.

2.3 Architecting Cooperation

All services in the peer-to-peer grid architecture are provisioned by workers. As a consequence, the utility a grid provides to its users is determined by the number of workers made available by participants and by the capacity of these workers. The larger the number and the higher the capacity of the workers made available by each participant, the higher the utility the grid can provide to all users. However, providing resources has a cost. Making resources available potentially increases energy and maintenance costs, as well as security risks. In some cases, and perhaps often, these costs may be unimportant: for example, if the additional maintenance costs needed in a site are relatively small or if the benefit of contributing for research done by others in the grid is perceived as valuable. However, other users may perceive the costs of contributing to the grid as non-negligible. For users who perceive the costs as considerable, the peer-to-peer grid is a social dilemma: not contributing and still benefiting from the grid is the most profitable alternative, but if all users act like this, the social outcome is worse than that of prevalent contributions.

OurGrid explicitly tackles this social dilemma, leveraging resource allocation to encourage contributions to the grid. All participants in OurGrid implement together a decentralized resource allocation mechanism called the *Network of Favors* (NoF) to reward participants who have contributed the most to the system. This reward is provided through a combination of prioritization and direct reciprocity, resulting in a scalable and robust mechanism in the context of peer-to-peer grids.

In the remainder of this section, we detail and discuss the design and properties of the NoF. For that, we first present an overview of the NoF considering a peer-to-peer grid where a single service computation is provided. After that, we discuss the underlying accounting mechanism used by the NoF that allows peers to autonomously estimate the work performed by the other peers in the grid. The section concludes with a discussion of how the NoF can be extended to address the fact that in practical deployments more than one service can be traded in a peer-to-peer grid.

2.3.1   Network of Favors

The Network of Favors (NoF) is a decentralized resource allocation mechanism through which participants in the peer-to-peer grid collectively encourage service provision. The core of this mechanism is that peers use a prioritization policy implemented through direct reciprocity when allocating its resources. We now detail how this mechanism is implemented for the main service shared in grids: computation. In later subsections we revisit this principle, considering multiple services (such as data transfer and storage) are shared.

Direct reciprocity means that each participant P_i reciprocates the behavior of each other participant P_j in the system had toward it in the past, and this is done according to a local record of their pair-wise interactions maintained by P_i. Because P_i uses only local information it observed directly to decide whether to allocate a resource to P_j or not, the NoF is robust against misinformation: a malicious peer cannot affect the decision by spreading any type of information. On the other hand, the allocation P_i will perform is as good as the evaluation it can make of its past interactions with its peers. We return to the matter of such precision in the next subsection. For the following discussion, assume P_i can account for the service it provides and receives from its peers.

From the perspective of the resource consumer P_j, a direct reciprocity mechanism makes contributing to the grid the best strategy if interactions between P_j and P_i happen frequently enough, so that P_i's decision has an impact on the resources available to P_j [Now06]. In the particular setting of peer-to-peer systems, Feldman et al. [FLSC04] show that an incentive mechanism based on direct reciprocity is not effective for large populations in a file-sharing system due to high churn and to the fact that not all peers have content that interests all others.

However, differently from a file-sharing system, in a peer-to-peer grid, interaction between any two peers is usually frequent. This is because both service requesters and providers interact with multiple and often many partners whenever they act in the system. Requesters typically participate in the system because they have demand for a large amount of service at once. This service is probably more than any participant in the grid can provide alone, which makes possible and likely that the consumer will interact with multiple providers. Furthermore, as we discuss later, service consumers can increase the performance of their applications in a heterogeneous grid replicating tasks on different resources [PCB03, SNCBL04]. Such replication potentially increases the number of providers a consumer interacts with. At the same time, a resource provider typically owns multiple resources, which can be split among multiple consumers at a given time. Finally, although multiple services may be provided in the grid, the number of services will be orders of magnitude smaller than the number of files in a file-sharing application. The reduction in the number of different services available, again, augments the chance of two peers interacting.

When effective, direct reciprocation has the advantage of simplicity over its alternatives in the design space. Two proposed alternatives to this approach that have received considerable attention from the community are reputation (sometimes termed indirect reciprocity) [KSGM03, VCS03, FLSC04] and market-based mechanisms [BV01, LRA+05, SNP+05]. In the former, a peer allocates its resources to a requester according to the aggregate opinion of its peers about the requester; in the latter, participants use a currency and either bargaining or auctions to determine to which requester to allocate resources. Both of these alternatives demand the implementation of more sophisticated and less robust underlying mechanisms than direct reciprocity.

To implement direct reciprocity, the participants of the peer-to-peer grid must divide their resources among requesters. Using the NoF, each participant always provides its available resources and prioritizes peers that simultaneously request the same resource based on its local record of past interactions with the requesters. The local record of participant P_i of its interactions with P_j takes the form of a balance, calculated as the value of service received minus the value of service provided, or zero if this value is negative. The balance is added with the value of service provided or received every time P_i interacts with P_j. The use of non-negative quantification for past interactions aims at discouraging whitewash attacks. A whitewash attack is performed when the attacker leaves the system and returns as a newcomer, for instance, to get rid of a bad reputation (a negative balance in the case of the NoF). In the presence of cheap identities and both positive and negative records of interactions, a participant with a negative balance in the system can always benefit from whitewashing. With non-negative records, participants cannot increase their records in the eyes of other peers. This use of non-negative balances was inspired by the work of Yamagishi and Matsuda [YM02].

The prioritization implemented in the NoF makes it so that whenever there is resource contention in the system, requesters that have contributed more in the past receive more service. This is less strict than other policies that could starve non-contributing participants. However, OurGrid chooses to implement prioritization on the grounds that it eases bootstrapping the system, and the absence of contention in the system can be taken as a sign of an acceptable level of service provision. Bootstrapping the face of allocation policies that deny service under certain conditions may lead to situations where all peers deny service to each other; this cannot happen through prioritization. The lack of rewards for cooperating participants only happens when the system is able to provide service to all requesters, in which case it is possible to punish non-cooperators only by decreasing the systems throughput, an alternative whose benefit is debatable.

When peers interact frequently, direct reciprocity and prioritization efficiently promote service provisioning in a peer-to-peer grid. For a detailed anal-

ysis of the conditions that affect the effectiveness of the NoF and other properties of this mechanism, we refer the reader to another publication [ABCM07].

2.3.2   Accounting for Computation

The local records of pair-wise interactions maintained by the participants
in the NoF must be accurate if the system is to provide accurate prioritization.
Performing such accounting robustly is not a trivial problem because there
is uncertainty about resource consumption from the consumer's side and a
potential benefit for the provider to misreport accounting information. This
section presents a solution for this scenario based on exploring a typical usage
pattern of grid participants to enable autonomous accounting of computation.

When providing a resource, providers benefit from forging the accounting
of its service provision because this leads to an increased credit in the local
records the other peers maintain for it. In the NoF, this translates into future
prioritization in resource allocation. In such a scenario, it is risky for a service
consumer to trust accounting information from service providers.

An alternative to using information from the providers is to estimate the
cost on the consumer side. The issue with this approach is that users typically
do not know how many resources a typical task in their application consumes. Studies with users of supercomputers have exposed that users cannot
predict accurately their applications' runtime even in a homogeneous platform [LSHS04]. To address this issue, OurGrid leverages the fact that participants in a peer-to-peer grid typically use the grid to augment a local infrastructure that cannot fully accommodate users' demand for computation. This
implies that often each application submitted to the grid will run partly on
resources that are local to the peer to which the application user is associated
and partly on the resources of the other peers that comprise the grid. The
progress made by the part of the application that runs on the local resources
can then be used by a participant to estimate the relative computing power
that the other peers are making available to run the application. The relative
power can be continuously updated as users run more applications on the grid,
moving toward more accurate estimates.

Santos and colleagues evaluated the efficiency of this autonomous accounting mechanism by comparing it against an optimal mechanism that is perfectly
informed [SAC+07]. Their results show that even when there is high heterogeneity among the resources that comprise the grid, as well as among the
processing demands of the individual tasks of the BoT applications, the accuracy of the autonomous accounting mechanism is only slightly worse than
the perfect accounting, with a small bias toward peers with faster resources.
In fact, this side effect may introduce an extra incentive for peers to provide
their most powerful resources to the grid.

2.3.3   Multiple-Service Peer-to-Peer Grids

The NoF was originally designed to share computation as the sole service
in the grid. However, in practice, applications have requirements for multiple

services. For example, some applications need as much data transfer, and thus network and storage services, as computation. Other applications need higher-level services such as datasets or software libraries in addition to raw computation.

Considering that multiple services are provided in the system complicates matters because participants may perceive differently the values and costs associated with different services. When choosing which peer contribution to reciprocate, a participant should then be interested not only in the probability that each peer will reciprocate, but also in how profitable a long-term relationship with that peer can be. This profitability, in turn, is a result of the services that a peer is likely to provide and request in the future.

The NoF has been enhanced to address this matter [MBA+06]. This extended mechanism, dubbed ExtNoF, performs nearly as well as utility-maximizing policies that have access to truthful information from other peers. It encourages the provision of multiple services by making it so that peers cluster with partners that lead to profitable relationships, still using only their local information about past interactions.

If participants had access to the perceived values and costs of all their peers, the utility-maximizing strategy would be to combine a selection policy with a prioritization policy. The former determines which peers can submit requests for services, while the latter determines how to arbitrate between requesters. With the information about preferences of others, a participant is able to calculate whether it is possibly profitable to maintain a long-term relationship with each other peer and who are the peers that return the most valuable services more often. However, because the information about the valuation of services may lead to prioritization if published, it becomes in the interest of participants to lie about this information. With this consideration, the NoF was extended without relying on such data. Instead, using the ExtNoF mechanisms, the participants recur to their own cost evaluations for evaluating the value of service provided by their peers. This valuation is then used for prioritization similarly to the NoF, and peers use no selection policy.

Mowbray and colleagues have simulated the effects of the NoF for the allocation of resources in a grid where computation and storage services are provided and participants have different levels of service demand, different costs, and various service availability [MBA+06]. They found that in spite of not using the selection policy or the information about peer preferences, the NoF performs similarly to the ideal policies, encouraging service provision in the grid even when the cost of providing service in the grid with idle resources approximates the benefit of receiving service from it when in high demand.

2.4   Worker Perspective

The autonomous and decentralized nature of a peer-to-peer desktop grid implies that two important concerns must be addressed by the grid middleware. First, the owners of the desktops need to unilaterally define when the desktops are available to run the grid workload, and when they are not. Second, there is a need to protect the desktops from any harm that could be caused by remote computations submitted by unknown, and therefore not trustworthy, peers that execute in the local desktops. In OurGrid, every desktop runs a *Worker* component that is in charge of implementing mechanisms to handle these problems.

In the current version of OurGrid, a single policy is used to define when desktops are made available to the grid. It is based on the monitoring of the activity of interactive users. In other words, whenever there is a period of time long enough without any activity in both the keyboard and the mouse, then the desktop is deemed to be idle, and made available to the peer-to-peer grid. Other policies based on a predefined schedule, or on the amount of resources that are being consumed by the applications that are running in the desktop, could also be used, alternatively or in combination.

On the other hand, OurGrid's mechanism for protecting workers leverages virtualization technologies and the fact that the tasks of BoT jobs do not need to communicate with each other. Every time a Worker is donated to execute grid tasks from a particular *Broker*, a fresh virtual machine is spawned by the worker to execute these tasks. If the worker is allocated to a different Broker, the virtual machine is recreated from scratch before starting the execution of the tasks of the new Broker. The virtual machine isolates the foreign code into a sandbox, where it can neither access local data nor use the network. The OurGrid Worker supports different virtualization technologies. During the OurGrid installation process, the site administrator chooses which worker version is more appropriate for the site.

2.5　Managing the Resources of Each Site

In OurGrid's architecture, workers are grouped in sites. In each site, a software component dubbed an *OurGrid Peer* is responsible for managing workers. These components are the peers in OurGrid's peer-to-peer network; workers are seen as resources a peer can provide to other peers in the network. Resources are requested and allocated among peers, in turn, upon demand from users, expressed through their *Broker* instances.

Users who are authorized to issue requests for service to a peer are named local users for that peer; similarly, this peer is seen as the local peer of the users. The responsibilities of an OurGrid peer are therefore to request and provide service to other peers, maintaining the accounting necessary for the Network of Favors and, efficiently allocate currently available resources, both local and obtained from other peers, among local users.

So that the local users of a given peer are never worse off participating in the grid than not, a local user always has priority in the allocation of its local peer's resources. More specifically, if a worker is currently allocated to another peer and there is a request for this worker from a local user, the non-local allocation will be preempted immediately.

Resource allocation among peers, in turn, is dictated by the Network of Favors. To implement prioritization, each peer periodically reviews its allocations and updates it accordingly to the local balance of requesting peers. This balance is also updated over time and is maintained by the peer. For most services provided by other peers, such as data transfer, accounting is done by the peer directly; some other services use information provided by the broker after it uses the received resources notably computation, as described in Section 2.3.

In its consumer role, a peer has two main responsibilities: to divide resources available locally and those obtained from other peers among local users, and to ensure resources are used efficiently from an energy perspective. Resource division is presently implemented through a policy that maintains an equanimous division of resources among users while minimizing preemptions when the allocation changes.

With regard to energy efficiency, it is important that resources that are not currently used to provide any service are put in low-consumption mode. Several popular operating systems, such as Microsoft Windows and Linux distributions, use a standard called Advanced Conguration and Power Interface (ACPI) to implement this kind of power management. In OurGrid, the peer uses ACPI technology to switch workers into sleeping mode when there is no demand for them, and brings them back when there is work to perform by sending a Wake-on-LAN message on the network. Two ACPI sleeping states meet OurGrid's requirements: standby (S3) and hibernate (S4).

Naturally, the procedure of sleeping and waking up worker machines impacts the response time of applications ran on the grid. Standby, or suspend-to-RAM, is a low-latency sleeping state, as the RAM is kept powered on. On the other hand, hibernate, or suspend-to-disk, consumes less power but requires a larger latency in order to return the machine to the operational state, as information needs to be retrieved from the disk. This wake-up cost introduces delays in the makespan of the jobs submitted to the grid.

Interestingly, simulation results [PB10] show that using standby and hibernate results in approximately the same savings in energy consumption in a peer-to-peer grid. Moreover, these savings are achieved without significant impact on application's makespan. As a result, the OurGrid peer presently puts worker machines in standby mode after a period of availability with no requests.

Given that machines are put to sleep, there is a final decision that must be made by the peer component of OurGrid: which machines to wake up first when there is a request that can be satisfied by multiple workers. Again, the OurGrid peer implements a policy derived from a simulation study. The results that guided implementation show that a heuristic based on waking up the most energy-efficient workers first allows for energy-efficient operation with minimal impact on the application's makespan [PB10].

2.6 Scheduling Applications on the Grid

At first glance, scheduling independent tasks on a desktop grid does not seem to be particularly challenging. As with most parallel scheduling problems, the goal is to keep all resources busy throughout the application execution, ensuring that the application as a whole finishes as fast as possible. Because tasks are independent, scheduling them reduces to bin-packing the tasks into the available desktops. Although this is an NP-Complete problem for the case of heterogeneous tasks and/or resources, there are good heuristics for the problem.

However, to run these heuristics, one must obtain accurate information about resource capacity and load, what can be a daunting endeavor in a large desktop grid. To make matters worse, often different tasks demand different amounts of resource capacities. In this case, such information must also be available for traditional, knowledge-based bin-packing schedulers. In fact, bin-packing schedulers need to know a priori, and accurately, the execution time of each task onto each resource that forms the desktop grid. Furthermore, bin-packing schedulers may display complexity that (although polynomial) is unacceptably high when there are many tasks and/or resources, a common case in large desktop grids.

2.6.1 Scheduling with No Information

A way to deal with the lack or the unreliability of the information is to use replication. In this approach, no information is used to guide task-to-resource assignments. Then, task replication can be used to mitigate the impact of unfortunate assignments. Task replication consists of dispatching multiple replicas of a single task. Naturally, when a task is replicated, the first replica that finishes is considered the valid execution of the task, and the other replicas are cancelled. Any changes in global state made by such other replicas is undone. In practice, this is often implemented by allowing only the finished replica to commit changes to global state. We term schedulers that employ task replication as *replication schedulers*. Note also that because replication copes with task-to-resource unfortunate assignments, scheduling decisions can be much simpler than in bin-packing schedulers (at random, for example), thus greatly reducing the complexity of the replication scheduler.

The replication approach minimizes the effects of the dynamic resource load, resource heterogeneity, and task heterogeneity, and does so without relying on information on resources or tasks. One way to think about replication is that it allows the trading of some additional resource consumption for the need of performance information about resources and tasks. On the other hand, replication schedulers demand more resources than bin-packing schedulers [PCB03].

2.6.1.1   Worqueue with Replication

The *OurGrid Broker* uses a very simple scheduling algorithm, named *Workqueue with Replication*, or WQR for short. As its name suggests, the scheduler is based on the very basic *Workqueue* scheduler, adapted to use replication.

Workqueue is a knowledge-free scheduler in the sense that it does not need any kind of information for task scheduling. Tasks are chosen in an arbitrary order in the "bag of tasks" and sent to the grid workers as soon as they become available. After the completion of a task, the worker sends back the results and the scheduler assigns a new task to the worker.

The idea behind this approach is that more tasks will be assigned to the fast workers while the slow workers will process a small load. The great advantage here is that Workqueue does not depend on performance information. Unfortunately, Workqueue does not attain performance comparable to schedulers based on full knowledge about the environment [PCB03]. The problem with Workqueue arises when a large task is allocated to a slow worker toward the end of the schedule. When this occurs, the completion of the whole application will be delayed until the execution of this task is finished.

The WQR algorithm uses task replication to cope with the heterogeneity of workers and tasks, and also with the dynamic variation of resource availability due to contention created by others users. The beginning of the algorithm execution is the same as Workqueue and continues the same until the bag of tasks becomes empty. At this time, in Workqueue, workers that finished their tasks would become idle during the remainder of the application execution. Using the replication approach, these workers are assigned to execute replicas of tasks that are still running. Tasks are replicated until a predefined maximum number of replicas is achieved. When a task finishes, its replicas are cancelled. With this approach, the performance of the application is increased in situations that tasks are delaying the completion execution because they were assigned to slow workers. When a task is replicated, there is a greater chance that a replica is assigned to a faster worker.

Note that the replication assumes that the tasks do not cause collateral effects (e.g., database accesses by independent tasks can generate inconsistencies). This assumption is reasonable when talking about desktop grid environments because it is not common to have this kind of application in these environments.

Paranhos and colleagues have conducted a number of simulation experiments to assess the performance of WQR when compared to bin-packing schedulers that were fed with complete and accurate information [PCB03]. Their results show that WQR is able to attain the same level of performance delivered by knowledge-based schedulers. As expected, this comes at the expense of using more computing cycles. The cycles' waste is considerably smaller for applications with small task sizes (in the range of 15 minutes to 1.5 hours), achieving less than 5% of waste. This percentage increases when the mean tasks sizes grows, but still the percentage of wasted cycles does not exceeds 40% for applications with fairly large tasks (around 7 hours). On the other hand, when the average of the task sizes is very large (more than 30 hours), the percentage of wasted CPU cycles can exceed 100%.

In their experiments, Paranhos and colleagues considered applications with a fixed total workload. Therefore, the larger the average task sizes, the smaller the number of tasks. The increase in cycles wasting when the average task sizes grow occurs due to the quick start of replication. When the application has small tasks, the replication only starts after executing many tasks of the application, and the fraction of time that the remaining tasks represent is small. When the application has large tasks, the replication starts much sooner, wasting more cycles. Nevertheless, with the settings evaluated, the percentage of wasting cycles can be maintained smaller than 50% if the replication level is set to be only $2x$. This, in turn, has just a minor impact on the performance of the applications.

2.6.1.2 System-Wide Impact of Replication

The fact that replication schedulers are very used in practice suggests that the advantages of task replication outweigh its disadvantages. Paranhos and colleagues performed a comprehensive analysis of replication scheduling and their system-wide impact [CBP+07]. They assess the effect of having multiple replication schedulers in the same system. This is a key issue because the extra consumption of resources promoted by replication schedulers raises severe worries about the system-wide performance of a distributed system with multiple, competing replication schedulers [KCC04]. Their study establishes the efficacy (the performance delivered to the application), efficiency (the amount of resources wasted), and emergent behavior (the system-wide behavior of a system with multiple schedulers) of replication schedulers.

Their results show that, on average, replication schedulers deliver good performance (on par with bin-packing schedulers) at the cost of a relatively modest extra consumption of resources. This is especially true when there are many more tasks than resources. For the opposite case (more resources than tasks), replication schedulers' performance degrades and the extra consumption of resources increases. When this is the case, one can improve efficiency (i.e., reduce the waste of resources) by limiting replication, a strategy that has little impact on the replication schedulers' efficacy (i.e., increase the applications' performance).

The case of more processors than tasks is also troublesome for the emergent behavior in a peer-to-peer desktop grid. However, a simple referee strategy (which can be locally and autonomously implemented by each grid resource) greatly improves matters [CBP+07]. The local referee strategy consists of having resources restrict themselves to serving one scheduler at a time. This strategy improves the emergent behavior of a system with competing replication schedulers in all scenarios investigated by Paranhos and colleagues, and is critical when there are more resources than tasks.

2.6.2   Other OurGrid Schedulers

WQR is currently the only scheduler that is distributed with the Our-Grid middleware. Nevertheless, a number of different schedulers have been proposed, prototyped, and evaluated. In the following subsections the most important of them are briefly discussed.

2.6.2.1   Data-Aware Schedulers

There are many scientific and enterprise applications that deal with a huge amount of data [AGM⁺90, SNTF⁺01]. In order to process large datasets, these applications typically need a high-performance computing infrastructure. Fortunately, because the data splitting procedure is easy and each data element can often be processed independently, these applications can typically be structured as BoT applications. However, because resources in a peer-to-peer desktop grid are typically accessed through wide area network links, the bandwidth limitation is an issue that must be considered when running these applications in such environments.

This is particularly relevant for those data-intensive BoT applications that present a data reutilization pattern. For these applications, the data reuse pattern can be exploited to achieve better performance. Data reutilization can be either among tasks of the same particular job or among a succession of job executions of the same or related applications. For instance, in the visualization process of quantum optics simulations results [SNTF⁺01] it is common to perform a sequence of executions of the same parallel visualization application, simply sweeping some arguments (e.g., zoom, view angle) and preserving a huge portion of the data input from the previous executions.

Targeting these applications, Santos-Neto and colleagues have proposed *Storage Affinity*, a heuristic for scheduling data-intensive BoT applications on peer-to-peer desktop grids [SNCBL04]. Storage Affinity takes into account the fact that input data is frequently reused. It tracks the location of data to produce schedules that avoid, as much as possible, large data transfers. The data location information, which can be gathered locally by the scheduler, is the only information that is required to perform the schedules. Therefore, similar to WQR, it also uses replication to reduce the effects of inefficient task-processor assignments.

Storage Affinity has been compared, through simulations, against XSufferage [CLZB00], a data-aware knowledge-based heuristic that uses information about the CPU and network loads, as well as the execution time of each task on each machine. Their results show that taking data transfer into account is mandatory to achieve efficient scheduling of data-intensive BoT applications. Further, they show that grid and application heterogeneity have little impact on the performance of the studied schedulers. On the other hand, the granularity of the tasks that comprise the application, given by their average tasks size, has an important impact on the performance of the two data-aware schedulers analyzed. Storage Affinity is outperformed by XSufferage only when application granularity is large. However, the granularity of data-intensive BoT applications can be easily reduced to levels that make Storage Affinity outperform XSufferage. In fact, independently of the heuristic used, the smaller the application granularity, the better the performance of the scheduler. In the favorable scenarios, Storage Affinity achieves a performance that is, on average, 42% better than XSufferage. The drawback of Storage Affinity is the waste of grid resources due to its replication strategy.

2.6.2.2 Scheduling with Partial Information

Replication schedulers can achieve performance that is comparable with that provided by bin-packing schedulers, yet without using any information about the environment. This comes at the cost of wasting computing resources. Nelson Jr and colleagues have further investigated this trade-off [NJAB08]. Their motivation comes from the fact that although it is hard to obtain accurate information in grid environments, it is not impossible to have it. In practice, some information can be gathered using services that collect resources and network information [MCC04, YSB+06] and publish them in a grid information service (GIS) [BCB+09, CFFK01].

The research question they investigated was "can a scheduler reduce the execution cost of BoT applications on a peer-to-peer grid, maintaining the same efficiency, by using whatever information that it is available?" To perform the study, they designed and evaluated through simulations a number of scheduling heuristics that are able to deal with the availability of partial information about both the grid and the application. The heuristics proposed try to combine the characteristics of both bin-packing and replication schedulers. The idea is to use whatever information is available to reduce the amount of replication required without impacting the application performance.

Their results show that the heuristics that try to adapt bin-packing schedulers to work with partial information have a performance worse than that of the replication heuristics. On the other hand, the heuristics that adapted the replication heuristics were very efficient in a large number of the scenarios evaluated, reaching a performance comparable to the performance of the WQR replication heuristic, but reducing the resource waste considerably. They concluded that the strategy based on the usage of any available information to control better the task replication process can greatly decrease the resource waste level, without impacting the application's performance.

2.7 Extending an OurGrid System with Public Cloud Resources

Some BoT applications are urgent, in the sense that the utility gain associated with their execution is greater if they finish earlier. Such kinds of applications can be associated with utility functions, which are typically monotonically decreasing functions of application runtime. In some cases, there is a deadline after which no utility gains are associated with the completion of the application, or even penalties are applied. A common example of an urgent BoT application is periodic weather forecast [ACW+05], which needs to be completed before the time for which the prediction is made. In general, e-science applications also fall into this class of applications, as the faster the experiments are processed, the faster is the research cycle completed.

If for one side, peer-to-peer desktop grids can be very cost-effective platforms for the execution of BoT applications, on the other side, their best-effort nature does not allow for any guarantees to be placed on the quality of service that can be attained. This might be a problem for BoT applications that need to meet a deadline, or, in general, whose utility decreases with the duration of its execution.

To address the demands of this type of application, Maciel and colleagues have proposed the use of a hybrid infrastructure in which computing power can be obtained from in-house dedicated resources, from cloud computing providers, and also from resources received from a best-effort peer-to-peer grid that trades the spare capacity of the in-house dedicated resources [JdFM+08], in such a way that more guarantees are provided on the quality of service delivered by the computing infrastructure.

Following the model proposed by Maciel and colleagues, Candeia and colleagues have proposed smart scheduling heuristics that can maximize the overall "profit" that is achieved by the execution of urgent BoT applications on OurGrid systems [CALB10]. In this setting, profit is defined as the difference between the perceived utility of completing the execution of the application in a given time frame, minus the associated cost to execute it. Intuitively, the more resources that are acquired from the cloud computing market, the sooner the application will finish and the higher will be the associated utility, thus positively impacting the profit. On the other hand, the higher will be the cost, thus negatively impacting the profit. So, the use of cloud computing resources needs to be judiciously planned.

In this business-driven approach, when an application is submitted, the set of tasks associated to the application is added to a processing queue in the scheduler. Then the scheduler starts requesting for workers from the peer-to-peer grid and the cloud computing providers. Each task of the application can be executed by a desktop in the grid or by a virtual machine instantiated in the cloud. Assuming that the cost of using grid resources is negligible, whenever possible, tasks are executed on the peer-to-peer grid. Any extra capacity required to improve the profitability of the execution of the application is purchased from the cloud providers. This strategy has been dubbed *cloud bursting*.

If the scheduler could anticipate the number of tasks that the peer-to-peer grid would be able to process in a given future interval, then it would be straightforward to calculate the amount of resources that would need to be purchased from the cloud, so that the profit of the execution is maximized.

Because this information is not available, the scheduler uses heuristics that try to approximate the optimum outcome.

The cloudburst scheduler works in turns of fixed durations, using a control loop that monitors the throughput of the peer-to-peer grid in each turn, that is the number of tasks completed in a turn. When an application is submitted for execution, the scheduler normally has no information about the current throughput of the peer-to-peer grid. Thus, for the first turn, the scheduler tries to obtain from the grid as many workers as the number of tasks it has to run. Then, the scheduler starts to compute the grid throughput. At the end of each turn, the scheduler computes the throughput of the grid for that turn and estimates the throughputs of future turns of the grid according to a certain heuristic. These can be approximately estimated based on the recent past throughput measured. The scheduler also knows the mean number of tasks a cloud instance is able to compute during a turn, the application demand, and its associated utility function. With such information, the scheduler calculates the profit it would achieve if it completed the application at the end of each future turn, considering the maximal number of instances that it can get from the cloud provider, and the estimated grid throughput for future turns. Then it decides to take the decision that leads the execution to complete at the end of the turn that maximizes the profit.

Candeia and colleagues have investigated several heuristics to predict the future throughput of a peer-to-peer desktop grid. Their initial results show that the approach is promising, but it is not straightforward to design heuristics that can achieve high efficiencies.

2.8   Open Challenges and Future Directions

Development of the OurGrid middleware and operation of the OurGrid Community are ongoing efforts since 2002 and 2004, respectively, and have so far moved from a proof-of-concept of the viability of a peer-to-peer approach for grid computing into a mature technology that has opened up a series of possibilities for grid users and researchers.

An overarching lesson in this path is certainly on the advantages of a simplicity-driven approach for the complex problem of providing a large-scale platform for distributed computing. Our experiences with designing incentive mechanisms, scheduling heuristics, and energy-efficient resource management all point to the efficacy of fundamentally simple and therefore robust and scalable solutions.

At the same time, efforts so far have allowed us to identify a number of challenges ahead. One of these challenges is extending the OurGrid middleware to better suit contributors with diverse motivations. The incentives provided by the Network of Favors often have a positive reaction among grid participants that both provide and use grid resources. This mechanism, however, does not resonate with a second class of participants that turns out to be more common than originally thought: altruistic contributors.

Over time, a number of institutions have joined OurGrid with the sole purpose of contributing resources, and a brief experiment with combining voluntary computing with the OurGrid Community suggests this approach is promising. Resource providers in these two use cases are not concerned with the return they can obtain through the Network of Favors, but rather with how their resources are used. For example, a domestic user contributing the idle time of his laptop may be more motivated to keep doing so depending on what applications run on his worker. Also, an institution that joined the grid to help a specific project may find it important that its resources are allocated to unknown peers only when this project does not need them.

We are currently researching how to incorporate mechanisms in the Our-Grid middleware that leverage a wider range of motivations for resource contribution, in addition to the Network of Failures. Interdisciplinary research on cooperation points to a number of design levers that can be considered in this context [Ben09].

A second challenge we are presently dealing with is the efficient use of virtual machine images. Virtualization enables any given worker to provide different images to different users, better addressing the individual needs of these users in terms of system configuration, for example. At the same time, however, this creates a problem for managing the images available in a site, to certify an image as the one requested by a user, and to efficiently distribute virtual machine images to workers.

Another continuing effort is the advancement of the OurGrid middleware to support more classes of applications. BoT applications have proven to be a popular type of parallel application, but researchers often need to execute applications that demand communication with external servers or synchronization among tasks. Because of this, we have started to address the expansion of the gamut of applications that are suitable for peer-to-peer grid computing.

Finally, in addition to the technological enterprises, we believe another valuable front for the development of peer-to-peer grid computing is the popularization of this approach among non-scientists. We contend that a number of "layman" applications such as video rendering and citizen activism could benefit from the computing power available in the grid, and we have been involved in increasing awareness of peer-to-peer grid computing among users of such applications, so as to further materialize the democratization made possible by this approach.

Bibliography

[ABCM07] Nazareno Andrade, Francisco Brasileiro, Walfredo Cirne, and Miranda Mowbray. Automatic grid assembly by promoting collaboration in peer-to-peer grids. *Journal of Parallel and Distributed Computing*, 67(8):957–966, 2007.

[ACK+02]   D. P. Anderson, J. Cobb, E. Korpela, M. Lebofsky, and D. Werthimer. SETI@home: an experiment in public-resource computing. *Communications of the ACM*, 45(11):56–61, Nov. 2002.

[ACW+05]   Eliane Arajo, Walfredo Cirne, Gustavo Wagner, Nigini Oliveira, Enio P. Souza, Carlos O. Galvo, and Eduardo Svio Martins. The seghidro experience: Using the grid to empower a hydrometerological scientific network. In *Proceedings of 1st IEEE Conference on e-Science and Grid Computing*, pages 64–71, 2005.

[AGM+90]   S. F. Altschul, W. Gish, W. Miller, E. W. Myers, and D. J. Lipman. Basic local alignment search tool. *Journal of Molecular Biology*, 1(215):403–410, 1990.

[And04]   David P. Anderson. BOINC: A system for public-resource computing and storage. In *Proceedings of the 5th IEEE/ACM International Workshop on Grid Computing*, GRID'04, pages 4–10, Washington, DC, USA, 2004. IEEE Computer Society.

[BCB+09]   Sujoy Basu, Lauro Costa, Francisco Brasileiro, Sujata Banerjee, Puneet Sharma, and Sung-Ju Lee. Nodewiz: Fault-tolerant grid information service. *Peer-to-Peer Networking and Applications*, 2009.

[Ben09]   Y. Benkler. *Government and Markets: Toward a New Theory of Regulation*. Cambridge University Press, 2009. Forthcoming.

[BV01]   Rajkumar Buyya and Sudharshan Vazhkudai. Compute power market: Towards a market-oriented grid. In *CCGRID*, pages 574–581, 2001.

[CAG+06]   E. Cavalcanti, L. Assis, M. Gaudencio, W. Cirne, F. Brasileiro, and R. Novaes. Sandboxing for a free-to-join grid with support for secure site-wide storage area. In *Proceedings of the First International Workshop on Virtualization Technology in Distributed Computing*, Tampa, FL, USA, November 2006.

[CALB10]   David Candeia, Ricardo Araujo, Raquel Vigolvino Lopes, and Francisco Vilar Brasileiro. Investigating business-driven cloudburst schedulers for e-science bag-of-tasks applications. In *CloudCom*, pages 343–350, 2010.

[CBA+06]   Walfredo Cirne, Francisco Brasileiro, Nazareno Andrade, Lauro Costa, Alisson Andrade, Reynaldo Novaes, and Miranda Mowbray. Labs of the world, unite!!! *Journal of Grid Computing*, 4(3):225–246, 2006.

[CBP+07] Walfredo Cirne, Francisco Brasileiro, Daniel Paranhos, Luís Fabrício W. Góes, and William Voorsluys. On the efficacy, efficiency and emergent behavior of task replication in large distributed systems. *Parallel Computing*, 33(3):213–234, 2007.

[CFFK01] K. Czajkowski, S. Fitzgerald, I. Foster, and C. Kesselman. Grid Information Services for Distributed Resource Sharing. In *Proceedings of the 10th IEEE Symposium on High-Performance Distributing Computing*, August 2001.

[CLZB00] H. Casanova, A. Legrand, D. Zagorodnov, and F. Berman. Heuristics for Scheduling Parameter Sweep Applications in Grid environments. In *Proceedings 9th Heterogeneous Computing Workshop*, pages 349–363, Cancun, Mexico, May 2000. IEEE Computer Socity Press. Describes Sufferage, max min and min min scheduling algorithms.

[FLSC04] Michal Feldman, Kevin Lai, Ion Stoica, and John Chuang. Robust incentive techniques for peer-to-peer networks. In *ACM Conference on Electronic Commerce*, pages 102–111, 2004.

[JdFM+08] Paulo Ditarso Maciel Jr., Flavio de Figueiredo, D. Maia, Francisco Vilar Brasileiro, and Alvaro Coelho. On the planning of a hybrid it infrastructure. In *NOMS*, pages 496–503, 2008.

[KCC04] Derrick Kondo, Andrew A. Chien, and Henri Casanova. Resource management for rapid application turnaround on enterprise desktop grids. In *Proceedings of the 2004 ACM/IEEE conference on Supercomputing*, SC'04, pages 17–, Washington, DC, USA, 2004. IEEE Computer Society.

[KSGM03] Sepandar D. Kamvar, Mario T. Schlosser, and Hector Garcia-Molina. The eigentrust algorithm for reputation management in p2p networks. In *WWW*, pages 640–651, 2003.

[LLM88] M. Litzkow, M. Livny, and M. Mutka. Condor - a hunter of idle workstations. In *Proceedings of the 8th International Conference of Distributed Computing Systems*, pages 104–111, San Jose, CA, USA, Jun. 1988. IEEE Computer Society.

[LRA+05] Kevin Lai, Lars Rasmusson, Eytan Adar, Li Zhang, and Bernardo A. Huberman. Tycoon: An implementation of a distributed, market-based resource allocation system. *Multiagent and Grid Systems*, 1(3):169–182, 2005.

[LSHS04] C. B. Lee, Y. Schwartzman, J. Hardy, and A. Snavely. Are user runtime estimates inherently inaccurate? In *10th Workshop on Job Scheduling Strategies for Parallel Processing*, June 2004.

[MBA+06] Miranda Mowbray, Francisco Vilar Brasileiro, Nazareno Andrade, Jaindson Santana, and Walfredo Cirne. A reciprocation-based economy for multiple services in peer-to-peer grids. In *Peer-to-Peer Computing*, pages 193–202, 2006.

[MCC04] Matthew Massie, Brent Chun, and David Culler. The Ganglia distributed monitoring system: Design, implementation, and experience. *Journal of Parallel Computing*, 30(7), July 2004.

[NJAB08] Nelson Nóbrega-Júnior, Leonardo Assis, and Francisco Brasileiro. Scheduling cpu-intensive grid applications using partial information. In *ICPP '08: Proceedings of the 37th International Conference on Parallel Processing*, pages 262–269, Washington, DC, USA, 2008. IEEE Computer Society.

[Now06] M Nowak. Five rules for the evolution of cooperation. *Science*, 314(5805):1560–1563, 2006.

[PB10] Lesandro Ponciano and Francisco Brasileiro. On the impact of energy-saving strategies in opportunistic grids. In *Energy Efficient Grids, Clouds and Clusters Workshop, Proceedings of the 11th ACM-IEEE International Conference on Grid Computing (Grid 2010),*, pages 282 – 289, Brussels, Belgium, 2010. ACM-IEEE.

[PCB03] D. Paranhos, W. Cirne, and F. V. Brasileiro. Trading Cycles for Information: Using Replication to Schedule Bag-of-Tasks Applications on Computational Grids. In *Euro-Par 2003: International Conference on Parallel and Distributed Computing*, volume 2790 of *Lecture Notes in Computer Science*, pages 169–180. Springer, January 2003.

[SAC+07] Robson Santos, Alisson Andrade, Walfredo Cirne, Francisco Brasileiro, and Nazareno Andrade. Relative autonomous accounting for peer-to-peer grids: Research articles. *Concurrent Computing : Practice and Experience*, 19:1937–1954, September 2007.

[SNCBL04] E. L. Santos-Neto, W. Cirne, F. V. Brasileiro, and A. Lima. Exploiting Replication and Data Reuse to Efficiently Schedule Data-intensive Applications on Grids. In *Proceedings of the 10th Workshop on Job Scheduling Strategies for Parallel Processing*, pages 123–135, New York, USA, June 2004.

[SNP+05] Jeffrey Shneidman, Chaki Ng, David C. Parkes, Alvin AuYoung, Alex C. Snoeren, Amin Vahdat, and Brent N. Chun. Why markets could (but don't currently) solve resource allocation problems in systems. In *HotOS*, 2005.

[SNTF+01] E. L. Santos-Neto, L. E. F. Tenório, E. J. S. Fonseca, S. B. Cavalcanti, and J. M. Hickmann. Parallel visualization of the optical pulse through a doped optical fiber. In *Proceedings of the Annual Meeting of the Division of Computational Physics*, Boston, MA, USA, 2001.

[VCS03]   V. Vishnumurthy, S. Chandrakumar, and E.G. Sirer. Karma: A secure economic framework for Peer-to-Peer resource sharing. In *Proceedings of P2PECON*, 2003.

[YM02]   T. Yamagishi and M. Matsuda. Improving the lemons market with a reputation system: An experimental study of internet auctioning. Technical report, University of Hokkaido, May 2002. http://joi.ito.com/archives/papers/Yamaghishi_ASQ1.pdf.

[YSB+06]   P. Yalagandula, P. Sharma, S. Banerjee, S. Basu, and S.-J. Lee. S^3: A scalable sensing service for monitoring large networked systems. In *Proceedings of SIGCOMM'06 Workshops*, September 2006.

Chapter 3

The XtremWebCH Volunteer Computing Platform

Nabil Abdennadher

University of Applied Sciences, Western Switzerland (HES-SO), Geneva, Switzerland

Marko Niinimaki

University of Applied Sciences, Western Switzerland (HES-SO), Geneva, Switzerland

Mohamed BenBelgacem

University of Geneva, Geneva, Switzerland

3.1 Introduction

The scene of distributed computing has been changing considerably since the early 2000s. As an example, Buyya et al. [BYV08] state in an influential paper that new computing paradigms include cluster computing, Grid computing, peer-to-peer (P2P) computing, service computing, market-oriented computing, and cloud computing. They estimate the value of the combination of these, a "utility/pervasive computing industry", to be one trillion $.

Here, we consider cluster computing as methods of distributing computation within one cluster. Well-known cluster computing systems include Condor [LLM88] and Portable Batch System [Hen95]. Grid systems, on the other hand, combine the resources of many clusters [TTL02]. Service computing, market-oriented computing, and cloud computing provide methods for on-demand provisioning of computing resources [BYV08].

Peer-to-peer can be considered as per Milojicic et al. in [M$^+$02] as systems and applications that employ distributed resources to perform a function in a decentralized manner. From that perspective, most distributed computing systems that are not based only on client server communication are P2P systems. Typical characteristics of most P2P systems include decentralization of data, scalability (due to the fact that peers can connect directly to each other instead of using a central server), anonymity of peers, and ad-hoc connectivity [M$^+$02].

Anderson [AF06] sees P2P computing as synonymous with volunteer computing or global computing. The ultimate problem of all new IT is to find users, and there volunteer computing has shown great promise. The pioneer volunteer computing project BOINC (Berkeley Open Infrastructure for Network Computing) had more than 330, 000 participating computers delivering at 535, 000 GFLOPS in 2006 [AF06]. The BOINC software has been used as the backbone for several Desktop Grid (DG) systems as well; these include the European FP7 project "Enabling Desktop Grid for e-Sciences" (EDGeS) [U$^+$09] and the SZTAKI [B$^+$07] platforms. Other notable volunteer computing software includes Bayanihan [Sar98] and XtremWeb [F$^+$01].

In what follows, we introduce a derivative of XtremWeb called XtremWebCH (XWCH). This volunteer/cluster computing software features a three-part architecture inspired by P2P systems. Like many modern distributed computing systems, XWCH has a rich API, by which many applications have been parallelized. XWCH has, likewise, been used as a back-end of the JOpera workflow management system [PA04]. An XWCH related project has also studied the feasibility of an economic schema of using and providing resources in volunteer computing [AAFM09].

The next section of this chapter describes the XWCH architecture and the programming model supported by it. Section 3.3 deals with a high-level API that proposes an intuitive programming model for XWCH. Section 3.4 details three applications ported and deployed on the XWCH platform. Section 3.5 presents a bridge allowing the migration of jobs from the Advanced Resource Connector (ARC) middleware to XWCH. In the final section we provide a short conclusion.

3.2 XWCH Architecture and Functionality

3.2.1 Architecture

An XWCH system is composed of four modules (Figure 3.1): coordinator, client, worker, and warehouse:

1. The *coordinator* is the central element in XWCH. It registers new workers and manages their availability. Once a client submits a job, the coordinator pre-assigns it to a worker and waits for a worker request's signal from the given worker. If a worker on which a job is running or to which a job is allocated dies, the coordinator re-assigns the job to another available worker. In order to manage jobs and workers, the coordinator keeps information related to submitted jobs and connected workers. The coordinator also provides a Web interface that can be used for administration and monitoring.

2. A *client* program sends "computation requests" to the coordinator using a specific API (Figure 3.1). A computation request is a job submission. Jobs can communicate with each other by exchanging files. A client program can, depending on his rights, create a computation module, add codes binaries (executables) to these modules, or just use the available modules to submit jobs. A module is a set of executables having the same functionality but compiled for different platforms (Operating Systems, CPUs). An XWCH job is associated to one module. Modules and jobs are explained in detail in Section 3.2.2.

3. *Workers* are computing nodes provided either by individuals or institutions. Workers are smallish Java daemons that periodically report themselves to the coordinator. The first step for a worker is to contact the coordinator for registration. During this step, the worker sends a set of information to the coordinator regarding its performance and characteristics. After that the worker is available for computation; it accepts jobs, retrieves input, computes jobs, and stores the results of the computation. Workers are assumed to be volatile, and they may not be able to communicate with each other, due to firewalls and NAT sub-networks. If two communicating jobs are running on two workers that can reach each other, the communication can take place directly. Otherwise, warehouses are used. Warehouses are "file servers" in which workers can load output data and/or upload input data.

4. *Warehouses* are used for storing data and binary executables. Once a worker finishes its calculation, it will send its result to one or multiple warehouses. A warehouse node acts as a repository or a file server. Workers contributing to the execution of the same application should reach at least one common warehouse.

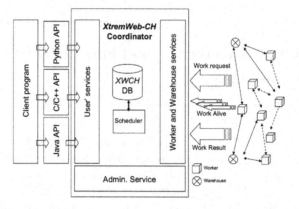

FIGURE 3.1: The XWCH architecture.

3.2.1.1 Communication Protocol

Communications between the coordinator and the workers are always initiated by the workers. Indeed, workers are very often behind NAT or firewalls and thus do not accept incoming connections. There are four types of signals that are sent by workers to the coordinator (Figure 3.1). These signals are:

1. WorkRegister: Sent to the coordinator when a new worker connects to the coordinator for the first time

2. WorkAlive: Sent periodically to the coordinator to inform that the worker is still alive (connected)

3. WorkRequest: Sent to the coordinator to ask for a job

4. WorkResult: Sent to the coordinator to provide results of a job

The protocol is the following:

- After it registers with its coordinator (Register Request), a worker receives the list of available warehouses.

- After a worker sends a Work Request for a new job, it receives as a reply, the locations of its binary executable and input data. If it can, it starts the execution of the job.

- When a worker finishes the execution of a job, it uploads its result in one or several warehouses. Thus, the result is stored both in the worker node and in the warehouse. The worker sends a WorkResult message to the coordinator with the locations (the IP of the worker itself and the IPs of the warehouses) of the job's results.

Job submission is supported by a flexible API, available for Java and C/C++ programs. Jobs are submitted to the coordinator by a "client node" which executes a client program that calls a users service (Figure 1). The number of jobs and their interconnection are not necessarily known in advance. Two other types of services are supported by the coordinator: Admin and worker & warehouses services. The three services are implemented using WSDL [Cer02].

All the components of XWCH are programmed mainly using Java, and their process memory sizes in a typical 32-bit Linux computer are shown below:

- Coordinator: 190 MB including the Glassfish Java container

- Worker: 40 MB

- Warehouse: 80 MB

The main characteristics of XWCH, compared to other volunteer computing middlewares such as XtremWeb or BOINC, are: dynamic job generation, data replication and control/monitoring of jobs execution.

3.2.1.2 Dynamic Job Generation

Applications are built using the XWCH API (see Section 3.2.2 for more details), either by a "native" programming language like Java, or by using a command-line-based tool that creates an application package using an application description file, and then submitting the application through the API. The first option naturally gives more control to the programmer. It allows developers to create jobs on-the-fly according to their needs, even if the number of jobs and their interconnections are not known in advance.

3.2.1.3 Data Replication

In many cases, the output file generated by a job J_0 is the input data of several successive jobs J_i ($i = 1$ to n). In order to avoid overloading one warehouse with parallel requests, the XWCH API allows developers to replicate the J_0 output file on different warehouses. Workers who execute jobs J_i ($i = 1$ to n) will receive the list of warehouses where the input data is available. They will then randomly select one of the warehouses in order to retrieve the requested file.

3.2.1.4 Controlling and Monitoring of Jobs

Due to security and performance concerns, users sometimes prefer managing "manually" the allocation of jobs to workers. Indeed, some applications require the allocation of a set of jobs to the same worker, or the allocation of a particular job to a specific worker. XWCH provides this kind of user control of allocation of jobs to workers even if this is against the spirit of grid and volunteer computing.

3.2.2 XWCH Programming Model

From the user's point of view, there are three levels of abstraction in XWCH:

1. Application: Set of jobs sent by the same client program they could be submitted to precedence constraints. An application is identified by an application identifier.

2. Job: Execution of a binary on a given worker. A job can have one or several input files and can generate one or several output files. The output files are compressed into one or several files according to the needs of the user. It is worth reminding here that the compressed output file of one job can be an input file of another job.

3. Module: Set of binary codes having, in general, the same source code. Each binary code targets a specific platform (Operating System, CPU). Once a module is created, it is available for use. Developers can use it as a building block to develop their applications. A module can be provided with binary executables either through a client application or by the Web interface.

On the other hand, there are three "types" of users in XWCH:

1. Administrators: They have all rights on the platform and can manage user rights.

2. Developers: They can update binaries belonging to a given module and can execute applications (client programs)

3. Users: They can only execute applications.

User management is supported by the Web interface. A client program drives the execution of an XWCH application by an API. The available methods could be classified into three main categories:

1. Application management methods: Set of methods used to create the "infrastructure" of the application, such as AddModule or AddModuleApplication (see Table 3.1).

2. Data management methods: Set of methods used to manage data on the platform, such as AddData (see Table 3.1).

3. Jobs management methods: Set of methods used to add or monitor jobs, such as GetJobStatus or AddJob (see Table 3.1)

The submission and execution of an XWCH application follow these main steps:

- Initialize connection with the XWCH coordinator.

- Create the application.

- Eventually create and initialize modules belonging to the application. It is also possible to use modules that are already available on the coordinator.

- Insert input data into the XWCH platform.

- Submit/monitor jobs.

- Download results.

- Finish the application.

In what follows, we give a brief summary of the main Java API functions (Table 3.1) that allow the user to create XWCH jobs according to his or her needs. The entire API documentation is available at the XWCH website http://www.xtremwebch.net.

TABLE 3.1: XWCH Functions

XWCHClient (java.lang.String serverendpoint, java.lang.String datafolder, java.lang.String clientID)	This method creates a connection with the coordinator. Serverendpoint refers to the URL of the user services in Figure 3.1.
AddApplication (java.lang.String appName)	This method adds an application to the coordinator. appName refers to the name given by the user. The method returns an applicationId.
AddModule (java.lang.String moduleName)	This method adds an empty "module" to the coordinator and returns its moduleId.
AddModuleApplication (java.lang.String moduleName, java.lang.String binaryzip, PlatformType)	Adds an executable binary file to a given module. This is "per platform" basis, i.e., one binary file can be added for each of the platform (MS Windows, MacOS, Linux, Solaris, etc.).
AddData (java.lang.String fileName)	This method is used to upload the input data of one application (one execution) into the XWCH system.
AddJob (java.lang.String jobname, java.lang.String appName, java.lang.String moduleName, java.lang.String commandLine, java.lang.String inputfiles, java.lang.String listoutputfiles, java.lang.String outfilename, java.lang.String flags)	This method submits a job to the coordinator. A job ID is returned. appName and moduleName refer to the application and the module to which the job belongs. commandLine is the command that invokes the binary with parameters. inputfiles is the set of input files. listouputfiles is a set of files lists; each list of this set will be compressed in one file. outfilename refers to the names of the compressed files. The flags parameter is explained in detail below.
GetJobStatus (java.lang.String JobID)	Gets the status of the job: submitted, assigned, running, killed, or finished.
GetJobFileOutName (java.lang.String JobID)	Gives the "references" (identifiers) of the output files of a given job.
GetJobResult (java.lang.String JobID, java.lang.String outfilename)	Gets the output file (outfilename parameter) of a given job.

The "flags" parameter of the AddJob method could be considered "fine tuning" options that can be used by the programmer according to his needs. The format of these flags are "OPTION : value" where OPTION is a predefined word and value is chosen by the user. Several options are supported by XWCH. We present here only two options, which are REPLICATION and SAME_WORKER:

- The "REPLICATION : x" option means that the output files of the concerned job will be replicated x times. This option can be used when the output file generated by a job J_0 is the input data of several successive jobs J_i $(i = 1$ to $n)$. In this case, in order to avoid overloading one warehouse by parallel requests of the same file, the REPLICATION option is used to replicate the output file of J_0 in different warehouses. Jobs $J_i(i = 1$ to $n)$ will randomly select one of the warehouses in order to retrieve the concerned file.

- The SAME_WORKER option is always followed by a job identifier: "SAME_WORKER, jobID." This option means that the job submitted by the AddJob method will be assigned to the worker that has already executed the job jobID. This option is useful when a set of jobs must execute on the same worker in order to avoid burdening the network with unnecessary data transfers.

An example of using the API to create and execute three communicating jobs: Job1, Job2, and Job3 are given in the code source below. Job2 and Job3 are using the output of Job1 as input data.

```
// Initialisation of the variables BinaryPath_Module1_win and BinaryPath_Module2_lin
// Initialisation of the connection
String InputFile = "input.zip"; //Input file of Job1
c = new XWCHClient(ServerAddress, ".", IdClient);
c.init();
String appid = c.AddApplication("Hello World application");
String ModuleId1 = c.AddModule("Module1");
String refWind = c.AddModuleApplication (ModuleId1,
                                BinaryPath_Module1_win,
                                PlateformEnumType.WINDOWS); //Windows binary
String ModuleId2 = c.AddModule("Module2");
String refLinux = c.AddModuleApplication(ModuleId2,
                                BinaryPath_Module2_lin,
                                PlateformEnumType.LINUX); //Linux binary


. . .

fileref input_ref = c.AddData(InputFile); //sends the input file to one of
                                //the available warehouses\index{warehouses}
String OutPutFile = c.GetUniqueID(); //Gets a unique file name: used to store the
                                //result of the job
String job1 = c.AddJob ("First Job", //Job description
                appid,    //Application identifier
                ModuleId1,  // Module identifier
                Job_CmdLine,  //Command line
                input_ref.toJobReference(), //List of input files. The syntax is:
                                //File1:md51;File2:md52; etc.
                "result_First_Job", // List of files to compress.
                                // In this example, there is one file
```

```
                    OutPutFile1, // name of compressed file
                    "");
//Wait until job1 ends
String status = "";
while (!status.equalsIgnoreCase("COMPLETE")) status = c.GetJobStatus(job1).toString();

// Retrieve the reference of the job1 output file
String in = c.GetJobFileOutName(j1);

String job2 = c.AddJob ("Second Job", appid, ModuleId2, Job2_CmdLine, in,
"result_Second_Job", OutPutFile2, "");

String job3 = c.AddJob ("Third Job", appid, ModuleId2, Job3_CmdLine, in,
"result_Third_Job", OutPutFile3, "");

//Wait until job2 and job3 end
String status = "";
while (!status.equalsIgnoreCase("COMPLETE")) status = c.GetJobStatus(job1).toString();
status = "";
while (!status.equalsIgnoreCase("COMPLETE")) status = c.GetJobStatus(job2).toString();

//retrieve the result\index{result}s of job2 and job3

GetJobResult (job2, file_out_id_job2);
GetJobResult (job3, file_out_id_job3);
```

This client program does not show all the features supported by XWCH. Nevertheless, it details how XWCH handles communication and precedence rules between jobs. Although this program does not show it, calls to XWCH services can of course take place in loops and tests controls. This means that the number of jobs and the structure of the graph representing the application are not known in advance. This "low-level" API provides complete control of the applications workflow. However, its use requires an advanced knowledge of its functions and their parameters. The purpose of the next section is to propose a simple, high-level, and intuitive programming model for XWCH.

3.3 XWCH High-Level API

In the process of writing distributed applications, we have noticed that the application resembles a workflow. The idea of supporting a directed acyclic graph (DAG) of task has been supported by XWCH from its first versions [AB05]. However, the process of implementing this support in parallel applications has normally required a lot of code. In our previous demonstrations of the XWCH API (see [BAN10]), we have been more concerned with complete control of the application than with making the workflow easy for programmers; in Section 3.2.2 we have seen that even a simple XWCH Java application written with the low-level API requires explicitly checking the availability of warehouses, packaging files, creating an application, and controlling its execution. In order to make future application development easier, we present a higher-level API that can be used alongside the low-level one.

While workflow frameworks have been available for some time (see [Fit91]), embedding workflow concepts into programming languages appears to be more recent [PDBH97], and supporting workflows in distributed parallel computing has become popular only in the 2000 [vLAHN02], [vLH05].

In the XWCH high-level API, the workflow is modeled as a directed acyclic graph, similar to Condor's DAGman ([The11]). As a simple example, we consider the traveling salesman application with three computing nodes running the same task with different parameters. Each of the nodes generates an output containing simply the shortest path and its length. A collector task gathers the results and returns the shortest among them.

The workflow can be modeled for XWCH as follows. The components form a set of nodes consisting of Salesman1, Salesman2, Salesman3, and Gather. Their dependencies are modeled as pairs {<Salesman1,Gather>, <Salesman2,Gather>, <Salesman3,Gather>}, implying that Salesman1, Salesman2, and Salesman3 must be run prior to Gather. An XWCH application is created in order to manage the workflow. Jobs are declared as follows:

PrepareJob (NodeName, Executable, InputFiles, Parameters, Output-Files, final), where

- NodeName refers to a name in the set of Nodes

- Executable is the name of the executable file to be run

- InputFiles is a list of input files

- Parameters is a list of runtime parameters for the executable

- OutputFiles is a list of output files

- Final is a Boolean that indicates if the job is the last job of the work-flow

The PrepareJob method adds jobs to an application. Our salesman example can therefore be written in Java as follows:

```
1. Nodes node\index{node}s = new Nodes("salesman1,salesman2,salesman3,Gather");
2. Dependencies dep = new Dependencies(nodes,"{salesman1,salesman2,salesman3:Gather};");
3. Application app = new Application("http://xtremwebch.hesge.ch:8080/", userid, "Salesman");
4. app.PrepareJob("job1", "tsp.sh", "tspdata.zip", "1 3 abcdefghijk out1", "out1", false);
5. app.PrepareJob("job2", "tsp.sh", "tspdata.zip", "2 3 abcdefghijk out2", "out2", false);
6. app.PrepareJob("job3", "tsp.sh", "tspdata.zip", "3 3 abcdefghijk out3", "out3", false);
7. app.PrepareJob("Gather", "gather.py", "out1 out2 out3", "out1 out2 out3", "shortestpath.txt"
            ,true);
8. app.XWCHRun(nodes, dep);
```

The nodes in the workflow are declared on line 1. Line 2 lists the dependencies in the set of nodes, namely that salesman1, salesman2, and salesman3 must be executed before Gather. The order of salesman1, salesman2, and salesman3 is not significant and they can be executed in parallel. The application is created in line 3, with the server contact URL, the identifier of the user and a name of the application. Lines 4 through 7 prepare the individual tasks related to the application. There, tspdata.zip contains the distances between cities and the fourth parameter "1 3 abcdefghijk out1" informs the executable that it is the first of three tasks computing the paths of eleven cities represented by the eleven characters string "abcdefghijk" and that it should write the result to file out1. In line 7, the gather job is declared to have executable "gather.py", input files out1, out2, and out3 (created by job1, job2, and job3), the same file names that were used as parameters, and an output file shortestpath.txt.

Inter-process communication tasks like Master/Slave (Master/Worker), however, require slightly a different approach. Obviously, we need to send the slave tasks to XWCH and let them broadcast their availability to the master (or to the platform that passes the information to the master). In practice, we support this in the API in the following way:

- The slave is submitted as an XWCH job and a job id is received;

- With the job id, execution information is queried, and the slave node's IP address is found in it;

- When all the slaves are in the RUNNING state, the master is invoked with the slaves' IP addresses as its parameters.

This simple method is embedded in a higher-level API, where the programmer does not need to know the technical implementation, as in the following matrix calculation (with one master and two slaves):

```
Nodes nodes = new Nodes ("slave1, slave2, master");
Dependencies dep = new Dependencies (nodes,"{slave1, slave2:master};");
Application app = new MasterWorkerApplication (url, uid, "Masterslave");
app.PrepareJob ("slave1", "slave1.sh", "zero.zip", "1", "slave1.out", false);
app.PrepareJob ("slave2", "slave2.sh", "zero.zip", "2", "slave2.out", false);
app.PrepareJob ("master", "master.sh", "zero.zip", "3", "master.out", true);
app.XWCHRun(nodes, dep);
```

As we see, the only significant difference in the API is the class MasterWorkerApplication instead of Application. This class overrides the XWCHRun method so that the tasks that are prior to master (identified by the last parameter as being "final") are executed in parallel and will need to reach the state RUNNING before the master is started with the slaves' IP addresses as its parameters.

We have compared the execution of the traveling salesman and master slave on XWCH with similar implementations on Condor. The computers where the test was run were single-CPU, single-core. The programs were executed Ubuntu 8.04 on three low-end computers (730 MHz Pentium3 with 256 KB RAM). The distributed computing platforms used were Condor 7.5 and XWCH version 2010-1-31. With both of them, default configurations were used, except that Condor execution nodes would always start a task, and never suspend it.

A non-parallel version of our traveling salesman implementation calculates the shortest path in 3 hours 4 minutes in a computer where no distributed computing daemons are running, and in about 3 hours 10 minutes with either Condor or XWCH running. When the search space is divided into three equal parts, the execution took 1 hour 7 minutes under Condor and 1 hour 6 minutes under XWCH.

For the Master/Slave problem, we used a simple matrix multiplication of a 100×100 long integer matrix written in C. The Condor implementation was done using the MW package [GKLY00], [GLY00]. With both Condor and XWCH, we had one master and two slave nodes (same computers as in the previous example). The speed of modern computers has made the distribution of such small tasks completely redundant; the multiplication of a 100x100 matrix in a single CPU takes about 3 seconds. However, as a benchmark test case, this type of task is often used [BBRR01]. With Condor, the execution took 1 minute 6 seconds, with XWCH 1 minute 50 seconds. In both cases, the overhead of the platform is large. With direct socket communication between the master and the slave, the task is executed in 5 seconds.

3.4 Applications

In this section, we discuss applications that utilize the XWCH platform. As discussed earlier, XWCH has a rich API for supporting functions like remote execution, data transfer, replication, and job control.

Applications are deployed and executed on an XtremWeb-CH platform with a number of connected workers ranging from 300 to 800. These machines are very heterogeneous in terms of performance and operating systems.

Workers are not dedicated to the XWCH platform; they are mainly used by students. These machines are administrated by the local IT supports of the participating institutions: HES-SO (hepia, Geneva), University of Geneva and University of Franche Conté (France). The administration policies carried out by some participating institutions often require nodes (PCs) to disconnect (outage during the nights, on weekends and/or holidays, etc.). Although the number of nodes in which XWCH is installed is more than $1,000$ PCs, the real number of connected workers is often less than 600.

Several applications are deployed on XWCH. This section details only three of them. For each of the three applications, a client program is written. This program submits jobs to the XWCH platform, monitors their execution, and uploads the output result.

3.4.1 NeuroWeb

NeuroWeb builds neuronal maps of brain activity using the input from noninvasive measurements [Abd09]. It is based on "extracting" the activity of the different regions of neurons from captors attached to the scalp of the patient. The reconstruction of dynamic neural maps requires the solution of a spatiotemporal inverse problem, which is an active field of research of great clinical importance. This technique can be used, for example, to localize an epileptic region for a remote surgery. Measurements of the internal cerebral activity around the scalp are obtained, for example, from magnetoencephalography (MEG).

Neuronal activities, called generators, are electromagnetic fields captured by the MEG captors. These fields are extremely noisy and distorted by the cell tissue. In essence, we have a lot of generators (about 60,000) but very few captors (about 200). The difficulty, and hence the computational requirement effort, is to construct a robust and efficient estimator for a neuronal map [AEB08].

The estimated images of brain activity are a solution to a high-dimensional, nondifferentiable optimization problem. The size of the solution space is extraordinarily large, about 107 and requires more than one day of intensive calculations. Finding the solution to this mathematical problem can be done efficiently using grid and parallel computing. The practitioners will then be able to virtually cut the cortex and identify sources of "abnormal" brain activities such as epileptic crisis, Parkinson, Alzheimer, etc.

The NeuroWeb application is assumed to produce a dynamic neural map, DNM, in which each row represents the electromagnetic activity of one region of neurons (generator). With 60,000 regions and 1,500 measures (assuming that the duration of the experiment is 1.5 second, and one measure is taken every 1 millisecond), the size of the DNM is greater than 240 MB. Moreover, other input data are required for the computation of the resulting DNM. These are the functional and anatomic data that are provided from IRM scanners.

The application starts by choosing a random map called DNM_0. At each step i, a new matrix DNM_i is constructed based upon DNM_{i-1}. The iterations (steps) stop when the DNM_i matrices can no longer be improved, that is, when convergence is attained. The DNM matrix is split into several blocks B_k. Each block B_k is processed by a daemon process DP_k that executes an iterative asynchronous algorithm. DP_k is created by an XWCH job and remains running after the end of this job. During its execution, a DP_k receives data from its neighbors (DP_{k-1} and DP_{k+1}) and integrates them to calculate a new "improved" version of the block B_k. Indeed, in a given block B, columns C_{l-1} and C_{l+1} are required to process column C_l. This means that DP_k and DP_{k-1} need to exchange a shared column, and so do DP_k and DP_{k+1}.

Data transfers among daemon processes are carried out by XWCH jobs. We denote by Job_{ik} the XWCH job that transfer data to DP_k during step i.

It is worth remembering here that the generation of a block B by a deamon process DP can take place even if no input data are received from the neighbors of DP. Indeed, DP is an iterative asynchronous algorithm, that continues improving its block B even if it does not receive any data from its neighbors.

On the other hand, a central daemon process (CDP) is deployed in order to gather blocks from all DPs and decide if convergence is reached or not. If the convergence is not yet reached, the CPS launches a new step. It can be deduced from what is presented above that:

- The number of steps in the workflow of NeuroWeb cannot be known before the execution;

- Daemons processes continue their execution and reside on the main memory of the XWCH workers until convergence is attained;

- During its execution, a given DP_k executes an iterative asynchronous algorithm and exchanges data with its neighbors DP_{k-1}, DP_{k+1} and with the CPS.

In this context, XWCH is used to:

- Create the daemon processes (CDP and DP_k) on the workers

- Feed the DPs with necessary data

- Feed the CDP with data to decide if convergence is reached or not

- Stop the CDP and the DP_i

These items are done by XWCH jobs created by a client program. The dataflow graph of NeuroWeb is represented in Figure 3.2.

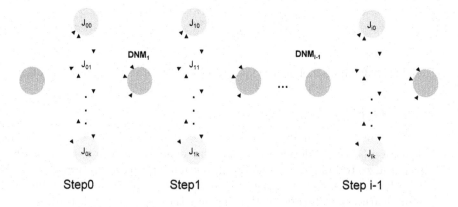

FIGURE 3.2: Dataflow in NeuroWeb

The NeuroWeb execution was run on the XWCH platform composed of more than 500 machines and five warehouses. This application is assumed to end when the convergence is reached. The convergence is described by a number called the "epsilon score". The central daemon process tries, on each iteration, to compare the neuronal correspondence of the intermediate matrix with the "epsilon score." Once the convergence is reached, the CDP ends the calculation.

Figure 3.3 illustrates the impact of the number of workers used for the calculation. We can see that with an "epsilon score" equal to $1e^{-3}$ and an initial matrix size equal to 43 Mb, the optimal needed resources for minimal execution time (12:25 minutes) is about thirty workers. Compared to local execution on a normal PC (3 GB of memory, Intel dual core: 2x2GHz) that took 59 minutes, the use of XWCH speeds up the execution time by a factor of 5. This grid experiment is an initial proof-of-concept to study the application resource requirements and convergence aspect. Future work will consist of processing big matrix that will need more resources. It is worth noting here that one of the main benefits that makes XWCH attractive for NeuroWeb application is the possibility to use more resources when needed: in case of a big matrix or simultaneous treatment of a set of brains.

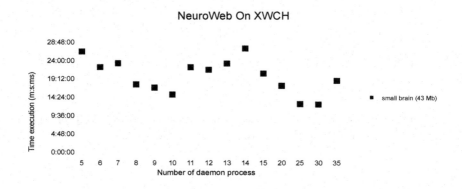

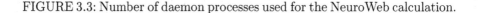

FIGURE 3.3: Number of daemon processes used for the NeuroWeb calculation.

3.4.2 GIFT

The Gnu Image Finding Tool, GIFT, is a content-based image indexing and retrieval package developed at the University of Geneva in the late 1990. GIFT utilizes techniques common from textual information retrieval and uses a very large collection of binary-valued features (global and local color and texture features) [SMMP00]. GIFT extracts these features and stores them in a file. In a typical desktop PC, the speed of this feature extraction is about one or two images per second (an earlier study with older hardware states four seconds [RLK03]). An inverted file database is created after the extraction. In such a database, all the features are listed, and this enables fast feature look-ups.

Content-based image retrieval aims at searching image databases for specific images that are similar to a given query image and has many applications in medical imaging [LGD+05]. However, hospitals produce large amounts of images, for instance, the University Hospitals of Geneva radiology department produced about 70,000 images per day in 2007 [NZD+08]. Feature extraction from such a large set of images would normally not be possible without parallelization.

We have used XWCH as a platform for simulating a large feature extraction task using the ImageCLEFmed [MKCJ+09] collection of about 50,000 images, as shown in Figure 3.4. The implementation was kept simple: the source images were packaged into ZIP files, each containing 1,000 images. The feature extraction binary and the package were sent to the coordinator nodes as XWCH jobs (1), and the resulting feature file was recovered as the output of each job (2). The entire process took 4 hours 53 minutes 58 seconds. The average of the execution times of a job was 16 minutes 36 seconds and the sum of the execution times is 51,316 seconds (ca. 14 hours). The figure of 51,316 seconds equals to executing the entire task on a single CPU [NBA10].

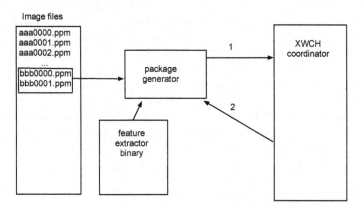

FIGURE 3.4: Running GIFT on XWCH.

GIFT is currently used in two XWCH-related collaborations. The first is based on CDS Invenio bibliographic content management sofware [MLRV10] developed at CERN. CDS Invenio has been integrated with GIFT to allow the user searching bibliographic records with image content based on image similarities. The feature extraction is carried out as an XWCH application.

In another collaboration, MUSIC, with the University of Applied Sciences of Luzern (Switzerland), artists create drawings with a SmartPen digital pen [Har09] and upload them in large-scale image storage. The digital pen records drawings and audio based on Anoto's (http://www.anoto.com) technology and the Frauenhofer IIS/AAC Audiocodec. Collaborative Web technologies are used to annotate the images, GIFT is used to study image similarities, and feature extraction is carried out using XWCH.

3.4.3 MetaPIGA

Phylogeny, that is, construction of evolutionary trees, is a very CPU-intensive task. Optimality-criterion-based phylogeny inference is an NP-hard combinatorial optimization task; the number of solutions increases explosively (factorials) with the number of taxa: 32 million different trees for 14 taxa, and 3×10^{84} trees (more than the number of atoms in the universe) for 55 taxa.

Though a brute-force optimization approach to the problem is impossible with large numbers of taxa, several Maximal Likelihood (ML) solutions have been proposed. Of these, genetic algorithms (GAs) are fast and accurate.

MetaPIGA [Ano] is a robust implementation of several stochastic heuristics for large phylogeny inference (under maximum likelihood), including a simulated annealing algorithm, a classical genetic algorithm, and the metapopulation genetic algorithm (metaGA) together with complex substitution models, discrete Gamma rate heterogeneity, and the possibility to partition data. MetaPIGA also implements the Likelihood Ratio Test, the Akaike Information Criterion, and the Bayesian Information Criterion for automated selection of substitution models that best fit the data. Heuristics and substitution models are highly customizable through manual batch files and command-line processing. MetaPIGA also offers an extensive graphical user interface for parameter setting, generating and running batch files, following run progress, and manipulating result trees. MetaPIGA uses standard formats for data sets and trees, is platform independent, runs in 32- and 64-bit systems, and takes advantage of multiprocessor and multicore computers.

MetaPIGA is a CPU time-consuming application. One big dataset needs in general 500 CPU hours. Assuming that 200 analyses are launched every year, the total number of CPU hours needed per year is equal to 100, 000.

MetaPIGA is exceptionally well suited for parallelization because:

- Several populations must be run in parallel and can therefore be sent to different machines.

- Multiple analysis must be performed to yield posterior probabilities of trees, and these can be sent to multi-core computers as well.

The main flexibility that makes this application attractive for the XWCH platform is the dynamic job generation aspect provided by XWCH's API.

The input data for each job is a dataset, while its output file represents P solutions (a solution tree for each population). A number of XWCH jobs (between 100 and 10, 000 jobs according to the data size) needs to be submitted to XWCH during the analysis generation process. An XWCH job generates randomly P populations composed of N individuals. Then, for each population, the best individual (under the maximum likelihood criterion) is kept while the other $N-1$ individuals are subjected to selection and mutation events. This process is repeated until a global optimum is found, that is, the current optimal solution can no longer be improved after m successive generations. The final population represents the output of the XWCH application.

When a job has been processed successfully, its output file is downloaded from XWCH and the result trees are added to the consensus tree on the client program. However, because of performance and reliability considerations, it has turned out to be more efficient to submit progressively bags of jobs on XWCH. One of the reasons is that the number of jobs to run cannot be known in advance. Moreover, a convenient stopping rule was developed in MetaPIGA to stop the generation of replicates when convergence in the consensus tree is detected; that is, the expected degree of confidence has been attempted on the final constructed tree.

The main aim of the parallel MetaPIGA is to accelerate the large-scale analysis. Here, we present measurements of the parallel version and compare it with a sequential execution on a single machine, a dedicated multi-core machine, and XWCH.

The experiments performed on XWCH started from a small population of fifteen individuals; a dataset composed of fifteen DNA sequences (1, 314 nucleotides each). For the analysis, we have launched the execution with a set of 1, 000 to 10, 000 jobs. The execution on the grid took about 50 minutes with around 300 performed jobs. This is more than 50 times faster than an execution on a single machine (Intel core 2 duo E8200 2.66GHz with 3GB memory). Moreover, in a 16-core machine, the execution of the same dataset took 180 minutes; here we can observe that even with a 16-core machine where the MetaPIGA jobs need 4 GB of memory, we do not have a speed-up of 16 compared to a single machine execution that needs only 250 MB. Indeed, if we increase the dataset size, each job will need about 1.5 GB for the analysis, that is, less than the typical amount of memory for a PC today. The analysis in the multi-core machine needs about 24 GB for a big dataset size. This is quite expensive and produces the results more slowly than the XWCH solution.

3.5   Integration with the ARC Grid Middleware

The NorduGrid Advanced Resource Connector (ARC) middleware [E+07] is an implementation of fundamental Grid services such as information services, resource discovery and monitoring, job submission and management, brokering, data management, and resource management. As the name suggests, it is often used to combine the resources of several geographically distributed institutes, to create grid collaboration. Today, these collaborations include SWEGRID (Sweden), Swiss National Grid Association (SwiNG), Material Sciences National Grid Infrastructure (M-grid) (Finland), NORGrid (Norway), and the Ukrainian Academic Grid. ARC can utilize computing resources provided by many cluster-computing softwares (so-called Local Resource Management Systems, LRMSs), including Condor, Torque, OpenPBS, and Sun Grid Engine [E+07].

This section describes a bridge between ARC and XWCH, effectively making XWCH a Local Resource Management System for ARC. In this context, the XWCH infrastructure is considered an ARC cluster. This bridge is used in a concrete case to connect the Swiss Multi-Science Computing Grid (SMSCG) [S+09] platform with an existing XWCH platform composed of hundreds of workers provided by different academic institutions. A map of SMSCG's Swiss clusters is shown in Figure 3.5.

ARC has three main components; computing servers, brokering clients and

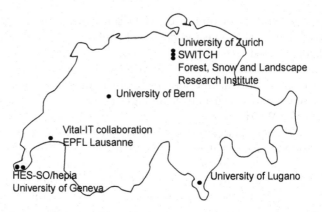

FIGURE 3.5: SMSCG infrastucture providers. HES-SO stands for University of Applied Sciences, West Switzerland. Hepia stands for Geneva School of Technology, Architecture and Landscape.

resource index servers, as shown in Figure 3.5 . The principal processes within a computing server are the information system and the grid job execution manager (earlier called GridManager [E$^+$07], currently A-REX [S$^+$10]).

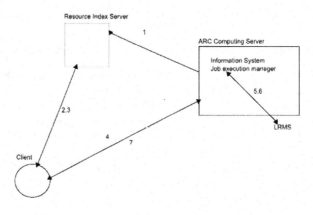

FIGURE 3.6: Basic ARC functionality

In practice (see Figure 3.6), the ARC computing servers send their characteristics (like number of CPUs, architecture, installed applications, or so-called runtime environments) to resource index servers (1). Using a resource index server, an ARC client's scheduler finds an ARC server that is available and suitable for executing its job (2, 3). The selection is most often done based on the maximum number of free CPU among the available servers. The job, typically consisting of executables and input files, is transferred from the client to the ARC server (4). The server's job execution manager calls the functions of the underlying LRMS to execute the job and to periodically observe its status (5, 6). The information system publishes this status so that the client can retrieve it. In a positive case, the job execution succeeds, and the job execution manager moves the output files, making them available to the client, and the client downloads them (7).

Our integration task of XWCH thus consisted of creating a set of scripts by which ARC's job execution manager can communicate with XWCH. For this purpose, an XWCH command-line interface was written. This interface contains high-level routines like "list the characteristics of all workers," "submit an XWCH application whose input files, executable, command line, and output files are described in file X," "retrieve the status of application Y" and "retrieve the output files of application Y." An example of a job description is shown in Figure 3.7. There, it is stated that this is a single task application (XWCHJob), not a part of a workflow. The name of the application is APP_CLI_TEST and it uses a module whose ID is shown. The application has input file data.zip, command line "run.sh parameters," and it will generate output that is put into a file modj.zip from all the files in the application directory (regex:.*). The application requires that the Linux operating system runs in the worker node. After the submission, the client will not wait for the application to finish (downloadresult=0); rather, the results will be retrieved using a separate client command-line call.

Examples of using the command line are shown in Figure 3.8. They include submitting an application (1), checking its status (2), retrieving its output (3), showing jobs (4), the platform status (5), and the available modules (6).

```
type = XWCHJob
xwserver = http://xtremwebch.hesge.ch:8080
client_ID = myIdIsSecret
applicationname = APP_CLI_TEST
moduleID = 17a201b7-36fa-418c-a13d-d3be1d951777
datainputfilename = data.zip
jobname = ModJob
workercommandline = run.sh parameters
outputfilename = modj.zip
listeoutputfiles = regex:.*
requirements = (os=LINUX)
downloadresult = 0
```

FIGURE 3.7: XWCH job description.

The ability to list modules and to use them in job descriptions has enabled us to implement ARC's idea of runtime environments (RTEs) in a natural and simple way. With ARC, one can define an RTE, that is, an application or a set of libraries needed by a job. This means that the users do not need to prepare the environment themselves (as this would take a lot of time) nor transfer it with their Grid task (as this would take a lot of bandwidth). Instead, if a job

```
1. xw\index{xw}ch -f job.xwch
2. xw\index{xw}ch -status 61097681-1943-47ed-9691-9f7844874ca5 -server http://xtremwebch.hesge.ch:8080
3. xw\index{xw}ch -d 61097681-1943-47ed-9691-9f7844874ca5;modj.zip -server http://xtremwebch.hesge.ch:8080
4. xw\index{xw}ch -display jobs -server http://xtremwebch.hesge.ch:8080

JOB_NAME         APP_NAME                       MODULE_NAME   STATUS
preparation job  distributed\index{distributed} sudoku solver SudokuPrepare  COMPLETE
computation job  distributed\index{distributed} sudoku solver Sudokuprocess  COMPLETE

5. xw\index{xw}ch -display machine\index{machine}s -server http://xtremwebch.hesge.ch:8080

NAME         OSTYPE_ARCH CPUTYPE CPUSPEED  RAM FREEMEM HD    FREEHD WANIP LANIP
unige146226  LINUX_32    Intel   2167Mhz x1 384  4      7666  5465   --    --
st01-b113-cao WINDOWS_32 Intel   2992Mhz x2 3328 594    152586 108027 --    --
unige146237  LINUX_32    Intel   2250Mhz x1 384  5      7666  5472   --    --

6. xw\index{xw}ch -display modules -server http://xtremwebch.hesge.ch:8080

MODULENAME    MODULE ID                              ARCH
SudokuPrepare 17a201b7-36fa-418c-a13d-d3be1d951777   LINUX_32
```

FIGURE 3.8: Using the command line client.

requires an RTE, it is written as a requirement in an ARC job description. Typically, in high-energy physics, large RTEs like ATLAS [The09] consist of hundreds of files and can take several gigabytes when installed. With a specific configuration option, an XWCH module is published to ARC as a runtime environment.

The ARC server "arcXWCH at HEPIA" shown at the top of Figure 3.9 accepts jobs and reports its status and resources to the SMSCG index server. It is rare for the servers shown in the SMSCG index server to have a varying number of CPUs, but this is the case with ARCXWCH. The number of CPUs reflects the number of worker PCs connected to the XWCH platform.

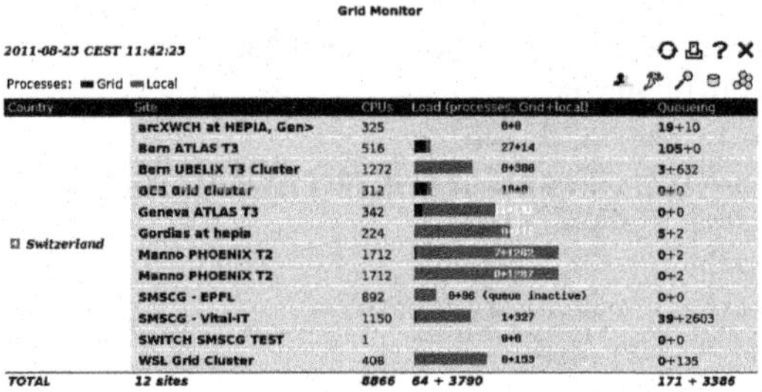

FIGURE 3.9: A list of clusters as seen through an ARC monitor.

3.6   Conclusion

The XtremWebCH system has been under constant development since 2005. During these years, time has been dedicated to make the system run on various platforms (Unix, Windows, Mac), to porting programs to the system, and also to make it accessible and easy to use. The users have extensive documentation of the features and architecture on the website, as well as several example applications. Though there are many desktop computing systems that are more feature-rich and better known than XtremWebCH, we aim to be the system with easiest setup and usage. For this purpose, we provide simple installation packages of all the components for system administrators, and a high-level API for programmers.

XtremWeb-CH is currently being bridged with several middlewares and platforms. Three projects funded by the Swiss confedaration and the European Commission (FP7) are in progress in order to connect XtremWeb-CH with grid and cloud platforms such as EDGI and Venus-C. One of the ideas consists of "creating" XWCH workers on cloud platforms when the resources required by the user are not available on the volunteer computing platform. In this context, the cloud is used as a "backup."

Bibliography

[AAFM09]   N. Abdennadher, A. Agrawal, E. Fragniere, and F. Moresino. Services pricing: A shared grid case study. In *IEEE International Conference on Service Operations, Logistics and Informatics*, Chicago, IL, USA, 2009.

[AB05]   N. Abdennadher and R. Boesch. Towards a peer-to-peer platform for high performance computing. In *Proc. HP-ASIA*, Beijing, China, 2005.

[Abd09]   N. Abdennadher. Using the volunteer computing xtremeweb-ch: Lessons and perspectives. In *Proc. Advances in Computer Science and Engineering*, Phuket, Thailand, 2009.

[AEB08]   N. Abdennadher, C. Evquoz, and C Bilat. Gridifying phylogeny and medical applications on the volunteer computing platform xtremweb-ch. In *Proc. HealthGrid'08*, Chicago, IL, USA, 2008.

[AF06]   D. Anderson and G. Fedak. The computational and storage potential of volunteer computing. In *IEEE International Symposium on Cluster Computing and the Grid*, Singapore, 2006.

[Ano]   Anonymous. Metapiga 2 - large phylogeny estimation. http://http://www.metapiga.org/.

[B+07]   Z. Balaton et al. Sztaki desktop grid: a modular and scalable way of building large computing grids. In *Proc. 21st IEEE International Parallel and Distributed Processing Symposium*, Long Beach, CA, 2007.

[BAN10]   M. Ben Belgacem, N. Abdennadher, and M. Niinimaki. Virtual ez grid: A volunteer computing infrastructure for scientific medical applications. *International Journal of Handheld Computing Research*, 2010. appacted, forthcoming.

[BBRR01]   O. Beaumont, V. Boudet, F. Rastello, and Y. Robert. Matrix multiplication on heterogeneous platforms. *IEEE Transactions on Parallel and Distributed Systems*, 12, 2001.

[BYV08]   R. Buyya, C. Yeo, and S. Venugopal. Market-oriented cloud computing: Vision, hype, and reality for delivering it services as computing utilities. In *Proc. 10th IEEE International Conference on High Performance Computing and Communications*, Dalian, China, 2008.

[Cer02]   E. Cerami. *Web Services Essentials*. O'Reilly, 2002.

[E+07]   M. Ellert et al. Advanced resource connector middleware for lightweight computational grids. *Future Generation Computer Systems*, 23(2), 2007.

[F+01]   G. Fedak et al. Xtremweb: a generic global computing system. In *Proc. First IEEE/ACM International Symposium on Cluster Computing and the Grid*, Brisbane, Australia, 2001.

[Fit91]   G. Fitzmaurice. Form-centered workflow automation using an agent framework. Master's thesis, Brown University, Providence, RI, USA, 1991.

[GKLY00]   J.-P. Goux, S. Kulkarni, J. Linderoth, and M. Yoder. An enabling framework for master-worker applications on the computational grid. In *9th IEEE International Symposium on High Performance Distributed Computing*, pages 43–50, Pittsburgh, USA, 2000.

[GLY00]   J.-P. Goux, J. Linderoth, and M. Yoder. Metacomputing and the master-worker paradigm. Preprint ANL/MCS-P792-0200, 2000.

[Har09]   J. Harboe. Drawing in a new dimension. Yearbook of the LUASA Art and Design, 2009. Luzern, Switzerland.

[Hen95]   R. Henderson. Job scheduling under the portable batch system. *Job Seheduling Strategies for Parallel Processing*, LNCS 949/1995, 1995.

[LGD+05]   T. Lehmann, M. Gulda, T. Deselaersb, D. Keyser, H. Schubert, K. Spitzera, H. Ney, and B. Wein. Automatic categorization of medical images for content-based retrieval and data-mining. *Computerized Medical Imaging and Graphics*, 29(2-3), 2005.

[LLM88]   M. J Litzkow, M. Livny, and M. W. Mutka. Condor-a hunter of idle workstations. In *Proceedings of the 8th International Conference of Distributed Computing Systems*, 1988.

[M⁺02]   D. S. Milojicic et al. Peer-to-peer computing. HPL Technical Report HPL-2002-57, 2002.

[MKCJ⁺09]   H. Mueller, J. Kalpathy-Cramer, C. E. Kahn Jr., W. Hatt, S. Bedrick, and W. Hersh. Overview of the imageclefmed 2008 medical image retrieval task. *Evaluating Systems for Multilingual and Multimodal Information Access*, LNCS 5706/2009, 2009.

[MLRV10]   L. Marian, Jean-Yves LeMeur, M. Rajman, and M. Vesely. Citation graph based ranking in invenio. *Research and Advanced Technologies for Libraries*, LNCS 6273/2010, 2010.

[NBA10]   M. Niinimaki, M. Ben Belcagem, and N. Abdennadher. Running gift image recognition software on xwch. Technical Report, HEPIA, 2010.

[NZD⁺08]   M. Niinimaki, X. Zhou, A. Depeursinge, A. Geissbuhler, and H. Mueller. Building a community grid for medical image analysis inside a hospital, a case study. In *Medical imaging on grids: achievements and perspectives, MICCAI Grid Workshop*, New York University, NY, USA, 2008.

[PA04]   C. Pautasso and G. Alonso. Jopera: A toolkit for efficient visual composition of web services. *International Journal of Electronic Commerce*, 9(2), 2004.

[PDBH97]   M. Papazoglou, A. Delis, A. Bouguettaya, and M. Haghjoo. Class library support for workflow environments and applications. *IEEE Transactions on Computers*, 46(6), 1997.

[RLK03]   M. Rummukainen, J. Laaksonen, and M. Koskela. An efficiency comparison of two content-based image retrieval systems, gift and picsom. *Image and Video Retrieval*, LNCS 2728/2003, 2003.

[S⁺09]   H. Stockinger et al. The swiss national grid association and its experince on a national grid infrastructure. In *Austrian Grid Symposion*, 2009.

[S⁺10]   O. Smirnova et al. Recent arc developments: Through modularity to interoperability. *Journal of Physics*, Conf. Ser. 219, 2010.

[Sar98]   L. Sarmenta. Bayanihan: Web-based volunteer computing using java, worldwide computing and its applications. *WWCA'98*, LNCS 1368/1998, 1998.

[SMMP00]   D. M. Squire, W. Mueller, H. Mueller, and T. Pun. Content-based query of image databases, inspirations from text retrieval: inverted files, frequency-based weights and relevance feedback. *Pattern Recognition Letters*, 21(13-14), 2000.

[The09]   The NorduGrid Project. Atlas data challenge on nordugrid, 2009. http://www.nordugrid.org/applications/atlas-dc2/.

[The11]   The Condor Project.   Condor dagman web page, 2011. http://www.cs.wisc.edu/condor/dagman/.

[TTL02]   D. Thain, T. Tannenbaum, and M. Livny. Condor and the grid. In Fran Berman, Geoffrey Fox, and Tony Hey, editors, *Grid Computing: Making the Global Infrastructure a Reality*. John Wiley & Sons Inc., 2002.

[U+09]   E. Urbah et al. Edges: Bridging egee to boinc and xtremweb. *Journal of Grid Computing*, 7(3), 2009.

[vLAHN02]   G. von Laszewski, K. Amin, S. Hampton, and S. Nijsure. Gridant white paper, 2002.

[vLH05]   G. von Laszewski and M. Hategan. Workflow concepts of the java cog kit. *Journal of Grid Computing*, 3:239–258, 2005.

Chapter 4

XtremWeb-HEP: Designing Desktop Grid for the EGEE Infrastructure

Oleg Lodygensky

IN2P3/CNRS, Orsay, France

Etienne Urbah

IN2P3/CNRS, Orsay, France

Simon Dadoun

IN2P3/CNRS, Orsay, France

Desktop grids such as XtremWeb [FGNC01], OurGrid and BOINC [And04], and service grids such as EGEE, ARC or UNI-CORE, are two different approaches for science communities to gather computing power from a large number of computing resources. Today, work is being done to combine these two Grid technologies in order to establish a seamless and vast grid resource pool.

In this chapter we introduce service grids, especially focusing on EGEE as the use case, and desktop grids focusing on XtremWeb-HEP and BOINC. Finally, we discuss standardization for grid inter-operability.

4.1   Introduction

XtremWeb-HEP (*XWHEP*) is a middleware permitting a distributed data processing infrastructure (*grid computing*) to deploy . XWHEP belongs to the so-called Cycle Stealing family that uses idle resources. Like some other grid middleware stacks, XWHEP uses remote resources (PCs, workstations, PDA, servers) connected to the Internet, or a pool of resources inside a LAN. Participants in an XWHEP infrastructure cooperate by providing their computing resources, such as processor, memory, and/or disk space. XWHEP is a lightweight grid middleware. It is a Free Software (GPL), Open Source, and non profit software to explore scientific issues and applications of global computing and peer to peer distributed systems. For example, XtremWeb-HEP allows a high school, a university or a company to setup and run a global computing or peer to peer distributed system for either a specific application or a range of applications.

The middleware is written in the Java 1.6 computing language.

XWHEP, based on XtremWeb 1.6.0 (http://www.xtremweb.net), introduces innovative features, specifically in the security field, that permit inter-grid sharings. The middleware is provided with the necessary services and tools to interconnect to other grid families, especially focusing on the EGEE middleware.

The next sections introduce the XWHEP middleware, detail the security issues, and describe inter-grid sharings.

4.2   Architecture

The XWHEP architecture contains three parts, as shown in Figure 4.1: one or a group of several servers is installed and maintained by system administrators to host centralized XWHEP services such as the "Scheduler" and the "Result Collector"; distributed parts deployed over the Internet: XWHEP workers are installed by citizens on their PCs to propose their computing resources to be aggregated within an XWHEP infrastructure; XWHEP clients are installed by XWHEP users (scientists, for example) on their PCs to interact with the infrastructure. The XWHEP client permits users to manage the platform as well as use distributed resources (submit jobs, retrieve results...).

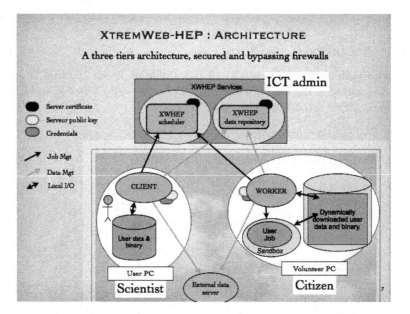

FIGURE 4.1: XtremWeb-HEP architecture.

This architecture corresponds to most well-known Global Computing Projects: The XWHEP central services permit the XWHEP administrator to manage registered applications. A client prepares the data that are needed to successfully compute jobs. These data may be stored in the XWHEP infrastructure or in any repository as soon as the data can be described by an URI and are accessible through the network. There is no limit on data size, but it is assumed that data should not exceed a few hundred megabytes. A client registers applications on the XWHEP infrastructure. A client prepares jobs (units of work) containing the reference of a registered application, optional parameters, optional references to additional files. Finally, a client submits jobs to the XWHEP Scheduler.

The workers contact a Scheduler to get jobs suitable for their architecture. In response, the Scheduler sends a suitable job description to a worker. For each file referenced by the job which is not present in the local cache of the worker yet, the worker fetches the file from the XWHEP Data Repository or from an External Data Server. To reduce bandwidth usage, the worker manages a local cache in least recently used (LRU) mode. Each time the computer is restarted, the worker automatically cleans this local cache. This ensures that the worker does not use too much local disk space, but the worker may then sometimes need to download the same file from run to run.

As soon as a job has finished on the worker side, the worker contacts the Result Collector to send the results.

4.3 Security Considerations

The middleware ensures security at different levels. In this chapter we detail security in the client server interaction. Security services and protocols in this context can be introduced as follows:

1. Secured communications

2. Authentication of server, client, and Worker

3. Authorizations

4. Access rights

5. Confinements

6. Data confidentiality

7. Data integrity

8. Data persistence

9. Result certification

The security in these different fields must be fulfilled for the inter-grid sharings.

4.3.1 Secured Communications

As sensitive data may be transferred in the network, the XtremWeb-HEP middleware systematically ensures communications privacy using TLS encryption. This ensures confidentiality using encryption and identifying communicating entities.

4.3.2 Authentication of Server, Client, and Worker

Server authentication is implemented by the usage of an electronic key pair: the server private key (see Figure 4.1) is securely stored on server side, and the server public key (see Figure 4.1) is enclosed inside the installation packages of the client and worker. This permits certification of the server identity: distributed XtremWeb-HEP components establish a session with a server only if this server proves that it owns the private key corresponding to the public key. On the other side, the server does not accept connection without its own public key.

Being connected to the server is not sufficient to execute a command on server side. Each connecting entity (client and worker) must present valid credentials so that the server can securely identify who is connecting. In XtremWeb-HEP, two credential types are accepted by the server:

- **Login/password.** Login/password is a key pair that uniquely defines an identity. Such identities must be inserted on the server side by the XtremWeb-HEP administrator. It is the administrator's responsibility to securely send the login/password pair to whom it may concern.

- **X509 certificate.** The XtremWeb-HEP middleware is able to use X509 certificates. This is an IETF standard ensuring certificate owner identity. It is out of our scope to detail certificate in this document. The middleware authenticates the communication initiator presenting an X509 certificate by verifying the certificate authority(CA) that signed the certificate. It is the XtremWeb-HEP administrator's responsibility to manage its list of known CA certificates.

The authentication process, if successful, defines an identity. This identity is used on the server side to execute commands sent by the distributed component (client, worker). An identity has several attributes: first name, last name, email, etc. An identity also has some critical attributes involved in security: credentials (login and password, or X509 certificate), user group membership, and authorization.

4.3.3   Authorizations

Being authenticated by the server is not sufficient to execute a command on the server side. Each authenticated entity has a usage level called *"user rights,"* that defines authorization. Authorization allows or denies execution of the requested action. Authorizations are defined as enumerations from least to most privileged: NONE, LISTJOB, INSERTJOB, GETJOB, DELETE-JOB, LISTDATA, INSERTDATA, GETDATA, DELETEDATA, LISTGROUP, INSERT-GROUP, GETGROUP, DELETEGROUP, LISTSESSION, INSERTSESSION, GETSES-SION, DELETESESSION, LISTHOST, GETHOST, LISTUSER, GETUSER, LISTUSER-GROUP, GETUSERGROUP, LISTAPP, GETAPP, STANDARD_USER, UPDATE-WORK, WORKER_USER, INSERTUSER, DELETEUSER, INSERTAPP, DELETEAPP, ADVANCED_USER, INSERTHOST, DELETEHOST, INSERTUSERGROUP, DELE-TEUSERGROUP, SUPER_USER.

The SUPER_USER authorization allows all access rights on all XtremWeb-HEP objects. In the XtremWeb-HEP middleware, we call the administrator an entity associated to a SUPER_USER authorization. The STANDARD_USER authorization allows one to do actions such as insert job, insert data, etc.

The general case is as follows: user right R allows to perform an action A if and only if:

$$0 < A \leq R$$

There is one specific case: WORKER_USER. Presenting an identity associated with this user's right, one can:

- Never list anything (list app, list data, etc.)

- Never insert anything (insert app, insert data, etc.)

- Never delete anything (delete app, delete data etc.)

- Ask for a pending job

- Send a heartbeat signal

- Retrieve an object by providing the object unique identifier (UID)

- Download data

- Upload result for the job it has computed (this is checked by the server)

4.3.4   Ownership

In XtremWeb-HEP, each object has an owner. By default, the owner has full access to his or her objects. Any user may have access to an object he or she does not own, if that object permits it. For example, it is possible to define private objects to ensure that nobody but the owner (and the administrator) can access them. It is also possible to restrict access to a user group only; any user belonging to a user group may have access to an object owned by a member of the same user group, if this object permits group access.

Accesses are defined by object access rights presented in the next section.

4.3.5   Access Rights

Being allowed to execute a command on a server is not sufficient to execute this command on an object managed by the server. All object (applications, data, jobs, etc.) are associated to access rights. This has been implemented *à la* Linux file system access rights. Access rights allow or deny access to the object it is associated with. XtremWeb-HEP assigns Read, Write (which includes "modify" and "delete"), and Execute access rights separately for the owner of the object, for members of the group the owner belongs to (if any), and for others.

Access rights of an object can be modified by its owner or by an XtremWeb-HEP administrator. It is not possible to increase access rights, except for the administrator.

There is one exception where access rights are fully bypassed so that access is always allowed: an XtremWeb-HEP administrator is always allowed to access (read, write, execute) any object. This exception permits one to modify access rights in any manner (even increasing).

Access rights are defined like this:

$U_R \ U_W \ U_E \ G_R \ G_W \ G_E \ O_R \ O_W \ O_E$

where

U_R = User (owner) read access rights

U_W = User (owner) write access rights

U_E = User (owner) execute access rights, if applicable

G_R = Group read access rights

G_W = Group write access rights

G_E = Group execute access rights, if applicable

O_R = Other read access rights

O_W = Other write access rights

O_E = Other execute access rights, if applicable

Examples in octal notation:

- *755* defines

    - Full accesses for the owner;

    - Read and execute accesses to the owner group, if any;

    - Read and execute accesses to others.

- *750* defines

    - Full accesses for the owner;

    - Read and execute accesses to the owner group, if any;

    - No access to others.

- *700* defines

    - Full accesses for the owner;

    - No access to the owner group, if any;

    - No access to others.

4.3.6 Confinements

The XtremWeb-HEP middleware manages user groups. Any user may belong to a user group. User groups aim to confine accesses so that members of the group are allowed to access objects defined within the group. These objects confined in a group are not accessible by users who do not belong to the user group.

Because XtremWeb-HEP assigns an authorization to each identity and access rights to each object, it effectively confines all objects it manages. Inside XtremWeb-HEP, a combined setup of authorizations and access rights is called a confinement.

All applications, all data, and all jobs are confined. But also all users, all clients, and all workers.

Confinements aim to restrict accesses so that distributed entities (clients and workers) are allowed or denied access (read, write, execute) to confined objects. The XtremWeb-HEP middleware helps to define as many confinements as possible combinations of authorizations and access rights.

In this section we present the three major confinements:

1. Public

2. Group

3. Private

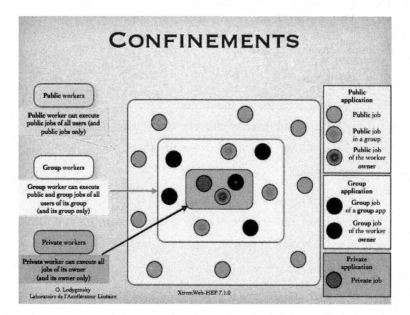

FIGURE 4.2: XtremWeb-HEP confinements.

Confinement paradigms are shown in Figure 4.2 representing the three majors confinement levels: public, group, and private. The following subsections detail these confinements.

4.3.6.1 Public Confinement

This is the default confinement. It is shown in Figure 4.2. In a public confinement, everything is public, all distributed entities (workers, clients) have read and execute accesses to public objects. The write access, if set, is reserved for the owner only. A public confinement is made of public objects:

- Public applications

- Public jobs

- Public data

- Public workers

- Public clients

- Public users

1. **Access rights for public confinement**
   Public access rights are defined to allow the owner full accesses, group read and execute accesses, as well as read and execute accesses to others. The octal value of the public access rights is 755.

2. **Management of a public confinement**

   The XtremWeb-HEP administrator is the only user allowed to manage a public confinement. Administrator authorization is the only one that permits to insert, delete, and modify public users, public workers and public applications.

3. **Resource aggregation within a public confinement**

   The XtremWeb-HEP administrator is the only user allowed to perform resource aggregation in a public confinement. To do so, the administrator must create an identity associated with WORKER_USER authorization.

   This identity must not belong to any user group.

   It is then the administrator's responsibility to prepare worker installation packages that will install the worker with this identity. All deployed workers will then connect to the server and present this identity. Such workers will be automatically public. There is no way to modify this. Public workers access public objects and execute public jobs only.

4. **Usage of a public confinement**

   Any user can read and execute public objects. Users who do not have administrator authorization cannot modify public objects they do not own. Users can send public data. Users can submit jobs referring public applications.

   Public applications must refer public data only. Jobs referring public application are automatically public. There is no way to modify this. Public jobs may refer public data only.

5. **Behavior of workers within a public confinement**

   All workers are allowed to access all public objects.

Worker confinement depends on the identity a worker presents when it connects to the XtremWeb-HEP server. A worker is public if this identity is associated with WORKER_USER authorization and does not belong to any user group. Public workers are not allowed to access non-public objects (group and private objects). This is why jobs referring public application are automatically public and must refer public data only.

4.3.6.2   Group Confinement

This confinement confines accesses within a group of users only. By default, all entities defined within a user group have read and execute accesses to group objects. A group confinement is made of group objects:

- Group applications

- Group jobs

- Group data

- Group workers

- Group clients

- Group users

1. **Access rights for group confinement**

   Group access rights are defined to allow owner full accesses, group read and execute accesses, and to deny accesses to others. The octal value of the group access rights is 750.

2. **Management of group confinement**

   The XtremWeb-HEP administrator is the only user allowed to create a group confinement. The administrator should delegate group management to a group administrator. A group administrator is an identity defined in a user group and associated to ADVANCED_USER authorization. A group administrator is allowed to insert, delete, and modify group object (users, workers, applications...).

3. **Resources aggregation within a group confinement**

   The group administrator is allowed to manage resource aggregation in a group confinement. To do so, the group administrator must create an identity within its group and associated with WORKER_USER authorization. It is then the group administrator's responsibility to prepare worker installation packages that will install a group worker with this identity. All deployed workers will then connect to the server and present this identity.

Such workers will automatically be group workers. There is no way to modify this. Group workers execute jobs owned by members of its group (this includes group and public jobs).

4. **Usage of a group confinement**

Any member of the group can read and execute group objects. Others cannot. Users who are not administrators of the group cannot modify group objects they do not own. Users can send group data. Users can submit jobs referring group applications.

Group applications must refer public data or data of its group only. Jobs referring group application are automatically group jobs. There is no way to modify this. Group jobs may refer any public data and data of the group.

5. **Behavior of worker within a group confinement**

All workers are not allowed to access group objects.

Worker confinement depends on the identity a worker presents when it connects to the XtremWeb-HEP server. A worker is a group worker if this identity is associated with WORKER_USER authorization and belongs to a user group. Group workers are allowed to access public objects and objects defined in the group. Group workers are not allowed to access private objects or objects defined in other groups. This is why jobs referring group application are automatically group jobs and must refer data accessible within the group (public data and data defined in the group).

4.3.6.3 Private Confinement

This confinement confines accesses to a single user only. All entities defined as private deny accesses to all users but the owner. A private confinement is made of private objects:

- Private applications

- Private jobs

- Private data

- Private workers

1. **Access rights for private confinement**

   Private access rights are defined to allow owner full accesses and to deny accesses to all others. The octal value of the private access rights is 700.

2. **Management of a private confinement**

   Any user can manage its own private confinement.

3. **Resource aggregation within a private confinement**

   Any user is allowed to manage its own private resource aggregation. A resource is private when it presents owner credentials not associated to WORKER_USER authorization (belonging or not belonging to any user group). It is then the user's responsibility to prepare worker installation packages that will install a private worker with its own identity. All deployed workers will then connect to the server and present this identity. Such workers will automatically be private workers. There is no way to modify this. Private workers execute jobs owned by its owner only. This includes private and group and public jobs.

4. **Usage of a private confinement**

   The owner is the only user allowed to access his or her private objects. Users can send private data. Users can submit jobs referring private applications.

   Private applications must refer data accessible by the application owner. Jobs referring private application are automatically private jobs. There is no way to modify this. Private jobs may refer any data accessible by its owner.

4.3.7   Data Confidentiality

Data may circulate over the network, from client to server, from server to workers, and back. To ensure data confidentiality, communications are secured as described in Section 4.3.1.

The middleware enforces data confidentiality according to the authorization and access rights mechanisms described above. The confinement ensures that:

- Private objects are sent to private clients and workers only

- Group objects are sent to group clients and workers only

- Public objects are freely accessible

XtremWeb-HEP provides absolutely no guarantee as soon as an object is sent to a distributed entity.

It is the user's responsibility to properly use confinement mechanisms to ensure confidentiality and restricted access to sensitive data. For example, X509 certificates are automatically private data. This ensures that this very sensitive data will never be downloaded by anyone but the owner. The middleware is written in such a way that workers can never access X509 certificates.

4.3.8 Data Integrity

The middleware ensures data integrity by checking data size and MD5 checksum on each access. It is the XtremWeb-HEP administrator's responsibility to ensure the integrity of data stored on server side.

4.3.9 Data Persistence

The XtremWeb-HEP middleware does not propose anything on this field.

It is the end-user's responsibility to ensure persistence of his or her data. For example, by using BitDiew2 P2P storage or EGI Storage Elements. The administrator may remove any data from the server without any formal notification to the data owner. It is the administrator's responsibility to ensure it does not remove any data that may lead to some problems regarding registered applications and submitted jobs.

4.3.10 Result Certification

The XtremWeb-HEP middleware does not propose anything for this field. It is the end-user responsibility to verify the results of his or her jobs.

4.4 Bridging XWHEP to Other Grids

There are various computing grids, also known as the Distributed Computing Infrastructure (DCI). Currently, these various grids are powered by different incompatible grid middleware stacks, and users of one grid have extreme difficulty getting access to resources managed by other grids. Inter-operation between several Service Grids has already been achieved by the joint work of several organizations, notably the Open Grid Forum (OGF), the Open Middleware Infrastructure Institute (OMII-Europe), and the Worldwide LHC Computing Grid (WLCG).

The Gliding-in approach to cluster resources spread in different Condor pools using the Global Computing system (XtremWeb) was first introduced in [LFC+]. The main principle consists of wrapping the XtremWeb worker as a regular Condor task and in submitting this task to the Condor pool. Once the worker is executed on a Condor resource, the worker pulls jobs from the DG server, executes the XtremWeb tasks and returns the result to the XtremWeb server. As a consequence, the Condor resources communicate directly with the XtremWeb server. Similar mechanisms are now commonly employed in Grid Computing. For example, Dirac uses a combination of push/pull mechanism to execute jobs on several Grid clusters. The generic approach on the Grid is called a pilot job. Instead of submitting jobs directly to the Grid meta-scheduler, this system submits so-called pilot jobs. When executed, the pilot job fetches jobs from an external job scheduler. The gliding-in or pilot job approach has several advantages. While simple, this mechanism efficiently balances the load between heterogeneous computing sites. It benefits from the fault tolerance provided by the DG server; if Grid nodes fail, then jobs are rescheduled to the next available resources. Finally, as the performance study of the Falkon system [RZD+07] shows, it gives better performances because series of jobs do not have to go through the meta-scheduler queue, which generally has long waiting times, and communication is direct between the worker running on the Computing Element (CE) and the DG server without intermediate agent such as the superworker. Grid security rules about multi-user pilot jobs require that the actual job owner must not be the pilot job owner, but the original job submitter. This job owner switching can be achieved using gLExec [SKV+07]. Inter-operation between SGs and DGs has already been explored, notably by the Lattice project [MBC08] at the University of Maryland (USA), the SZTAKI Desktop Grid [Bal07] (Budapest, Hungary), the Superlink project at Technion (Haifa, Israel), and the Clemson University (South Carolina, USA).

An important grid type is the Service Grid (SG), aggregating geographically distributed cluster-like resources as a coordinated federation of independently managed computing sites. Inside Service Grids, the job servers do not wait for computing resources to pull waiting jobs, but the job servers broker incoming jobs and push them to computing resources assessed as adequate. The most notable SG middleware stacks are Globus, gLite, ARC, Unicore, Naregi, and Genesis.

4.4.1 Xtremweb-HEP Plugin of the 3G Bridge

In order to bridge these various grids powered by different middleware stacks (see Figure 4.3), the EDGeS [UKF+09] and EDGI projects funded by the European Commission have designed, implemented, and put in production a Generic Grid-to-Grid Bridge, shortened as 3G Bridge and shown in figure 4.3. The XWHEP team has contributed the Xtremweb-HEP plugin of the 3G Bridge. An infrastructure using this plugin currently permits users of gLite, ARC, and Unicore to transparently take advantage of resources managed by desktop grids powered by XWHEP.

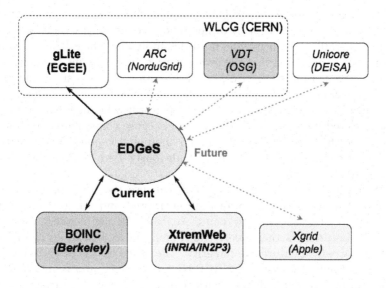

FIGURE 4.3: Grid middleware stacks bridged by the 3G Bridge.

4.4.2   XWHEP Bridge to gLite

XWHEP allows users to use not only volunteer, but also gLite Service Grid (SG) resources aggregating and using cluster-like resources, by proposing an innovative resource-sharing technology: The XWHEP bridge aggregates SG resources in a Desktop Grid (DG) deployment by submitting to the SG a pilot job (PJ) containing an XWHEP worker. XWHEP implements the DG to gLite bridge so that, under certain circumstances, jobs submitted to an XWHEP scheduler can be run by both types of resources.

4.4.2.1   Bridging Security for XWHEP Job Submission to gLite

The main concern of the XWHEP bridge is security. The XWHEP bridge must conform to SG security requirements, as well as logging and bookkeeping. To achieve these goals, XWHEP introduces unique DG features: authorization and access rights. These must be understood as well as those well known in the clustering environment. Any user willing to use the platform must have valid credentials, which define and ensure its identity as well as its authorizations. Those last allow (or deny) possible actions. For example, a user may be allowed to submit jobs, but not permitted to insert applications, etc. If the possibility of an action relies on user authorizations, the success of an action depends on object access rights.

The security in SG is ensured by X509 certificates, which must be signed by a known certificate authority(CA). These certificates are very sensitive data that must be securely kept by owner. They are designed with a long term life time (generally 12 months) and must definitely never be duplicated. They are secured by a password that the owner must be the only one to know. As grids delegate credentials to services, grid users can generate a short lifetime X509 certificate (generally 12 hours) known as an X509 proxy (or simply proxy). User delegation to grid services may be performed by migrating a proxy; this is considered an acceptable compromise because communications are always encrypted thus ensuring proxy integrity and point-to-point authentication. On the other hand, if a proxy was to be maliciously used, its short lifetime would then limit the misusage.

Following these concepts, XWHEP can use the X509 proxy, as any SG service, because communications are encrypted and services authentication ensured by the usage of key pairs. An DG user can connect using such a proxy. The proxy is then safely transferred and copied in an XWHEP data repository, which ensures its integrity. The usage of this proxy is confined to the owner only, thanks to XWHEPAR as explained above: in XWHEP, the proxy is private data, defined with a private access. This ensures that the proxy will never be sent to any volunteer computing resource. Finally, logging and bookkeeping are ensured on both grids (DG and SG) because the same X509 proxy is used.

A user submitting a job to XWHEP using a proxy gains potential access to an SG resource. The resource that will finally run the job is still unpredictable. It depends on the resource usage. The job will be sent to the first available one, which may be a DG or an SG one. To ensure its security and integrity, the X509 proxy will never be downloaded by any XWHEP resource, neither a volunteer one, nor a pilot job one.

4.4.2.2   Architecture of the XWHEP Bridge to gLite

The XWHEP bridge is the only distributed part of XWHEP that uses the X509 proxy. It is a daemon (a program that loops forever) that must be run on a machine where both XWHEP and SG client middleware are installed. Grid administrators must ensure the security and the integrity of the XWHEP bridge, because it performs some critical actions. The XWHEP bridge periodically looks for XWHEP jobs submitted with a valid X509 proxy. For each job, the XWHEP bridge downloads the proxy and uses it to perform SG actions necessary to start a pilot job. If the proxy is not valid, SG actions fail. The XWHEP bridge does not consider this an error because the job has then no chance to be computed on any SG resources, but only on DG ones (and the XWHEP bridge does not have to report this as an error because it does not manage the DG itself). If the proxy is valid, the XWHEP bridge will monitor the pilot job and report the status until completion or error. To ensure the pilot job will compute the proxy owner job, the XWHEP bridge configures it as a private resource: an XWHEP resource that can only compute the job for a given user.

Figure 4.4 shows the overall infrastructure: XWHEP and SG and the XWHEP bridge.

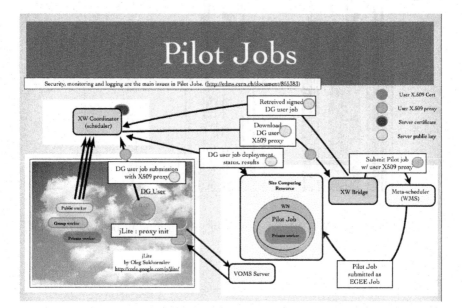

FIGURE 4.4: XtremWeb-HEP pilot jobs.

4.4.2.3 Usage of SG Resources by the XWHEP Bridge

One can easily understand that there may be resource overbooking. Because there are delays in the submission path, it may happen that the XWHEP bridge submits a pilot job that will start up after all jobs have been computed by other resources. Our experience shows that gLite resources are effectively used at an average rate of 75% by our XWHEP bridge. Figure 4.5 shows resources usage: One line shows the amount of jobs submitted with an X509 proxy, hence able to be computed by any gLite resources. Another line shows the total amount of completed jobs, independently of their submission type (with or without an X509 proxy), hence computed by XWHEP and gLite resources. Yet another line shows job failures (data unavailable, etc).

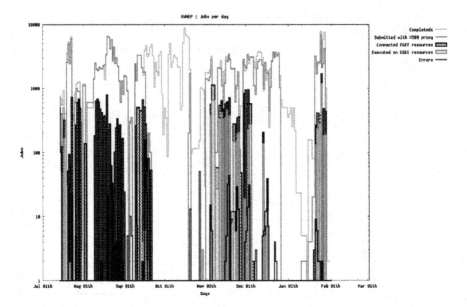

FIGURE 4.5: XWHEP: Snapshots of job running in both DG and SG.

4.5   Acknowledgments

The EDGeS (Enabling Desktop Grids for e-Science) project receives community funding from the European Commission within the Research Infrastructures initiative of FP7 (grant agreement Number 211727).

Bibliography

[And04]   David P. Anderson. Boinc: A system for public-resource computing and storage. In *GRID*, pages 4–10, 2004.

[Bal07]   Balaton, Z., Gombas, G., Kacsuk, P., Kornafeld, A.,Kovacs, J., Marosi, A.C., Vida, G., Podhorszki, N.,Kiss, T. Sztaki desktop grid: a modular and scalable way of building large computing grids. In *the 21st International Parallel and Distributed Processing Symposium*, Long Beach, CA, USA, 2007.

[FGNC01] Gilles Fedak, Cecile Germain, Vincent Neri, and Franck Cappello. XtremWeb: A Generic Global Computing Platform. In

CCGRID'2001 Special Session Global Computing on Personal Devices, 2001.

[LFC+] Oleg Lodygensky, Gilles Fedak, Franck Cappello, Vincent Neri, Miron Livny, and Douglas Thain. Xtremweb & condor : Sharing resources between internet connected condor pools. In *Proceedings of CCGRID'2003 Special Session Global Computing on Personal Devices*, number 2003, Tokyo, Japan. IEEE/ACM.

[MBC08] Daniel S. Myers, Adam L. Bazinet, and Michael P. Cummings. Expanding the reach of Grid computing: combining Globus- and BOINC-based systems. In A. Zomaya, editor, *Grids for Bioinformatics and Computational Biology*, Book Series on Parallel and Distributed Computing, pages 71–85. John Wiley & Sons, New York, 2008.

[RZD+07] Ioan Raicu, Yong Zhao, Catalin Dumitrescu, Ian Foster, and Mike Wilde. *Falkon: A Fast and Light-Weight Task Execution Framework.* 2007.

[SKV+07] Sfiligoi1, O Koeroo2, G Venekamp2, D Yocum1, D Groep, and D Petravick. Addressing the pilot security problem with gLExec. *Technical Report* FERMILAB-PUB-07-483-CD, Fermi National Laboratory, 2007.

[UKF+09] Etienne Urbah, Peter Kacsuk, Zoltan Farkas, Gilles Fedak, Gabor Kecskemeti, Oleg Lodygensky, Attila Marosi, Zoltan Balaton, Gabriel Caillat, Gabor Gombas, Adam Kornafeld, Jozsef Kovacs, Haiwu He, and Robert Lovas. Edges: Bridging egee to BOINC and xtremweb. *Journal of Grid Computing*, 7(3):335–354, September 2009.

Chapter 5

A Volunteer Computing Platform Experience for Neuromuscular Disease Problems

Nicolas Bard

CNRS, Lyon, France

Viktors Bertis

IBM Systems & Technology Group, Austin, Texas, USA

Raphaël Bolze

LIP Laboratory, Lyon, France

Frédéric Desprez

INRIA, Lyon, France

5.1 Introduction

Large-scale applications in many different fields can now be executed on platforms based on Internet-connected PCs volunteered by their owners. These are usually loosely coupled applications like search for extraterrestrial intelligence [ACK+02] or bioinformatics applications [TAKI06] that can benefit from a large number of relatively slow machines. The scheduling of their tasks is also quite easy and their performance comparable with those of large supercomputers or datacenters. However, this "simplicity" has a cost. Faults often occurs that make strong fault tolerant algorithms mandatory. These algorithms usually use replication of tasks over different nodes.

The World Community Grid's mission is to create a large public computing grid to help projects that benefit humanity [Gri]. World Community Grid initially built its grid platform using a commercial solution from UNIVA-UD [UD] (stopped since 2008) and now uses the BOINC open-source solution. The BOINC system [ACA06] is one of the most well known systems designed to build large-scale volunteer computing systems. Its simple and robust design allows the development of large-scale systems for several important applications all over the world.

The use of such platforms for embarassingly parallel applications is not new and several papers have presented results of efficient port of these application over large scale platforms. In [TAKI06], the authors present their experience of the use of P@H/BOINC system for testing structure prediction algorithms based on conformational sampling. The authors show how the use of a Desktop grid allows them to improve the prediction of protein structure. In [UTS+03], clustered processors are put within the United Device MetaProcessor to allow task and data parallelism to be used in the parallel CHARMM application. Their target application is protein folding and the aim is also to improve the quality of the folding calculations. Kondo et al. [KFC+07] use a cluster equivalence metric to quantify the utility of the desktop grid relatively to a dedicated cluster. Their study is based on a enterprise desktop grid with small number of hosts (around 220).

In this chapter we present our experience in porting a docking application, which predicts protein-protein interactions, to the World Community Grid. We are especially interested in performance estimation and the way we can predict the execution time for larger problems on such a large scale grid. We also compared the performance obtained on World Community Grid with the ones obtained on a dedicated grid. Our work should be thus taken as a practical example of how to port an application onto a Desktop Grid and how much performance and improvements can be obtained compared to a dedicated grid.

The rest of this chapter is organized as follows. The next section presents our target application, that is the "Help Cure Muscular Dystrophy project". Section 5.3 gives details about World Community Grid which we used to solve our problem. Section 5.4 gives details about the way the application is sliced into workunits of a given size that allow the best performance onthe World Community Grid. Then we describe the execution of the whole application over the grid in Section 5.5. Finally, before the conclusion, Section 5.6 presents the performance comparison of the Desktop Grid Platform with a dedicated grid for this specific application.

5.2 Help Cure Muscular Dystrophy Project (HCMD)

This project has been carried out in the framework on the Décrypthon program[1], set up by the CNRS (Centre National de la Recherche Scientifique), AFM (French Muscular Distrophy Association), and IBM. This project proposes to use the power offered by grid computing for the detection of protein-protein interactions. It directly addresses these problems by setting the goal of screening a database containing thousands of proteins for functional sites involved in binding to other protein targets. Information obtained on the structure of macromolecular complexes is important, not only to identify functionally important partners, but also to determine how such interactions will be perturbed by natural or engineered site mutations in either of the interacting partners, or as the result of exogenous molecules, and, notably, pharmacophores. A database of such information would be of significant medical interest because, while it now becomes feasible to design a small molecule to inhibit or enhance the binding of a given macromolecule to a given partner, it is much more difficult to know how the same small molecule could directly or indirectly influence other existing interactions.

The development of a docking algorithm to predict protein-protein interactions exploits knowledge on the location of binding sites. The basic algorithm uses a reduced model of the interacting macromolecules and a simplified interaction energy. Optimal interaction geometries will be searched for using multiple energy minimizations with a regular array of starting positions and orientations. Later on, knowledge of binding sites will greatly reduce the costs of the search. It provides the basis for real-case predictions of functionally significant partners. Further information on this project can be found on the website at World Community Grid [WCG].

---

[1]http://www.decrypthon.fr/.

5.2.1 MAXDo Program

The MAXDo program (Molecular Association via Cross Docking simulations) has been developed by Sacquin-Mora et al. [SMCL08] for the systematic study of molecular interaction within a protein database. The aim of the docking process is to find the best way to associate two proteins in order to form a protein-protein complex (and see whether these two proteins are likely to interact, should they ever meet in a biological system). The quality of the protein-protein interaction can be evaluated through an interaction energy (expressed in kcal.mol^{-1}), which is the sum of two contributions; a Lennard-Jones term (E_{lj}), and an electrostatic term (E_{elec}), which depends on the electric charges that are located all over the protein. The more negative the sum of these two contributions, the stronger the protein-protein interaction. The MAXDo program uses a reduced protein model developed by M. Zacharias [Zac03]. The proteins are rigid and the minimization of the interaction energy is computed according six variables: the space coordinates x, y, z of the mass center of the ligand and the orientation of the ligand α, β, γ. Those degrees of liberty are concatenated into two parameters: i_{sep} and i_{rot}, which respectively represents the *starting position* (x, y, z) of the mobile protein p_2 (also called the ligand) with respect to the fixed protein p_1 (called the receptor) and the *starting orientation* (α, β, γ) of protein p_2 relative to the fixed protein p_1. Then, minimizing the interaction energy between two proteins for a set of initial positions and orientations of the ligand gives a map of the interaction energy for the protein couple.

More formally, the MAXDo program is defined for one couple of proteins (p_1, p_2) as the computation of the interaction energy between them:

$$E_{tot}(i_{sep}, i_{rot}, p_1, p_2)$$

where i_{sep} is the starting position and i_{rot} is the starting orientation of the ligand with respect to the receptor. So the map of the interaction is obtained by computing the set of docking energy:

$$\forall i_{sep} \in [1..N_{sep}(p_1)], \forall i_{rot} \in [1..N_{rot}], E_{tot}(i_{sep}, i_{rot}, p_1, p_2)$$

Notice that the number of starting positions between the two proteins depends on the receptor p_1 and this number is directly linked with the size and shape of the protein. Finally, the total number of MAXDo program's instances that must be launched for a set of proteins P is defined by this quantity:

$$\forall (p_1, p_2) \in P, \forall i_{sep} \in [1..N_{sep}(p_1)], \forall i_{rot} \in [1..N_{rot}], E_{tot}(i_{sep}, i_{rot}, p_1, p_2)$$

The first Phase of the HCMD project targets 168 proteins. The number of rotations[2] has been fixed by the scientists at $N_{rot} = 21$, and the starting positions are evaluated by another program for each protein. The

---

[2]In fact the number of *starting orientation* is 210: 21 couples (α, β) for 10 values of γ.

MAXDo program is not symmetric, that is $if p_1 \neq p_2, E_{tot}(i_{sep}, i_{rot}, p_1, p_2) \neq E_{tot}(i_{sep}, i_{rot}, p_2, p_1)$. These 168 proteins have been selected because they are all known to take part in at least one identified protein-protein complex and they cover a wide range of protein structures and functions without redundancy [MWP+05].

The second Phase of the HCMD project targets 2,246 proteins and the docking of 2,466,753 couples of them. The behavior of MAXDo has changed after the end of the Phase I.

It is now symmetric,

$$\forall (p_1, p_2) \in P, MAXDo(p1, p2) = MAXDo(p2, p1)$$

so the number of necessary dockings is approximatively reduced by half.

A few steps of preprocessing were necessary before launching the application on the grid: executing the Joint Evolutionary Trees (JET) preprocessing on the 2,246 choosen protein files and generating the list of postions that will have to be effectively computed for each of the 2,466,753 couples of proteins. The JET step was executed on the dedicated Decrypthon university grid. JET is a java program that uses external databanks to produce a modified version of the protein files, in order to reduce the investigation space. The necessity to use such large databanks prompted us to launch these jobs on a dedicated grid instead of the desktop grid. Then the position generation step was launched on grid'5000, as the required computer time was not large enough to port it on the desktop grid. The result was a total of 137,652,178,995 conformations to compute, stored in 2,466,753 files, for a total of 401.77 GigaBytes compressed.

These 2,246 proteins were chosen because they are involved or suspected to be involved in neuromuscular dystrophy.

This project follows a first study on six proteins that was performed on the dedicated grid of the Decrypthon project. This study argues that preliminary work showed that the docking program required a lot of CPU time and produced promising scientific results [SMCL08], and can take advantage of desktop computing.

5.3   World Community Grid: A Volunteer Grid

Launched November 16, 2004, their sponsor, IBM, provides the technical infrastructure, expertise, and hosting to support the World Community Grid. The scientific projects are proposed by scientists from universities or non-profit organizations and are reviewed by prominent scientists and officials from leading public and private organizations. The reviews identify those projects with the best potential to benefit from technology of the World Community Grid and make important progress toward humanitarian goals [Gri]. This type of grid encourages public awareness of current scientific research, it catalyzes global communities centered around scientific interest, and it gives the public a measure of influence over the directions of humanitarian scientific progress.

5.3.1   Desktop Grid Description

World Community Grid is a volunteer grid, also known as a Desktop grid, or distributed grid. Basically this grid is composed of several servers that host a database of computing work (data + programs). The volunteers have to register to the Web server at the World Community Grid and download an agent that will be responsible to contact the grid servers to request workunits to process. A workunit consists of input data and programming code that runs in the background, at lowest priority, on the volunteer's computer for a number of hours. The agent is in charge of monitoring and controlling workunit computations. The agent connects to the server to get a new workunit, then it launches the program with the specific parameters corresponding to the piece of work they have been selected for. After the computing work is finished, the computing device sends back the results to the World Community Grid and asks for an another workunit. The user can configure the agent to use only the idle time of the device, or to launch the workunit only when the screensaver is active or to continuously work for World Community Grid projects.

The World Community Grid used two different types of middleware systems in order to provide the infrastructure of the Desktop grid: the BOINC system [And04] and the UNIVA-UD's middleware Grid MP [UD]. There are more than 575,000 subscribed members and more than 1,880,000 declared devices. Over 500,000 CPU-years of processing time has been contributed. The website at World Community Grid provides some basic information about the status and the global statistics of their grid [tea].

We introduced this paradigm of *virtual full-time processors*: ratio of the total CPU time produced during a period by the length of this period. With this notion we answer the question: "How many processors do we need to generate 10 years of CPU time in 1 day?". If in 1 day, 10 years of CPU time is consumed, it is equivalent to at least 3,650 processors that compute full time for 1 day. This notion of *virtual full-time processors* does not say anything about the processor. We do not have any information about the power of the grid and we only know the minimum number of processors needed to generate the CPU time. This notion is easier to understand than the total CPU time and it gives a better idea of how large the World Community Grid is. We will use this paradigm to compare World Community Grid with a dedicated grid.

According to the statistics available from the World Community Grid, the graph of the number of *virtual full-time processors* that participate in the World Community Grid can be plotted. Figure 5.1 shows the evolution of the number of *virtual full-time processors* that participate in the project hosted by the World Community Grid since the beginning of the grid. The website at World Community Grid provides the total CPU time generated by days. We converted this value into *virtual full-time processors*. We notice that the number of *virtual full-time processors* globally increases. The curve is not regular during the weekend there are less processors than during the week. There are some periods where the number of processors went down;

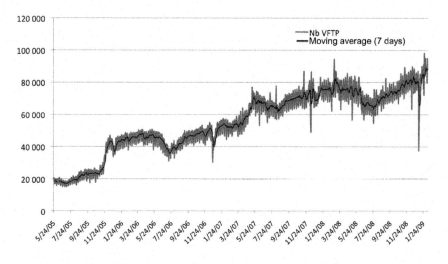

FIGURE 5.1: Virtual full-time processors of World Community Grid.

Christmas holidays and summer time, etc. The box and whiskers plot given in Figure 5.2 provides a synthetic view of statistics data of *virtual full-time processors* and it shows the growth of the platform over each year.

5.3.2 Needs and Requirements

Scientific projects must meet three basic technological requirements, to ensure benefits from World Community Grid computing power (these requirements are given in the documents called "Request for proposals" [ab]). Projects should have a need for at least many tens of millions of CPU hours of computation to proceed. The computer software algorithms required to accomplish the computations should be such that they can be subdivided into many smaller independent computations. And finally, if very large amounts of data are required, there should also be a way to partition the data into sufficiently small units corresponding to the computations.

These constraints are directly related to the nature of the World Community Grid. The computers that comprise the membership of the World Community Grid are usually simple desktop machines. Desktop computers have more and more computing and storage capacities, fast Internet connection, but the work distributed to the volunteer computer has to need reasonable memory space and data transfer time. Empirically, the team at World Community Grid has determined a workunit should last around 10 hours. This value can be considered a human factor; it represents the time a volunteer would expect to accomplish a workunit and it is a good value to monitor the progression of the work. This value is also constrained by the capacity of the servers at World Community Grid to distribute the work to volunteer devices. It determines the rate of transactions with World Community Grid servers.

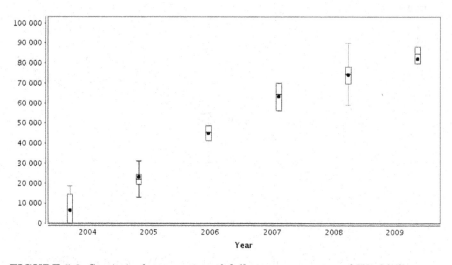

FIGURE 5.2: Statistic data on virtual full-time processors of World Community Grid since 2004.

An interesting study on performance issues of a BOINC task server has been done by the BOINC team [AKW05].

5.4 Workunits Preparation

As mentioned in the requirements for the World Community Grid, the work should be partitioned into "small pieces" of work that ideally take 10 hours to complete. In order to be able to achieve this goal, we must estimate the computing time needed by the MAXDo program.

5.4.1 Phase 1

5.4.1.1 Analysis of MAXDo Program Behavior

In order to launch the HCMD project on the World Community Grid, we have to model the behavior of the MAXDo program. The first evaluation was made to determine the parameters $N_{\text{sep}}(p)$ for the 168 proteins. This parameter is the number of starting positions of a ligand around a given receptor. Figure 5.3 gives the distribution of the number of starting positions. It shows that most of the proteins have less than 3,000 starting positions to compute. One of them has more than 8,000.

Then important properties of the MAXDo program computing time $ct(i_{\text{sep}}, i_{\text{rot}}, p_1, p_2)$ were established:

1. The MAXDo program has a reproducible computing time.

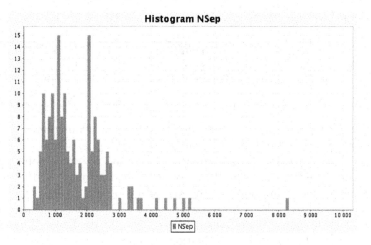

FIGURE 5.3: NsepMax distribution.

2. For one couple of proteins p_1, p_2, if the parameter i_{sep} is fixed, then the MAXDo program is linear in the parameter i_{rot} (Figure 5.4a):

$$\forall(p_1, p_2) \in P, \forall i_{rot}, i_{sep} \text{ fixed }, \exists a, b \in \mathbb{R}, ct(i_{sep}, i_{rot}, p_1, p_2)$$
$$= a * ct(i_{sep}, 1, p_1, p_2) + b$$

3. For one couple of proteins p_1, p_2, if the parameter i_{rot} is fixed, then the MAXDo program is linear in the parameter i_{sep} (Figure 5.4b):

$$\forall(p_1, p_2) \in P, \forall i_{sep}, i_{rot} \text{ fixed }, \exists a, b \in \mathbb{R}, ct(i_{sep}, i_{sep}, p_1, p_2)$$
$$= a * ct(1, i_{rot}, p_1, p_2) + b$$

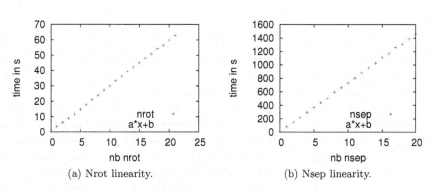

(a) Nrot linearity.

(b) Nsep linearity.

FIGURE 5.4: Parameter linearity.

The linear property was checked over 400 random couples of proteins. The correlation coefficient is always around 0.99. For the sake of simplicity, we

TABLE 5.1: Statistic Values of the Computation Time Matrix in Seconds.

Average	Standard Deviation	Min	Max	Median
671	968,04	6	46347	384

decided to assume that the computing time is a linear function of the number of *starting orientations* or the number of *starting positions* ($b = 0$). This means that we only need one point to determine the slope value (a) of the linear function for each couple. With these three properties, the number of computing time evaluations for the MAXDo program is highly reduced. It is only necessary to evaluate for a fixed number of N_{sep} and N_{rot} the computing time for each couple of proteins in the set P. The cardinal of the target set P is 168, so the number of evaluations is $168^2 = 28,224$. We launched the MAXDo program on four clusters with similar nodes (i.e., dual Opteron 246 at 2 GHz) on the Grid'5000 [BCC$^+$06] platform; 640 processors were used for this experiment during one day. This experimental run gives us the complete matrix M_{ct} of computing time, where the entry $ct_{i,j}$ represents the computing time for the couple of proteins (p_i, p_j). This 168^2 run consumed more than 73 days of CPU time; Table 5.1 represents the statistical value of the computation matrix M_{ct}. The distribution of the computing time is extremely disparate; there are ten proteins that represent 30% of the total processing time.

The matrix M_{ct} and N_{sep} table offers the possibility to evaluate the total CPU time needed on the reference processor. It needs more than 14 centuries and 88 years of CPU time on a single 2GHz Opteron processor, to be precise 1,488:237:19:45:54 (y:d:h:m:s). This quantity is represented by the following formula:

$$\sum_{p_1, p_2 \in P} N_{sep}(p_1) * 21 * ct_{iter}(p_1, p_2) \tag{5.1}$$

where $ct_{iter}(p_1, p_2)$ represents the entry value in M_{ct} for the couple of proteins (p_1, p_2), that is, time needed for the MAXDo program to run $E_{tot}(1, 1, p_2, p_1)$, and $N_{sep}(p_1)$ represents the value in the N_{sep} table, which gives the number of starting positions around the protein p_1.

At this time, some observations can be pointed out. The total CPU time needed on the reference CPU is huge, almost 15 centuries. The MAXDo program is embarrassingly parallel. Each step can be computed independently of every other, and there are a huge number of steps (49,481,544 workunits can be generated). The data needed for the MAXDo program is small: the 2 proteins files + program + parameters (no more than 2 MB). These observations show that the MAXDo program is a perfect candidate for a desktop grid such as the World Community Grid.

5.4.1.2   Workunit Packaging

The objective is now to slice the whole work represented by the formula 5.1 into small pieces of work that lasts approximately h hours ($h \simeq 10$). The constraints are the following. For a given couple of proteins (p_1, p_2), only the number of the N_{sep} to compute can change, and it must remain in the interval $[1..N_{\text{sep}}(p_1)]$. The number of orientations N_{rot} is fixed at 21. The workunit is defined for one couple of proteins, that is, we cannot build a workunit with three proteins p_1, p_2, p_3 in order to compute on a piece of the work of couple (p_1, p_2), then (p_2, p_3), and so on.

The two previous constraints are technical because it will demand unnecessary additional work to merge result files. So the problem is to find the parameter n_{sep}, which is the number of separation points to compute in one workunit for one couple of proteins:

$$\forall (p_1, p_2) \in P, \text{ find } n_{\text{sep}} \in [1..N_{\text{sep}}(p_1)],$$

$$\text{if} \quad \left\lfloor \frac{h}{21 * ct_{\text{iter}}(p_1, p_2)} \right\rfloor \leq 1, \qquad n_{\text{sep}} = 1$$

$$\text{if} \quad \left\lfloor \frac{h}{21 * ct_{\text{iter}}(p_1, p_2)} \right\rfloor \geq N_{\text{sep}}(p_1), \quad n_{\text{sep}} = N_{\text{sep}}(p_1)$$

$$\text{else} \qquad n_{\text{sep}} = \left\lfloor \frac{h}{21 * ct_{\text{iter}}(p_1, p_2)} \right\rfloor$$

There are several methods to build workunits, and we can have sub-goals such as to decrease the number of small workunits or minimize the number of workunits. It also depends on the softness of the h parameter. Figure 5.5 shows some examples of the workunits distribution generated with $h = 10$ hours and $h = 4$ hours. Indeed, the number of workunits increases when the workunit execution time wanted decreases.

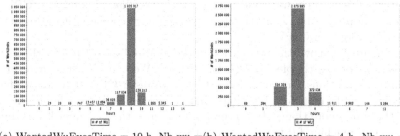

(a) WantedWuExecTime = 10 h, Nb wu = 1,364,476

(b) WantedWuExecTime = 4 h, Nb wu = 3,599,937

FIGURE 5.5: Examples of workunit distribution.

5.4.2   Phase 2

5.4.2.1   Analysis of MAXDo Program Behavior

The version of MAXDo for Phase 2 is very similar to the version for Phase 1, except for two points:

MAXDo is now symmetric, and the required computer time is no longer linear with the number of position to compute. We found the presence of randomly distributed irregular positions that can be up to 400 times more computer time costly than normal positions. We ran benchmarks to determine this fact.

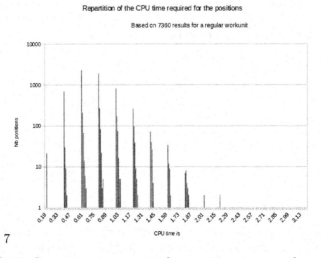

FIGURE 5.6: Computer time–required repartition on a regular workunit.

Figure 5.6 shows the repartition of the required CPU time for the 7,630 first positions of the couple of proteins "1RKC_B-2K2R_A" from the first batch. This is what we could expect after the Phase 1, and this CPU time would be predictible. Unfortunately, we also have couples with positions that require a lot more CPU time.

Figure 5.7 shows the repartition of the required CPU time for the 213 first positions of the couple "1RKC_A-2O72_A." We can see that there are two kinds of positions, "normal" ones that requires a CPU time lower than 30 seconds, and "monster" positions that require more than 1 hour (and most of the time, more than 4,100 seconds). Of course the benchmark was done on the same machine for these two couples.

5.4.2.2   Workunit Packaging

We ran a very basic kind of benchmark on every couple of proteins : for each of them we computed the first 9 positions. With this we made a raw evaluation of the number of workunits we required to split the couple into,

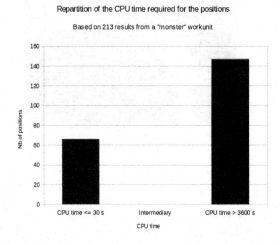

FIGURE 5.7: Computer time–required repartition on a dificult workunit.

based on the number of positions in the couple. This first benchmark was run on grid'5000, as the Decrypthon university grid was not powerful enough to do it in a reasonable time. We left a "timetable" file with required data to generate workunits, the list of the positions to compute for each couple, and parameters such as $Nsep$ for each protein available to World Community Grid staff, to generate the workunits. Then, the World Community Grid sends the workunits to the voluntary contributors.

After 6 hours of processing, if the workunit progress was at least 60% completed, it is allowed to continue, up to 4 more hours. If the 60% was not reached, or if the workunit was not finished after the 10 hours, it is stopped. The positions that were not computed are split again to generate "children" workunits, which would finish in 6 hours, on an average computer. For example, if the first workunit had 20,000 positions, and we could only get results for positions 1 through 5000 after 6 hours, four new workunits could be generated 5001–8750, 8751–12500, 12501–16250 and 16251–20000. Sometimes, these workunits needed to be split again, creating "grandchildren," "great-grandchildren," and even "great-great-grand children workunits."

5.4.3 Porting to World Community Grid

The MAXDo program has been packaged into a program with a screensaver and modified in order to monitor the progression of the program. This job has been done by the technical team at World Community Grid. Figure 5.8 shows the screensaver of the HCMD project. Several pieces of information are shown in the MAXDo agent: the name and the graphic of the two proteins that are currently being docked, the value of the docking energies, the current progress of the docking program, and the links to the website of the different partners in the HCMD project.

FIGURE 5.8: HCMD screensaver.

Furthermore, the technical team adds a checkpoint feature to the MAXDo program. The MAXDo program can be stopped at any time and restarted from the last checkpoint. This feature is essential; as the volunteers can stop or kill the process at any time, checkpoints are essential to preserve computations that have been carried out. The checkpoint occurs only between *starting positions*. If the program is stopped during the computation of one starting position, the MAXDo program must be relaunched from this position.

5.5 HCMD Project Launched on World Community Grid

5.5.1 Phase 1

5.5.1.1 Computing Phases

The previous section explained how the MAXDo program and the target set of proteins were tuned to fit the constraints of World Community Grid. The computation Phase was launched on December, 19, 2006. World Community Grid decided to launch the workunit of one protein after another. They also decided to first launch the protein that required less computing time. This choice was motivated by the fact that it can be easier to detect the failures at the beginning of the project when results return quickly from the volunteers. Furthermore, the World Community Grid is dynamic; there are always new members that join the grid with brand-new machines. So these new faster devices can work on more time-consuming workunits.

Figure 5.9a shows the number of *virtual full-time processors* that participated in the HCMD project. This graphic is generated with the data provided by World Community Grid. They give us the CPU time in years consumed each week. With this value, we estimate the number of *virtual full-time processors* that participate in HCMD. Then the following remarks can be given for Figure 5.9a:

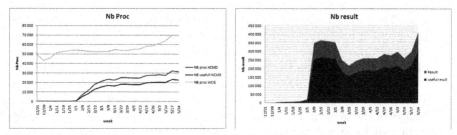

(a) Number of *virtual full-time processors* dur-(b) Number of results received per week during
ing the HCMD project. the HCMD project.

FIGURE 5.9: HCMD project on World Community Grid.

1. During the project, the number of processors that participated in World
   Community Grid always increased. The average number of processors
   available is 54,947 and the average number of processors dedicated to
   the HCMD project is 16,450.

2. Three different periods can be distinguished for the HCMD project.

   (a) During the 2 first months of the project, there are just a few pro-
       cessors that participated to the HCMD project. The priority of this
       project was very low compared to the others. This Phase can be
       assimilated to the *control period.*

   (b) In February, the number of processors that participated in the
       HCMD project increased. At the end of February, 45% of World
       Community Grid's devices participated in the HCMD project. This
       Phase can be named *project prioritization.*

   (c) From March to the end of the HCMD project: 4 months, the part
       of the processors that participated in the HCMD project stayed
       constant. As the number of processors in World Community Grid
       increased, the number of processors dedicated to the project in-
       creased. The Phase can be called the *full power working phase.*
       During this phase, 26,248 processors were used, on average.

Figure 5.9b shows the number of results received during the project. The
number of results is not necessarily linked with the number of processors
(Figure 5.9a), that is, the CPU time consumed, because each result does
not represent the same amount of work. There are two areas in Figure 5.9b;
only 73% are useful results. The World Community Grid system sends more
than one copy of each workunit to the volunteers. This is called *redundant
computing.* This mechanism allows World Community Grid to identify and
reject erroneous results. Redundant computing was used in two cases: the
result returned is not correct or the workunit sent to a volunteer reached the
timeout. Because the grid is composed of volunteers, we can expect that there

are some people who do not connect to the grid servers for a long period of time. Then when the agent reconnects and sends back the result to the servers, this result is taken into account even if the result has already been computed by some other device. The redundancy factor for all projects is 1.37; it is obtained by comparing the number of computing results disclosed by World Community Grid (5,418,010) with the number of effective results received from World Community Grid (3,936,010). This factor was not constant during all Phases of the project. It was higher at the beginning, because the results were compared to each other to be validated, but later we provided a method to validate the results by checking the values returned in the result file (there are some specific boundary conditions on each value).

5.5.1.2   Result Processing and Verification

During the project, World Community Grid sent results that were calculated by the volunteers to a storage server in France. Then we were put in charge of validating those results. The output of the MAXDo program is a simple text file that on each line gives the coordinate of the ligand and its orientation, and then the interaction energies values. World Community Grid sent us the results when one protein had been docked with the 168 others. Each time we received the results, we validated those results with three different checks: check if there are the correct number of files, check if there are the correct number of lines in the files, and check if the values in the file are within a valid range. Then when the files were checked, we merged result files in order to have one result file for one couple of proteins. All these result files represent 123 GB of text files (45 GB compressed) and there are 168^2 files.

In addition to these controls, we provide the graphics shown in Figure 5.10, which represents the progression of the project. The proteins are on the X axis, and the Y axis represents the cumulative percentage of computation. The part (on the left) is the percentage that has been computed, the part (on the right) is the not yet computed part. This graphic effectively shows that the time needed for each protein is different. For example, on the 05-02-07 (i.e., 5 months after the beginning of the project), 85% of the proteins were docked, but this represents only 47% of the 1488:237:19:45:54 (y:d:h:m:s) total computation.

5.5.2   Phase 2

5.5.2.1   Computing Phases

Phase 2 of the Help Cure Muscular dystrophy project was not completed when this chapter was written. Figure 5.11 shows the evolution of the virtual full time processors available for all of projects on the World Community Grid and for the HCMD2 project since its beginning.

As seen in Figure 5.12, the number of other projects on the World Community Grid since the beginning of the second stage of Help Cure Muscular Dystrophy varied from 5 to 8: four running projects have finished while

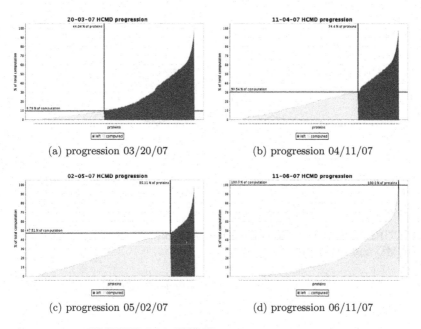

(a) progression 03/20/07   (b) progression 04/11/07

(c) progression 05/02/07   (d) progression 06/11/07

FIGURE 5.10: HCMD project progression.

HCMD2 was running (*Discovering dengue drugs together, Nutritious rice for the world, The clean energy project,* and *Inluenza antiviral drug search*), three other projects were started after the launch of HCMD2 (*Discover dengue drugs together phase II, The clean energy project phase II,* and *Computing for clean water*), and four projects were running before and are still active (*Fight aids at home, Help conquer cancer, Help fight children cancer, human proteome folding phase II*).

Figure 5.13 shows the part of the grid dedicated to HCMD2. With an average of 17.33%, HCMD2 is one of the most popular projects.

5.5.2.2 Result Processing and Verification

The result files were sent by World Community Grid to a Decrypthon server in Rouen (CRIHAN). For each "batch" of 1,000 couples there are a certain number of compressed files (usually from 3 to 200); these archive files are the gzip-compressed result files of the processing of the workunits, which are then concatenated with tar. The first step in the process is a sorting phase. Each of the archive tar-files is extracted, and the workunit results are moved to a folder corresponding to its corresponding couple. During this sorting step, we process a basic validation of the result file: we check that the number of results is correct by counting the number of lines. At this stage we are also processing a statistics-collecting step, to generate charts for the volunteers and the Internet users.

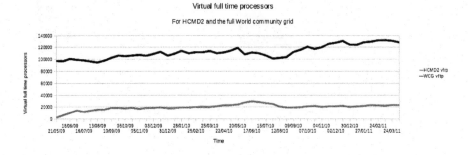

FIGURE 5.11: Repartition of World Community Grid computer time between HCMD and other projects.

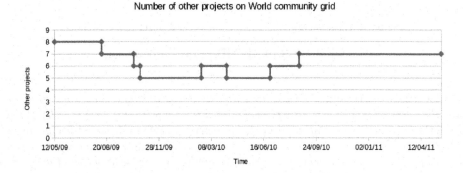

FIGURE 5.12: Other projects running on World Community Grid.

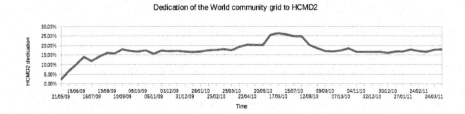

FIGURE 5.13: CPU time allocated to HCMD phase 2 on World Community Grid.

Then we need a compression step. A workunit result file contains a line for each position computed by the worker node; each line contains the coordinates of the position and the computed energy values. Thus, two values are added by World Community Grid, for checking reasons and to facilitate checkpointing. The compression step will extract the gzipped workunit result files, remove these two added columns, shorten some World Community Grid annotations in the result files, concatenate all the result files of the same couple, and then compress it at a better rate with bzip2. Usually the compression step reduces by 76% the uncompressed file, and by 30% compared to the size of the original gzip archives. Using lzma was also proposed, with a rate of 81% on our result files, but the compression time would have been too long. Now, to keep a backup of the result file, the results are stored on three different servers, and we plan to expand our storing capacity to make sure we can receive and process all the results.

5.6 Comparative Performance with a Dedicated Grid Phase I

World Community Grid's website reports that the total CPU time consumed by the HCMD project is 8,082:275:17:15:44 (y:d:h:m:s). This quantity is 5.43 times higher than the estimated value with the reference 2GHz Opteron processor on the Grid'5000 platform. However, this total CPU time includes the redundant results (the redundancy factor is 1.37; see Section 5.5). So if we avoid the redundant computation, the speed reduction factor is 3.96.

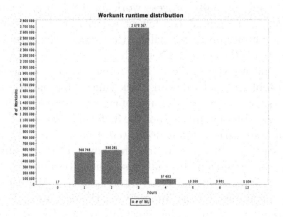

FIGURE 5.14: Real workunit distribution.

Furthermore, the distribution of real workunits that were sent to the volunteers reported in Figure 5.14 confirmed the speed reduction factor. This distribution shows that most workunits were tuned to take between 3 and 4 hours of CPU time (average is 3 hours 18 min 47s), according to the evalu-

TABLE 5.2: Equivalence between *Virtual Full-Time Processors* of World Community Grid and Processors of Dedicated Grid.

Grid	Whole Period	Full Power Working Phase
World Community Grid	16,450	26,248
Dedicated Grid	3,029	4,833

ation matrix obtained by the experiment described in Section 5.4.1.1 on the reference processor Opteron at 2 GHz. For the whole project, the average CPU time obtained on a World Community Grid device is around 13 hours (obtained from the total number of results and the total CPU time). This confirms the speed reduction factor of 3.96 (13 hours / 3.96 = 3h15): the average workunit computing time obtained with the packaging is approximately 4 times less than the real average execution time.

Several arguments can be given to explain the speed reduction factor. The first comment is that there are some limitations on the way the UD agent accounts for runtime. In particular, the UD agent measures wall clock time rather than actual process execution time. This comes into play in a couple of ways. First of all, World Community Grid has set the work for the UD agent to run, at most, at 60% of CPU time. This can only be changed by downloading a separate utility from World Community Grid's website to change this setting. This means that a computer using the UD agent with the 60% CPU throttle that runs a workunit for 8 hours of "wall clock" time will, at most, only actually process work for 4.8 hours. Second, because the research application runs at the lowest priority on the computer, any other use of the computer's processor will further reduce the actual amount of time the research runs. This can further reduce the actual CPU time. In essence, it would not be unexpected that the research application actually ran for less than 50% of the elapsed wall clock time. This means that the runtime reported overstates the actual amount of time the research was running. These constraints can explain about half of the 3.96 value. In addition, the devices on World Community Grid are not dedicated, that is, the volunteers share their computer and the agent program may be stopped for any number of reasons, even a power interruption. In that case, the program is relaunched by the agent at the last checkpoint; these interruptions consumed a large part of the additional computing time. In addition, the devices on World Community Grid are slower (on average) than an Opteron 2 GHz, and screensaver itself can add to CPU usage in varying degrees, depending on the platform and speed of the machine. Nevertheless, even if the performance of a volunteer's device is quite low compared to the power of one processor of a dedicated grid, this weakness is balanced by the huge number of *virtual full-time processors* of this kind of grid.

Table 5.2 represents the equivalence[3] between the average number of *virtual full-time processors* that were consumed during the HCMD project and the number of processors that would be necessary on a dedicated grid such as Grid'5000. Two distinct periods are shown; the *whole period*, that is, from the beginning to the end of the project, that is, 6 months or the *full power working phase*, which is only the 4 months when the HCMD project had a high priority on World Community Grid.

To be specific, during the prior week before this chapter was written, World Community Grid received 1,435 years of runtime, or an average of 74,825 days of runtime per day. This equates to 74,825 *virtual full-time processors*. Using the factor 3.96 determined above, this suggests that World Community Grid at that time provided at least the equivalent of 18,895 Opteron 2 GHz processors of Grid'5000. Since 2008, World Community Grid has used only BOINC middleware. Consequently, this is a low estimate of World Community Grid's computing power because BOINC measures runtime more accurately than GridMP did. We will explore this idea further in a future work.

5.7 Conclusion

Large-scale execution of applications on volunteer grids are no longer research projects. Several applications have been successfully ported on large-scale platforms available around the world. Among them, BOINC is one of the most utilized systems due to its performance and ease of use.

Our target docking application fits nicely on such a platform. The HCMD project was launched on December 19, 2006, and finished on June 11, 2007. It took 26 weeks to complete. During this time, 168 proteins were docked 2 by 2, which generates 123 GB of data and consumed more than 80 centuries of CPU time.

In this chapter we described all the steps that were needed to be able to launch this large-scale execution for our bioinformatics application. We showed the benefits of using a volunteer grid, but also described the limitations of this kind of grid compared to a dedicated one. The run of this project required 5.43 times more CPU time than expected. We have introduced the notion of *virtual full-time processors* that characterizes a volunteer grid against a dedicated one. We showed that the virtual full-time processor power of World Community Grid is around 4 times slower than a 2 GHz Opteron processor. We proposed a first estimation with several assumptions for the second phase of the project, which will be refined as soon as the analysis of the first phase results is completed. Additionally in Phase 1 of the HCMD project, the MAXDo program was only run on the United Devices agent, but in Phase 2 the program only runs on the BOINC agent. There are some

---

[3]This comparison must be taken carefully, because it supposed that the dedicated grid was optimally used.

differences between the way the two middleware systems account for runtime that may introduce differences in what represents a virtual full-time processor. Another way to approach the number of *virtual full-time processors* would have been to base the estimate on the number of points awarded instead of execution time. Points represent the amount of work done by computer to compute a result and are based on the runtime for that result multiplied by a weight factor determined by running a benchmark on the agent. This approach should reduce the differences between each platform and therefore be more middleware independent. This approach should also allow us to observe the trend toward more powerful processors in desktop computers.

Phase 2 of the HCMD project started in spring 2009. How the behavior of the World Community Grid volunteer devices have evolved and how it benefits to the HCMD project can clearly be seen: the number of *virtual full-time processors* available for HCMD remained higher than 20,000, whereas the World Community Grid had more projects running concurrently.

Finally, this project is a perfect example of how the specificities of two different kinds of grid (dedicated and volunteer) can be used in order to solve a bioinformatics project that required a huge amount of computation power.

Acknowledgments

This work was carried out in the framework of the Décrypthon program, set up by CNRS, AFM, and IBM. The cross-docking computations were performed on World Community Grid, and evaluations were performed on the Grid'5000 and the Décrypthon university grid. The authors wish to thank Sophie Sacquin-Mora from the IBCP, Alessandra Carbone from UPMC, Richard Lavery from IBPC, all members of the team at World Community Grid, and also World Community Grid volunteers who made this work possible.

This chapter is an update and extension of [BBDR09].

Bibliography

[ab]   World Community Grid advisory board. World Community Grid request for proposals. http://www.worldcommunitygrid.org/bg/rfp.pdf.

[ACA06]   D.P. Anderson, C. Christensen, and B. Allen. Designing a Runtime System for Volunteer Computing. In *Proceedings of the Supercomputing Conference*, Tampa, Florida, USA, 2006.

[ACK⁺02] D.P. Anderson, J. Cobb, E. Korpela, M. Lebofsky, and D. Werthimer. SETI@home: An Experiment in Public-Resource Computing. *Communications of the ACM*, 45(11), 2002.

[AKW05]   D.P. Anderson, E. Korpela, and R. Walton. High-performance task distribution for volunteer computing. In *E-SCIENCE '05: Proceedings of the First International Conference on e-Science and Grid Computing*, pages 196–203, Washington, DC, USA, 2005. IEEE Computer Society.

[And04]   D.P. Anderson. BOINC: A System for Public Resource Computing and Storage. In *Proceedings on the Fifth IEEE/ACM International Workshop on Grid Computing (CGRID04)*, 2004.

[BBDR09]   Viktors Bertis, Raphal Bolze, Frdric Desprez, and Kevin Reed. From Dedicated Grid to Volunteer Grid: Large Scale Execution of a Bioinformatics Application. *Journal of Grid Computing*, 7:463–478, 2009. 10.1007/s10723-009-9130-7.

[BCC+06]   R. Bolze, F. Cappello, E. Caron, M. Daydé, F. Desprez, E. Jeannot, Y. Jégou, S. Lantéri, J. Leduc, N. Melab, G. Mornet, R. Namyst, P. Primet, B. Quetier, O. Richard, E.-G. Talbi, and I. Touche. Grid'5000: a large scale and highly reconfigurable experimental grid testbed. *International Journal of High Performance Computing Applications*, 20(4):481–494, November 2006.

[Gri]   World Community Grid. World Community Grid web site. http://www.worldcommunitygrid.org.

[KFC+07]   D. Kondo, G. Fedak, F. Cappello, A. A. Chien, and H. Casanova. Characterizing resource availability in enterprise desktop grids. *Future Gener. Comput. Syst.*, 23(7):888–903, 2007.

[MWP+05]   J. Mintseris, K. Wiehe, B. Pierce, R. Anderson, R. Chen, J. Janin, and Z. Weng. Protein-protein docking benchmark 2.0: An update. *Proteins: Structure, Function, and Bioinformatics*, 60(2):214–216, 2005.

[SMCL08]   S. Sacquin-Mora, A. Carbone, and R. Lavery. Identification of protein interaction partners and protein-protein interaction sites. *Journal of Molecular Biology*, 382(5):1276 – 1289, 2008.

[TAKI06]   M. Taufer, C. An, A. Kerstens, and C.L. Brooks III. Predictor@ Home: A Protein Structure Prediction Supercomputer Based on Public-Resource Computing. *IEEE Transactions on Parallel and Distributed Systems*, 2006.

[tea]   WCG team. World Community Grid global statistics page. http://www.worldcommunitygrid.org/stat/viewGlobal.do.

[UD]   Univa UD. Univa UD, PCs grid solution: Grid MP. http://www.univaud.com/hpc/products/grid-mp/.

[UTS⁺03]   B. Uk, M. Taufer, T. Stricker, G. Settanni, A. Cavalli, and A. Caflisch. Combining Task- and Data Parallelism to Speed up Protein Folding on a Desktop Grid Platform - Is efficient protein folding possible with CHARMM on the United Devices MetaProcessor. In *Proc. of the IEEE International Symposium on Cluster Computing and the Grid (CCGrid '03)*, May 2003.

[WCG]   HCMD team World Community Grid. Help Cure Muscular Dystrophy description. http://www.worldcommunitygrid.org/research/hcmd2/overview.do.

[Zac03]   M. Zacharias. Protein-protein docking with a reduced protein model accounting for side-chain flexibility. *Protein Sci*, 12(6):1271–1282, 2003.

Chapter 6

How to Work with XtremWeb, Condor, BOINC on Top of BonjourGrid

Christophe Cérin

University of Paris, Villetaneuse, France

Heithem Abbes

University of Tunis, Tunis, Tunisia

Walid Saad

University of Tunis, Tunis, Tunisia

6.1   Introduction

This chapter is about the practical issues in using some popular Desktop Grid middleware, namely XtremWeb, BOINC, and Condor. Instead of presenting one by one how to use each Desktop Grid middleware, we introduce a system able to coordinate multiple instances of XtremWeb, BOINC, Condor, namely BonjourGrid. All the practival work that we introduce in this chapter is built on top of BonjourGrid.

The chapter is organized as follows. Section 6.2 introduces BonjourGrid, its philosophy and features. In Section 6.3, we explain how to install BonjourGrid on your system. You need an .iso file that is managed through VirtualBox[1] in our case. VirtualBox is a popular virtualization product for enterprise as well as home use and easy to install and configure on top of Windows or Linux. In the remainder of the chapter we do not provide examples showing how to manage multiple instances (one XW instance with one Condor instance with one BOINC instance), but for the sake of simplicity we explain how to use an XW instance on top of BonjourGrid (Section 6.4), then a BOINC instance on top of BonjourGrid (Section 6.4.1), and finally an instance of Condor on top of BonjourGrid (Section 6.4.2). In each case we show a distinct application. And last, Section 6.5 is a glossary.

6.2   BonjourGrid System

How to federate all the users and machines of all the BOINC, Condor, and XtremWeb projects? If you believe in volonteer computing and want to share more than one project, then BonjourGrid may help. We have proposed a

---

[1]See: http://www.virtualbox.org/.

novel approach, called BonjourGrid [ACJ08, ACJ09], to orchestrate multiple instances of Institutional Desktop Grid middleware. It is our way to remove the risk of bottleneck and failure, and to guarantee the continuity of services in a distributed manner. Indeed, BonjourGrid can create a specific environment for each user based on a given computing system of his choice, such as XtremWeb, Condor, or BOINC.

Many Desktop Grid systems have been developed using a centralized model in mind. These infrastructures run in a dynamic environment and the number of resources may increase dynamically. Hence, the need for decentralization is becoming increasingly important. BonjourGrid [ACJ08, ACJ09] is a novel decentralized approach to Desktop Grid systems. Its main objective is to provide a decentralized infrastructure of multi-coordinators, using the services offered by a publish/subscribe system. Unlike a classical Desktop Grid, BonjourGrid can create, on demand, a dynamic and decentralized execution environment for each user, based on existing computing systems such as XtremWeb [FGNC01], BOINC [And04], or Condor [LLM88], to run any kind of application, without the intervention of a system administrator.

The motivation is to facilitate access for the user to a large variety of projects, either for becoming a slave or for deploying a coordinator corresponding to his own project. Again, if you believe in volonteer computing, you may be interested in participating in CosmologyHome, but also CelsHome (two BOINC projects) and also in the Condor pool at the university of Madison. BonjourGrid addresses a lack for a target audience and it provides facilities to do it easily, we mean without the ability to describe the process in a separate way.

6.2.1 Overview of BonjourGrid System

The principle of the proposed approach is to create, dynamically and in a decentralized way, a specific execution environment for each user to execute any type of application without any system administrator intervention. An environment does not affect another one if it fails.

Let us give more details. Each user, behind a desktop machine in his office, can submit an application. It is important to note that BonjourGrid can handle bags of tasks and distributed applications with precedences between tasks. BonjourGrid deploys a master (coordinator), locally on the user machine, and requests for participants (workers). Negotiations to select them should now take place. Using a publish/subscribe infrastructure, each machine publishes its state (idle, worker, or master) when changes occur as well as information about its local load or its use cost, in order to provide useful metrics for the selection of participants. Under these assumptions, the master can select a subset of workers nodes according to a strategy that could balance the "power" of the node and the "price" of its use.

The master and the set of selected workers build the Computing Element (CE) that will execute, manage, and control the user application. When a CE finishes the application, its master becomes free, returns in idle state, and

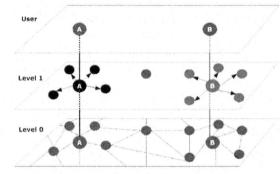

FIGURE 6.1: BonjourGrid abstract layers.

releases all workers to return also to the idle state. When no application is submitted, all machines are in the idle state.

The key idea of BonjourGrid is to rely on existing Institutional Desktop Grid middleware, and to orchestrate and coordinate multiple instances, that is, multiple CEs, through a publish/subscribe system (see Figure 6.1). Each CE will be owned by the user who has started the master on his machine. Then this CE is responsible for the execution of one or many applications for the same user. As shown in Figure 6.1, in the user level, a user A (resp. B) deploys his application on his machine and the execution seems to be local. Level 1 (middleware layer) shows that, actually, a CE with 4 (resp. 5) workers has been dynamically created, specifically for the user A (resp. B). Level 0 shows that all machines are interconnected and under the availability of any user.

To realize this approach, we compose BonjourGrid in three fundamental parts: (a) a fully decentralized resources discovery layer, based on Bonjour protocol [SC05]; (b) a CE, using DG middlewares such as XtremWeb (XW), Condor or BOINC, which executes and manages the various tasks of applications; (c) a fully decentralized protocol of coordination between (a) and (b) to manage and control all resources, services, and CEs. In the current work, we use XW as a computing system. Future works will deal with other systems such as BOINC and Condor, where we intend to take into account the security issue (for instance, applications signature for BOINC).

In a previous work [ACDJ08], we tackled the following questions: Can a system based on publish/subscribe protocols be "powerful"? Can it be scalable? What is the response time to publish a service? What is the discovery time of a service? We carried out experiments on various protocols such as Bonjour protocol, on the GdX node of Grid5000[2] platform using more than 300 machines (AMD Opterons connected with a 1GB/s network). Measure-

---

[2]https://www.grid5000.fr.

ments show that Bonjour is reliable and very powerful in resources discovery. Indeed, Bonjour discovers more than 300 services published simultaneously in less than 2 seconds with 0% loss. These evaluations support and justify our choice to use Bonjour as a basis for the discovery protocol in the Bonjour-Grid system. Moreover, we believe that using an existing and tested protocol for the industry (Bonjour is available on Macintosh machines) to evaluate its usability in the desktop grid environment is, in itself, of great value and very useful for the community.

6.2.2 Service-Oriented Architecture for Building a Computing Element

Each machine in BonjourGrid can have one of three states (Idle, Worker or Coordinator). Each state is associated to a service: *IdleService* for idle machine, *WorkerService* for worker, and *CoordinatorService* for coordinator. When a machine changes its state, it publishes the corresponding service to notify its new state after having deactivated the old one.

6.2.2.1 From Idle to Coordinator State

In the remainder, we assume that we have to coordinate XW instances. We suppose that each user keeps his machine connected to the desktop grid during the execution of his application in order to avoid managing the fault-tolerance problem between coordinators. The fault tolerance-issue has been studied in another work.

A user submits his application through the local user interface of Bonjour-Grid (installed on his machine). The user machine changes, now, its state into the coordinator state in order to initiate the construction phase of a new CE. If this machine is a worker for another application, BonjourGrid waits until it finishes the task in progress and then launches the coordinator in order to not degrade machine performance.

The coordinator starts by running a discovery program (or Browser) on idle machines. The discovered machines are recorded in an IdleMachineDict dictionary. From this data structure, the coordinator selects machines that fit the application requirements and creates a new MyWorkersDict dictionary. The coordinator continues the research, that is, its browser remains listening on idle machines until the size of MyWorkerDict reaches the number of required machines (i.e., the size of the CE). Thereafter, the coordinator stops the browser. Now the coordinator checks if selected workers accept to work for it. Indeed, the coordinator publishes for each selected worker, a new "RequestWorker" service using the "Worker Name" as service type. We note that the browser program listens on service type and not on service names.

The confirmation of participation of worker "A" to the "C" coordinator relies on the fact that "A" publishes a new "MyConfirmation" service putting as service type the name of the "C" coordinator. Only coordinator "C" runs a browser program on services of this type. Hence, if the browser discovers new registration of a service of this type, it means that there is a new idle machine that has accepted to work for it. If the confirmations number does not reach the size of the CE, the coordinator launches again the browser on idle machines, but now with the number of missing machines only.

It is possible that more than one coordinator selects the same worker "A" and publish a service "WorkerRequest" with the same service type (PTR: WorkerA)[3]. Thus, there may be contention in the access to the worker "A". BonjourGrid provides efficient and simple feedback on that. Indeed, an idle machine "A" confirms its participation to the first service "RequestWorker" discovered (i.e., the first coordinator), knowing that the protocol Bonjour guarantees that the browser does not discover more than one service at the same time.

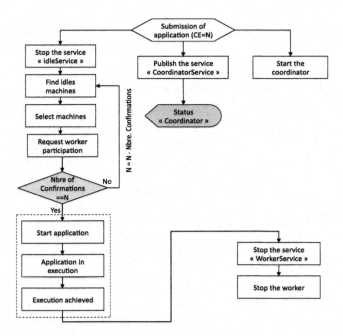

FIGURE 6.2: Steps to transform an idle machine into a coordinator

In the middleware layer, BonjourGrid establishes the connection between XW-Coordinator and XW-Workers that have accepted to participate. Indeed, when an idle machine publishes the *CoordinatorService*, it launches also the XW-Coordinator. This XW-Coordinator remains listening to new connections of XW-Workers. In the same way, when an idle machine browses the "Request-Worker" of a coordinator and publishes the service "MyConfirmation", it also starts the XW-Worker with the IP address of the coordinator (the address is stored in the TXT record of the "RequestWorker" service).

---

[3]Bonjour manages this problem and can distinguish the two services.

Since the first connection of an XW-Worker, the XW-Coordinator submits the first task of the application and remains ready for other connections from other XW-Workers. If the coordinator does not find available idle machines, the application finishes its execution with only connected XW-Workers before reaching the required size of the CE. When the application finishes its execution, the coordinator deactivates its *CoordinatorService* service and stops XW-Coordinator. It returns to its initial state by publishing the *IdleService* service again. Figure 6.2 illustrates the steps of the protocol from idle state to coordinator state.

6.2.2.2 From Idle to Worker State

Figure 6.3 presents states diagram to describe the steps taken by a host to change its state from idle to worker. Indeed, when a host joins the BonjourGrid system, it is recorded with the initial idle state. This host remains listening to notifications of coordinators that need its participation. When the machine discovers the participation request, it answers this request by publishing a "MyConfirmation" service as described above.

Thereafter, the machine stops the browser, removes the *IdleService* service, and publishes the *WorkerService* one to announce its new state: it is no longer free, and now it becomes a worker. The worker runs the XW-Worker program

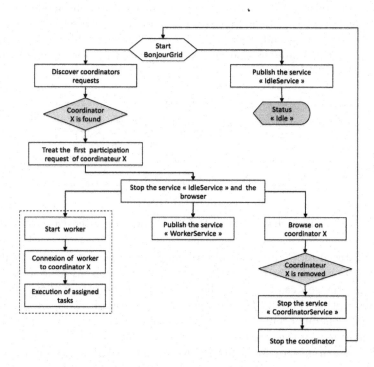

FIGURE 6.3: Steps to transform an idle machine into a worker.

with the IP address of the coordinator. The worker remains in possession of
the coordinator while this one is alive. It runs a browser in order to listen to
new events from its coordinator, especially its "death." When the coordinator
is stopped, the worker removes the *WorkerService* and stops the XW-Worker
process. Finally, the machine turns over to the initial state by publishing the
IdleService again. In this way, the coordinator does not need to contact its
workers to liberate them.

6.2.3   Implementation

BonjourGrid is based on Bonjour, the Apple protocol implementing
the Publish/Subscribe paradigm. BonjourGrid is entirely implemented using
Python. We have used the Bonjour-py package that offers a Python interface
to interact with Bonjour protocol. When a machine joins our system, information about its characteristics (i.e., MHZ, CPU, RAM, Hostname, IP address,
average load) are automatically collected and stored in a Python dictionary.
This information is useful for the selection of suitable machines that match
the application requirements. For instance, one would like to select workers
according to the CPU metric only; another one would like to select workers
that have been being idle for a long time. In our system, we can easily plug
in a new policy based on other selection criteria.

The Publish/Subscribe paradigm is an asynchronous mode for communicating between entities. Some users, namely the subscribers or clients or
consumers, express and record their interests under the form or subscriptions,
and are notified later by another event produced by other users, namely the
producers [EFGK03].

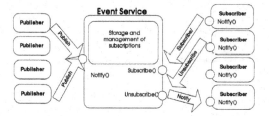

FIGURE 6.4: The Pub-Sub paradigm [EFGK03].

As stated on Figure 6.4, subscribers record their interest by a call to the
subscribe() operation inside the event service management system, without
knowing the source of events. The *unsubscribe()* operation allows one to stop
a subscription. The *notify()*(or *publish()*) operation is called by *publishers* in
order to generate events that will be propagated to subscribers and such event
are managed by the event service management system too. Each subscriber
will receive a notification for every event that conforms to its interest.

This communication mode is thus multipoint, anonymous, and implicit.
It is a multipoint mode (one-to-many or many-to-many) because events are

sent to the set of clients that have declared an interest in the topic. It is an anonymous mode because the provider does not know the identity of clients. It is an implicit mode because the clients are determined by the subscriptions, and not explicitly by the providers.

It is also known that this asynchronous communicating mode allows spatial decoupling (the interacting entities do not know each other) and time decoupling (the interacting entities do not need to participate at the same time). This total decoupling between the production and consumption of services increases the scalability by eliminating many sorts of explicit dependencies between participating entities. Eliminating dependencies reduces the coordination needs and consequently the synchronizations between entities. These advantages make the communicating infrastructure well suited to the management of distributed systems and simplifies the development of a middleware for the coordination of DGs.

6.3   Installing BonjourGrid

BonjourGrid sources are available on SourceForge[4] or on a server[5] at Centre de Calcul El Khawarizmi in Tunisia.

Figure 6.5 depicts what the top level of the distribution contains. We find two directories. The first one is related to "how to" and the folder is depicted in Figure 6.6. The folder also contains a Linux image if you prefer, and the files are swap.img.bz2, disk.img.bz2, bgrid.cfg.bz2. The second folder is related to the building of BonjourGrid from sources and the folder's content is depicted on Figure 6.7. The folder contains sub-folders: one contains the BonjourGrid sources, one contains BOINC, and one contains a Condor implementation able to run on top of BonjourGrid.

Each folder comes with documentation explaining what the user must to do with the files. It is important to note that the user/reader has two possibilities. Either he is working with the virtual machine containing the BonjourGrid system or he has to install (see Section 6.3.2) the BonjourGrid system from the sources on his physical machine. With the virtual machine in place, the user can start any number of BonjourGrid systems; then he decides (through a script) who plays the role of coordinator and who plays the role of workers.

6.3.1   How to Create and Set Up Your Virtual Desktop Grid Based on BonjourGrid

This manual will help you to easily create your own virtual grid based on BonjourGrid. Through this guide, you can create a grid with some virtual machines in order to understand the concept of BonjourGrid and perform research as well. It will be easy to set up a Desktop Grid based on Bonjour-

---

[4]http://sourceforge.net/projects/bonjourgrid/.
[5]http://grid6.cck.rnu.tn/BonjourGrid/.

Home Package:

FIGURE 6.5: The root files of the BonjourGrid distribution.

Grid in laboratories, research units, companies, and also universities using our manual "How to create your own grid based on BonjourGrid". You download the BG-VM package containing virtual machine with the kernel BonjourGrid and scripts necessary for the establishment of the grid.

The content of the package BG-VM is:

- Virtual Machine Disk (BonjourGrid Virtual Machine): This directory contains the BonjourGrid virtual machine:

  - bgrid: This folder contains the image of the machine named bgrid, created with XEN and containing the kernel of BonjourGrid (disk.img and swap.img).

  - bgrid.cfg: A configuration file for the machine bgrid on XEN.

- Supervisor-Machine (Header or front Machine): This directory contains the scripts necessary for a supervisor machine:

  - Create-CE.sh: To create a computing element (1 master + N workers). This script is started on one node and it spawns the coordinator and the workers (see below for the full list of parameters).

  - Start-BG.sh: To start the module coordinator on a "free" machine.

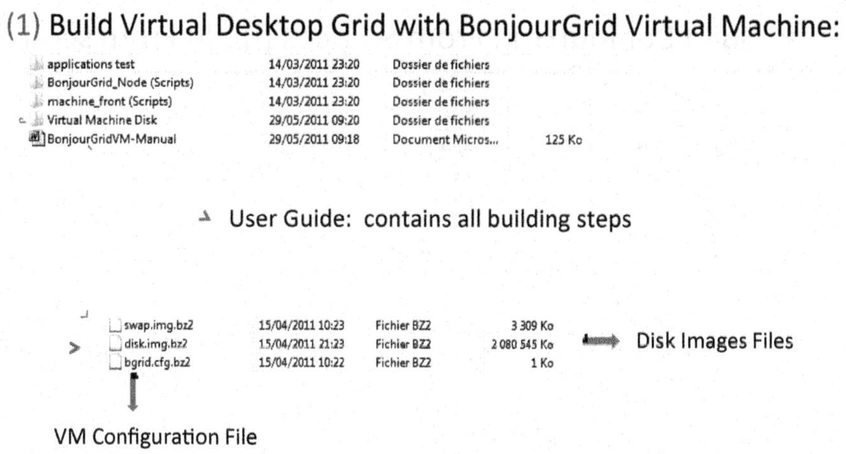

(1) Build Virtual Desktop Grid with BonjourGrid Virtual Machine:

applications test	14/03/2011 23:20	Dossier de fichiers	
BonjourGrid_Node (Scripts)	14/03/2011 23:20	Dossier de fichiers	
machine_front (Scripts)	14/03/2011 23:20	Dossier de fichiers	
Virtual Machine Disk	29/05/2011 09:20	Dossier de fichiers	
BonjourGridVM-Manual	29/05/2011 09:18	Document Micros...	125 Ko

↳ User Guide: contains all building steps

swap.img.bz2	15/04/2011 10:23	Fichier BZ2	3 309 Ko	
disk.img.bz2	15/04/2011 21:23	Fichier BZ2	2 080 545 Ko	**Disk Images Files**
bgrid.cfg.bz2	15/04/2011 10:22	Fichier BZ2	1 Ko	

VM Configuration File

FIGURE 6.6: Using a disk image

- Get-app-result.sh: To retrieve the results of the computing element.
- Get-app-status.sh: To follow the progress of the application.
- Get-host-status.sh: To view the status of each machine.
- Get-CES-status.sh: To view the status of the computing elements.
- Set_command.sh: To make the scripts as commands.

6.3.1.1 Preparing the Environment for the Grid

Before you begin creating your grid, you have to install XEN. It is an open-source virtualization platform that is both flexible and powerful. To download and install XEN check the manual of XEN. You also need to install ssh, which is a secure communication protocol. The protocol requires an exchange of encryption keys at the beginning of connection. Subsequently, all the frames are encrypted. The keys must be installed on your machine (it is already installed on the BG-VM).

To build statistics on grid usage, the virtual machine contains the Ganglia monitoring system. It provides a Web interface to monitor the grid, so you need a Web server to use it.

Installing and Configuring Xen on a Debian is easy:

```
apt-get update apt-get install xenman xen-hypervisor-3.2-1
xen-linux-system-2.6.26-2-xen-686 xen-tools libc6-xen xen-utils-3.2-1
```

From now on, you can reboot your new kernel. It should behave the same way as the old one. You get hypervisor assets and a nucleus on domain 0.

Your machine starts on the XEN kernel. Edit the following file to configure your new machine and follow the steps:

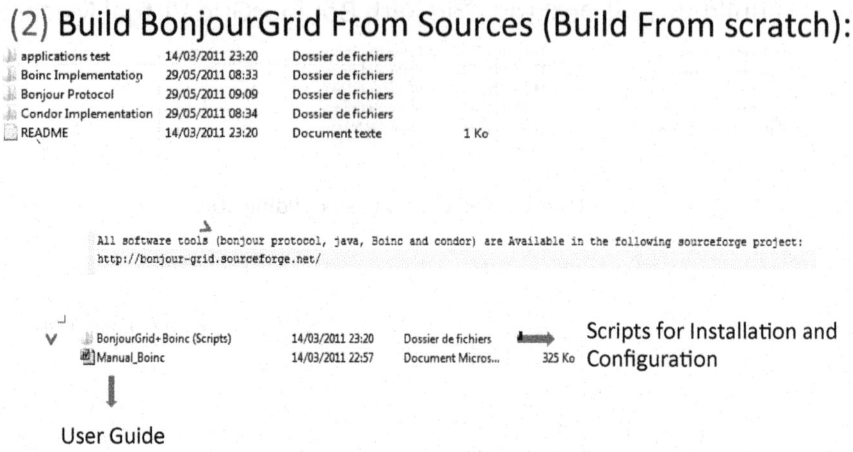

(2) Build BonjourGrid From Sources (Build From scratch):

applications test	14/03/2011 23:20	Dossier de fichiers	
Boinc Implementation	29/05/2011 08:33	Dossier de fichiers	
Bonjour Protocol	29/05/2011 09:09	Dossier de fichiers	
Condor Implementation	29/05/2011 08:34	Dossier de fichiers	
README	14/03/2011 23:20	Document texte	1 Ko

All software tools (bonjour protocol, java, Boinc and condor) are Available in the following sourceforge project:
http://bonjour-grid.sourceforge.net/

BonjourGrid+Boinc (Scripts)	14/03/2011 23:20	Dossier de fichiers	➡	Scripts for Installation and
Manual_Boinc	14/03/2011 22:57	Document Micros...	325 Ko	Configuration

User Guide

FIGURE 6.7: Building from scratch and other usage.

```
Edited file: xend-config.sxp ==> vim /etc/xen/xend-config.sxp

Make sure you have these lines:

60 ##
61 # To bridge network traffic, like this:
62 #
63 # dom0: fake eth0 -$>$ vif0.0 -+
64 #
65 # bridge -> real eth0 -> the network <= this one
66 #
67 # domU: fake eth0 -$>$ vifN.0 -+
68 #
69 # use
70 #
71 # (network-script network-bridge) <= and this one
```

Uncomment the lines here for example No. 65 and No. 71:

```
60 ##
61 # To bridge network traffic, like this:
62 #
63 # dom0: fake eth0 -$>$ vif0.0 -+
64 #
65 bridge -> real eth0 -> the network <= as above
66 #
67 # domU: fake eth0 -$>$ vifN.0 -+
68 #
69 # use
```

70 #
71 (network-script network-bridge) <= and as

Restart the service xend: /etc/init.d/xend restart

Edited file: xen-tools.conf (in /etc/xen-tools/xen-tools.conf)

Replace and uncomment line 27 (dir = /home/xen) by your home name:

23 # New instances will be stored in subdirectories named after their
24 # hostnames.
25 #
26 ##
27 dir = /home/xen

Add your kernel boot line No. 28 and No. 29:

28 kernel = /boot/vmlinuz-2.6.26-2-xen-686}
29 initrd = /boot/initrd.img-2.6.26-2-xen-686}

Create the machine\index{machine}s of the grid

Copy the folder bgrid, located in the directory BG-VM, under
/home/xen/domains.

Copy the file bgrid.cfg, located in the directory BG-VM under
etc/xen.

Copy the folder machine\index{machine}front under your home directory.

Open a terminal as a root account and run the script: add-machine.sh
with the following command: # sh add-machine.sh New_Machine_Name

This script makes a copy of bgrid folder with the name of the new
machine and copy also the configuration file located in /etc/xen/
with the name of the new machine.

Still in the terminal window and always placed under the header machine
folder, run the script: config-machine.sh with the following command:
sh config-machine.sh Machine_Name. This script automatically edit
/etc/xen/Machine_Name.cfg according to desired parameters. If you
want to edit manually you can do so using the command:
vim /etc/xen/Machine_Name.cfg
(You must change the parameters: disk, name, path)

Repeat these two commands (# sh add-machine.sh New_Machine_Name,
sh config-machine.sh Machine_Name) to create as many machines as you
want. So far, the grid's machines have been created; they must be
launched to configure each of them in the grid.

6.3.1.2   Configure a Virtual Machine for the Grid

The first thing to do is that you must start the virtual machine with the following command (replace machine_name with the name of your machine):

```
# xm create /etc/xen/machine_name.cfg
```

```
After launching your machine, you connect as root (see Part 6:
confidential information of a machine type). Under the root directory
you will find a script called addmac.sh that allows you to change the
MAC address of the machine according to your choice and will be
recorded:
```

```
# sh addmac.sh
```

```
You must configure the network interface of the virtual machine (eth0)
via the command:
```

```
# vim /etc/network/interfaces
```

```
Change the parameters of the interface eth0 of your choice (address,
net mask, broadcast, network, gateway,)
```

```
After the network configuration of the machine, we must change the
name of the host simply with the command:
```

```
# sh set-name.sh Name (Or you edit the file /etc/hostname)
```

```
To ensure communication between machines in grid with domain names,
you must add the IP addresses and machine names in the file /etc/hosts:
```

```
# vim /etc/hosts
```

```
After completing the above steps you must restart your machine to
take into account the changes:
```

```
# reboot
```

```
Restart the console of the machine with the command:
```

```
# xm console hostname
```

```
Now test the communication from the machine with the other elements
of the grid using the ping command.
```

6.3.1.3   Secure Communication within the Grid

SSH ensures that transactions are secure in the grid. We must create two keys for the machine: a public and a private key via the following command:

```
# ssh-keygen
```

Copy the public key to users (boincadm, condor, and root) of other
machines using ssh-copy-id:

```
# ssh-copy-id boincadm@machineip
# ssh-copy-id condor@machineip
# ssh-copy-id root@machineip
```

(For each machine, you should execute these three commands with the
same IP address of a machine in the grid, then repeat with the address
of another machine from the grid) is for the use of machine resources
in a tunnel, without having to authenticate each time.

Note: This step must also be done on the home machine.

6.3.1.4   Configure the Supervisor Machine (Home Machine, Front Machine)

The file "listnode" in the directory of the home-machine must contain the
IP addresses of every machine in the grid. You can edit it with this command:

```
# vim listnode
```

The scripts for the handling and the administration of the grid are:

create-CE.sh: This script is used to create computing element in passing
the necessary parameters:
```
# sh create-CE.sh Appdir middleware nbreofworker listeNode appname
nbjob filename
```
where
Appdir is the directory where the application is stored
middleware is either Condor or BOINC or Xtremweb
nbreofworker is the number of workers that you need
listeNode is the list of machines where we will run
      the aaplication and the BonjourGrid system
appname is the name of the application
nbjob is the number of jobs in the workflow
filename is the workflow

get-app-statut.sh: connects to the master of the computing element
to follow the progress of the application

```
# sh get-app-status.sh Appname Coordinator Middleware
```

get-app-result.sh: connects to the master of the element of calculation
and asks for the results of the application on this element.

```
# sh get-app result.sh Appname Coordinator Directory Middleware
```

```
get-host-status.sh: to view the status of each machine in the grid.
```

```
# sh get-host-status.sh listNodeFile
```

```
get-CESS status.sh: Shows the status of computing elements.
```

```
# sh get-EC-status.sh listNodeFile
```

```
Confidential Information of a virtual machine: The passwords for users
of the BG-VM
```

In the BG-VM (BonjourGrid Virtual Machine) we have three users:

1. Root: The root. His password is bonjourgrid.

2. Condor: a simple user. It is necessary for the operation of the Condor middleware. His password is also bonjourgrid.

3. Boincadm: a simple user, it is necessary for the operation of the BOINC middleware. His password is bonjourgrid too.

6.3.1.5 Configuring mySQL

The database in the VM-BG contains two users:

1. Root: the root of the database his password :bonjougrid

2. Boincadm: without password

6.3.2 Installing BonjourGrid from Sources

In the forthcomming paragraphs we explain how to install from sources the BonjourGrid system on the user's machine. We need to install several packages first, then configure the system. This approach is quite different from the previous one consisting of running the BonjourGrid system inside a virtual machine.

6.3.2.1 Installation and Configuration Package mDNSResponder-107.5

The installation procedure and configuration is as follows:

- Decompress the archive mDNSResponder-107.5 with this command:
  ```
  # tar xvfz mDNSResponder-107.5.tar.gz
  ```

- Move to the archive directory decompressed mDNSResponder-107.5/ mDNSPosix/:
  ```
  # cd mDNSResponder-107.5/mDNSPosix/
  ```

- Specify the target platform using the command:

  # make os=linux

- Specify the location of the JDK with the command:

  # make os=linux JDK=$JAVA_HOME Java

  Replace $JAVA_HOME with the location of the JDK.

- Proceed to the installation via the command:

  # make os=linux install (this command will be run in root mode)

- Copy the libraries required for the operation of implementing Bonjour in the $JAVA_HOME/jre/lib/i386/ so that they are supported by Java through the following three commands:

  # cp build/prod/libdns_sd.so $JAVA_HOME/jre/lib/i386/

  # cp build/prod/libjdns_sd.so $JAVA_HOME/jre/lib/i386/

  # cp build/prod/dns_sd.jar $JAVA_HOME/jre/lib/ext/

  Thus, at the end of this stage of installation and configuration of Bonjour, you can start the service via the command:

  # /etc/init.d/mdns start

6.3.2.2   Installation of Bonjour-py-0.1 (Programming Interface for Python)

The installation procedure is as follows:

- Decompress the archive bonjour-py-0.1.tar.gz with this command:

  # tar xvfz bonjour-py-0.1.tar.gz

- Move to the directory bonjour-py-0.1 with this command:

  # cd bonjour-py-0.1

- Run the installation script via this command:

  # python setup.py install

6.3.2.3   Installation of BOINC

The procedure is as follows:

STEP1: Basic settings:

- Create an account boincadm:

  # useradd -m -s /bin/bash boincadm

  # usermod -G boincadm www-data

- Log on with the account boincadm
- Copy the directory BonjourGrid+Boinc into /home/boincadm/
  ```
  # cp -r BonjourGrid+Boinc /home/boincadm
  ```

STEP 2: Configuration and installation of libraries to work with BOINC:

- As root or with command sudo:
  ```
  # cd /home/boincadm/BonjourGrid+Boinc
  # sh install-api-BOINC.sh
  ```

STEP 3: Installing the BOINC environment. On each node of the Grid:

- Installing BOINC Master: Copy the BOINC server package into /home/boincadm/BonjourGrid+Boinc
  ```
  # cp boinc-server.tar.gz /home/boincadm/BonjourGrid+Boinc/
  # cd /home/boincadm/BonjourGrid+Boinc/BGB/BG-master/
  # sh boinc-server-install.sh
  ```
- Installing BOINC Worker: Copy the BOINC client package into /home/boincadm/BonjourGrid+Boinc
  ```
  # cp boinc_5.8.16_i686-pc-linux-gnu.sh /home/boincadm
  /BonjourGrid+Boinc
  # cd /home/boincadm/BonjourGrid+Boinc/BGB/BG-worker/
  # sh boinc-worker-install.sh
  ```

STEP 4: Create a BOINC project:

```
# cd /home/boincadm/BonjourGrid+Boinc/BGB/BG-master/
# sh boinc-create-project.sh url_project projectname
```

Upon request, script will automatically create and configure a boinc project. Usage: `# sh boinc-create-project.sh http://127.0.0.1/bonjourgrid` or `# sh boinc-create-project.sh http://Grid01.localdomain bonjourgrid`

6.3.2.4   Installation of Condor

The procedure is as follows:

Basic settings:

- Create a condor account:
  ```
  # useradd -m -s /bin/bash condor
  ```
- Log on with the condor account
- Copy the directory BonjourGrid+CONDOR into /home/condor/
  ```
  # cp -r BonjourGrid+CONDOR /home/condor
  ```

- Copy the condor package into /home/condor/ BonjourGrid+CONDOR/:
  # cp condor-7.2.0-linux-x86-debian40.tar.gz
  /home/condor/BonjourGrid+CONDOR/ – Beware that you may download the tarball containing the latest version and not the source files.

STEP 2: Installing the Condor environment. On each node of the Grid:

- Condor CentralManager (Submit and manager machines):
  # cd /home/condor/BonjourGrid+CONDOR/BGC/BG-master/
  # ./condor-manager-install.sh
- Condor Execution machine:
  # cd /home/condor/BonjourGrid+CONDOR/BGC/BG-worker/
  # ./condor-worker-install.sh NameCoordinator (Ex: Name-Coordinator = 192.168.1.1 or Grid01.localdomain)

6.3.2.5   Installation of XtremWeb

The procedure is as follows:

STEP 1: Basic settings:

- Create an xwch account:
  # useradd -m -s /bin/bash xwch

  – Log on with the xwch account
  – Copy the directory BonjourGrid+XW into /home/xwch/
    # cp -r BonjourGrid+XW/home/xwch

STEP 2: Configuration and installation phpMyadmin:

- Download phpMyAdmin from this link http://www.phpmyadmin.net
- Copy the phpMyAdmin directory into /var/www/html/
- Test phpMyAdmin: http://localhost/phpMyAdmin or http://IPadress/phpMyAdmin .

STEP 3: Installing the XW environment. On each node of the Grid:

- Installing XW Master(Coordinator machine): Copy the XW server package into /home/xwch/BonjourGrid+XW
  # cp master.tar.gz /home/xwch /BonjourGrid+XW/
  # cd /home/ xwch /BonjourGrid+ xwch /BGB/BG-master/
  # sh create-xw-DataBase.sh
  # sh xw-server-install.sh
  # sh create -xw-WebSite.sh

- Installing XW Worker: Copy the XW client package into /home-/xwch/BonjourGrid+XW

  ```
  # cp xwchworker-debian5.tar.gz /home/xwch/BonjourGrid+XW/
  ```

  ```
  # cp xwchworker-debian5.tar.gz /home/xwch /BonjourGrid+XW/
  ```

  ```
  # cd /home/xwch /BonjourGrid+XW/BGB/BG-worker/
  ```

  ```
  # sh xw-worker-install.sh
  ```

6.4   Deploying an Application with BonjourGrid Middleware

6.4.1   Deploying an Application with BOINC as the Desktop Grid Middleware

6.4.1.1   Gridifying Application

In this section we describe the UPPERCASE application. It is a distributed application according to the SPMD model (Simple Program Multiple Data). It is a program written in C++ that takes as parameters a text file, consisting of lowercase letters, and CPU time (in seconds) planned to convert the file to uppercase. First, the user must create two XML file. The first describes the work unit (*workunit template file*, Figure 6.8). The second specifies the results (*result template file*, Figure 6.9).

```
<file_info>
<number>0</number>
</file_info>
<workunit>
<file_ref>
<file_number>0</file_number>
<open_name>in</open_name>
</file_ref>
<command_line>-cpu_time 10
</command_line>
</workunit>
```

FIGURE 6.8: The workunit template file.

In BonjourGrid, to deploy a BOINC application, the user must prepare a tree describing the various points mentioned previously (see Figure 6.10).

```
<file_info>
<name><OUTFILE_0/></name>
<generated_locally/>
<upload_when_present/>
<max_nbytes>5000000</max_nbytes>
<url><UPLOAD_URL/></url>
</file_info>
<result>
<file_ref>
<file_name><OUTFILE_0/>
</file_name>
<open_name>out</open_name>
</file_ref>
</result>
```

FIGURE 6.9: The workunit result file.

6.4.1.2 Deploying Application

For running BOINC applications on top of BonjourGrid, you need to follow this step:

STEP 1: Start BOINC and BonjourGrid environments:

- Master:
  `cd /home/boincadm/projects/Name_Project/`.
  Example: `cd /home/boincadm/projects/bonjourgrid`
  `sh startmaster.sh nbreCPU AppDir submitTime appname nbjob`
  nbreCPU: Number of execution machines in the computing element.
  `AppDir`: The path of folder containing all the applications to deploy. By default it is the apps folder that is already created under `/home/boincadm/projects/bonjourgrid/`
  Example: `sh startmaster.sh 10 apps 22:39:18 UPPERCASE 100`
  The scripts will build a computing element composed of a coordinator and execution machines (workers).
  In parallel, this script will launch the bonjour protocol and will deploy the boinc server (one boinc project):
  `sh boinc-server-conf.sh projectName URLPriject @IP`

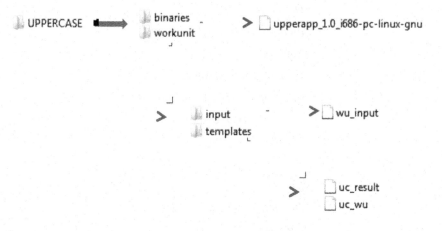

FIGURE 6.10: BOINC application.

Example: `sh boinc-server-conf.sh bonjourgrid`
`http://192.168.1.1 192.168.1.1`

- Application deployment:

  `python deploy-app.py appname inputFile projectname`

  Example: `python deploy-app.py UPPERCASE.zip`
  `UPPERCASE bonjourgrid`

  `cd /home/boincadm/projects/bonjourgrid/`
  `&& ./bin/update_versions`

- Worker

  `cd /home/boincadm/BonjourGrid+Boinc/BOINC/`

  `sh startworker.sh`

  In parallel, this script will launch the bonjour protocol and will deploy the boinc client:

  `sh boinc-worker-conf.sh urlproject projectname hostname`

  Examples: `sh boinc-worker-conf.sh`
  `http://Grid01.localdomain /bonjourgrid/bonjourgrid Grid02`
  or `sh boinc-worker-conf.sh http://192.168.1.1/bonjourgrid/`
  `bonjourgrid Grid02`

6.4.2 Deploying an Application with Condor as the Desktop Grid Middleware

6.4.2.1 Gridifying Application

In this section we describe the PI-Montecarlo application. This code[6] implements the calculation of pi by the Monte Carlo method. We activate n $(n > 1)$ processes that each performs *niter / n* calculations (niter is the number of iterations used to estimate pi). A shared memory is implemented to recover the level of the parent process as calculated by the son. Before deploying the application, the user must create a submit description file. It describes all information concerning this application (binaries, arguments, logs and results, matchmaking, etc.). See Figure 6.11.

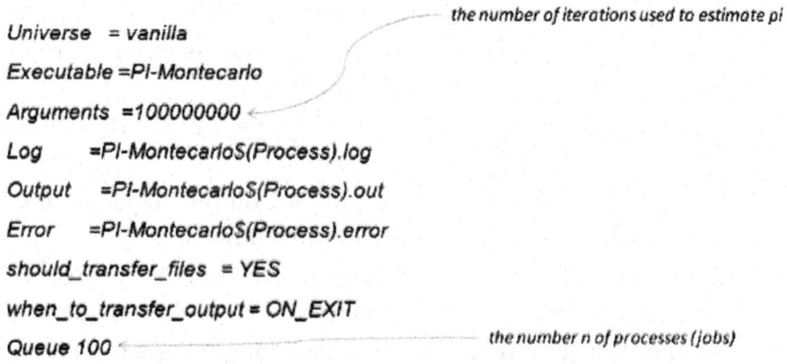

```
Universe   = vanilla                              the number of iterations used to estimate pi

Executable =PI-Montecarlo

Arguments  =100000000

Log        =PI-Montecarlo$(Process).log

Output     =PI-Montecarlo$(Process).out

Error      =PI-Montecarlo$(Process).error

should_transfer_files = YES

when_to_transfer_output = ON_EXIT

Queue 100                                         the number n of processes (jobs)
```

FIGURE 6.11: The Condor job submit file.

In BonjourGrid, to deploy a Condor application, the user must prepare a tree describing the various points mentioned above (see Figure 6.12).

6.4.2.2 Deploying Application

For running Condor applications on top of BonjourGrid, you need to follow this step:

STEP 1: Start Condor and BonjourGrid environments:

- Master

  `cd /home/condor/BonjourGrid+CONDOR/BGC/BG-master/`

  `sh startmaster.sh nbreCPU AppDir submitTime appname nbjob`

  nbreCPU: Number of execution machines in the Computing element.

---

[6]http://grid6.cck.rnu.tn/BonjourGrid/Build-BonjourGrid-from-Sources/applications-test/PI-Montecarlo/.

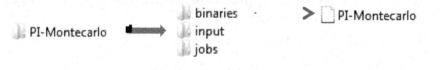

FIGURE 6.12: Condor application.

AppDir: The path of folder containing all the applications to deploy. By default it is the apps folder, which is already created under /home/condor/BonjourGrid_Condor/

Example: `sh startmaster.sh 10 apps 22:39:18 PI-Montecarlo 100`

For each user, this script will build a Computing element composed of a coordinator and n execution machines.

In parallel, this script will launch the bonjour protocol and will deploy the condor central manager (one pool manager).

`./condor-manager-conf.sh HostManagerName domain CollectorName @IP`

Example: `./condor-manager-conf.sh Grid01 localdomain Bgrid 192.168.1.1`

- Application deployment:

  `cd /home/condor/BonjourGrid+CONDOR/BGC/BG-master/`
  `python deploy-app.py appname inputFile`
  where:

  1. `appname` is the .zip file containing the application (binaries, inputs, submit file...) and it must be located in the same place as the `deploy-app.py` script;
  2. `inputFile` is the prefix name of the previous argument. In our example we will have `python deploy-app.py Monte-Carlo.zip Monte-Carlo`. We also use relative paths and not absolute paths.

- Worker:

  `cd /home/condor/BonjourGrid+CONDOR/BGC/BG-worker/`
  `sh startworker.sh`

  In parallel, this script will launch the bonjour protocol and deploy the condor client.

```
./condor-wotker-conf.sh HostManagerName domain
CollectorName @IP
```
Example: `./condor-wotker-conf.sh Grid01 localdomain
Bgrid 192.168.1.2`

6.4.3 Deploying an Application with XtremWebCH as the Desktop Grid Middleware

6.4.3.1 Gridifying Application

In this section we describe the gridification of the PHYLIP application. PHYLIP (the PHYLogeny Inference Package) is a package of programs for inferring phylogenies (evolutionary trees). Developed during the 1980s, PHYLIP is one of the most widely distributed phylogeny packages. It has been used to build the largest number of published trees. The package is available free over the Internet, and is written to work on as many different kinds of computer systems as possible. The binary and source code (in C) are distributed. In particular, already-compiled executables are available for Windows, MacOS, and Linux systems.

Five modules were ported on XWCH: *Seqboot, Dnadist, Fitch-Margoliash, Neighbor-Joining,* and *Consensus* (Figure 6.13). Input data for these modules are nucleotide sequence data (DNA and RNA) coded with an alphabet of the four nucleotides Adenine, Guanine, Cytosine, and Thymine. Each nucleotide is denoted by its first letter: A, G, C and T. Every nucleotide sequence belonging to the input data is a leaf node of the evolutionary tree to be constructed.

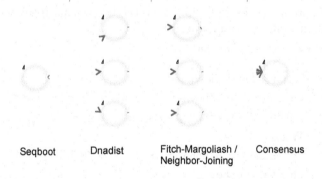

Seqboot	Dnadist	Fitch-Margoliash / Neighbor-Joining	Consensus

FIGURE 6.13: Gridification of PHYLIP application.

In BonjourGrid, to deploy a BOINC application, the user must prepare a tree describing the various points mentioned above (Figure 6.14).

6.4.3.2 Deploying Application

For running Condor applications on top of BonjourGrid, you need to follow this step:

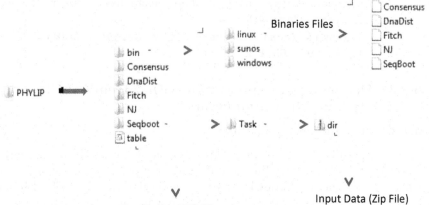

FIGURE 6.14: XtremWebCH application.

STEP 1: Start XW and BonjourGrid environment

- Master:

  cd /home/xwch/BonjourGrid+XW/BGXW/BG-master/

  sh startmaster.sh nbreCPU AppDir submitTime appname nbjobs

  For each user, this script will build a Computing element composed of a coordinator and *n* execution machines.

  In parallel, this script will launch the bonjour protocol and will deploy the xtremweb coordinator.

  sh xw-server-conf.sh @IP

  - Application deployment:

    cd /home/xwch/BonjourGrid+XW/BGXW/BG-master/

    python deploy-app.py appname inputFile

  - Worker:

    cd /home/xwch/BonjourGrid+XW/BGXW/BG-worker/

    sh startworker.sh

    In parallel, this script will launch the bonjour protocol and deploy the XW client.

    sh xw-wotker-conf.sh CoordinatorIP

6.5 Glossary

Grid Middleware: A Grid Middleware Component is a piece of software integrated in one or more grid middleware distributions that provides a service, implements protocols and algorithms with the objective of allowing users access to the grid's distributed resources.

Middleware Coordination and Collaboration: Activities that facilitate the inter-relationship between different middleware. At the grid level, such as EGEE, coordination means standards such as XACML, SAML, SRM, GLUE, BES, JSDL and See also the Open Grid forum website.

Desktop Grid: Cycle stealing, or shared computing, creates a "grid" from the unused resources in a network of participants (whether worldwide or internal to an organization). Usually this technique is used to make use of instruction cycles on desktop computers that would otherwise be wasted at night, or during lunch. Volunteer computing is a type of distributed system in which computer owners donate their computing resources (such as processing power and storage) to one or more projects.

Bibliography

[ACDJ08] Heithem Abbes, Christophe Cérin, Jean-Christophe Dubacq, and Mohamed Jemni. étude de performance des systèmes de découverte de ressources. *CoRR*, abs/0804.4590:–, 2008.

[ACJ08]   Heithem Abbes, Christophe Cérin, and Mohamed Jemni. Bonjourgrid as a decentralised job scheduler. In *APSCC*, pages 89–94. IEEE, 2008.

[ACJ09]   Heithem Abbes, Christophe Cérin, and Mohamed Jemni. Bonjourgrid: Orchestration of multi-instances of grid middlewares on institutional desktop grids. In *IPDPS*, pages 1–8. IEEE, 2009.

[And04]   David P. Anderson. BOINC: A system for public-resource computing and storage. In Rajkumar Buyya, editor, *GRID*, pages 4–10. IEEE Computer Society, 2004.

[EFGK03] Patrick Th. Eugster, Pascal Felber, Rachid Guerraoui, and Anne-Marie Kermarrec. The many faces of publish/subscribe. *ACM Comput. Surv.*, 35(2):114–131, 2003.

[FGNC01] Gilles Fedak, Cécile Germain, Vincent Néri, and Franck Cappello. Xtremweb: A generic global computing system. In *CCGRID*, pages 582–587. IEEE Computer Society, 2001.

[LLM88]   Michael Litzkow, Miron Livny, and Matt Mutka. Condor - a hunter
           of idle workstations. In *ICDCS*, pages 104–111. IEEE Computer
           Society, 1988.

[SC05]   Daniel H Steinberg and Stuart Cheshire. *Zero Configuration Net-
           working: The Definitive Guide.* O'Reilly, 2005.

Chapter 7

How to Work with PastryGrid

Christophe Cérin

University of Paris, Villetaneuse, France

Heithem Abbes

University of Tunis, Tunis, Tunisia

7.1 Introduction

This chapter is a chapter about the practical issues of using a particular Desktop Grid middleware, namely PastryGrid [ACJ08, ACJM10]. This middleware differs from other popular Desktop Grid middleware in the sense that it is built with the Peer-to-Peer philosophy in mind. The implementation is done on top of a popular P2P system, namely Pastry, and hence its name. PastryGrid is able to execute task graphs with precedences in a fully distributed manner. During the execution of the task graph, the choice of the machines that will execute the next tasks is unknown but it is set on the fly, after the completion of each task. This feature relies on a concept of Rendez-vous point (RDV) and we manage one RDV per application. To be clear, the PastryGrid overlay network is able to manage multiple applications at the same time. The most important fact for a machine is to join the overlay network. After that, a machine will become a worker for another machine, or a RDV point or will play yet another role for mastering the fault tolerance of the application.

In peer-to-peer distributed systems, each participant (peer) knows a set of logical neighbors to which it can directly send and receive messages. These sets of neighbors form a topology that we call an *overlay* network. The choice of overlay network is important for obtaining performance and tradeoffs are made between the number of neighbors, the diameter of the graph, the time for an unicast or a broadcast in the worst case. People consider $\log n$ neighbors among n peers and a broadcast made in $\log n$ hops is a good compromise. This is the case with Pastry,[1] which is a structured or regular overlay, network meaning that all the nodes are equivalent and can use the same routing and fault-handling algorithms.

Roussopoulos et al. [RBR+04] define peer-to-peer systems as those respecting the following criteria:

Self-organization: The network automatically adapts itself to peers' arrival or departure without compromising the integrity of the system. Peers exploit local information provided by its neighbors to organize the network;

Distribution: There is no centralized control to manage peers' behavior;

Scalability: Issues like bottleneck, overloaded node,s and single-point-of-failures are avoided, allowing the system to scale up to millions of peers.

Distributed Hash Tables (DHT) provide the means to build scalable, fault-tolerant- and load-balanced distributed applications on top of large-scale P2P systems. DHT is a distributed data structure that holds pairs $< key, data >$ in a completely distributed manner by mapping each key-data to a node of the overlay network.

To this end, a DHT exploits hashing techniques, which provide uniform distribution of data and can locate objects (resources) in a small number of hops (i.e., $\log n$) among n peers. Finding data is done through the hashing of the name of the data, which gives, typically, a 128-bit key. If the node requesting the data has the computed 128-bit identifier, we have found the data. Otherwise, we route the request to the neighbor whose node identifier is the closest to the requested 128-bit identifier.

To form the address of the next hop, the 128-bit key is divided into chunks with each chunk being b bits long, yielding a numbering system with base 2^b. The routing table contains the address of the closest known peer for each possible chunk and this results in the storage of $2^b - 1$ neighbors. Then, it is not difficult to observe that if we choose $2^b - 1 = \log n$, then a logarithmic number of hops will be traversed in the worst case to retrieve any data.

The organization of the chapter is as follows. In Section 7.2 we introduce

---

[1]http://www.freepastry.org/

the PastryGrid system, its philosophy and features. In section 7.3 we explain how to install PastryGrid on your system. You need the Pastry substrate, which is available under .jar files, and of course you need a Java environment. In Section 7.4 we investigate how to deploy and run a toy application in order to exemplify all the mecanisms that a user needs to know for mastering PastryGrid. We also provide a glossary in Section 7.5.

7.2 PastryGrid System

In the recent past, we have proposed a system called PastryGrid for the management of institutional DG (i.e, in the same institution) and user applications over a fully decentralized P2P network. Our system has the particularity of executing not only Bag of Tasks (BoT) but also distributed application with tasks precedences (DA) in a decentralized manner. The basic idea of *PastryGrid* is that each user application creates its own autonomous and decentralized execution environment. Each application holds, in a dynamic way, a site, called a *Rendez-Vous* Point (RDV), to deposit or recover data. Moreover, the RDV is not a central element; it is not a static machine because it varies from one application to another one and, most important, other machines do not know its site (neither its address nor its name). Concerning the running phase, machines that contribute to the execution of one or many tasks of an application may contribute also to the management and control (i.e., allocation, assignment, and scheduling of tasks) of other tasks of the same application. As detailed below, there is no specific element that coordinates the execution steps of applications.

The originality of PastryGrid consists of running distributed applications for tasks with precedences in a decentralized way. A Distributed Application (DA) is a combination of one or more modules. A module is a set of tasks, using the same binary file and, generally, different input files. Tasks of the same module can be carried out in a parallel way. To simplify the implementation, we consider in this work that tasks of the same module need the same machine requirements for their execution. Figure 7.1 illustrates an example of a DA with four modules and ten tasks.

We use the following notations (see also Figure 7.2 for an illustration of the concepts):

1. Mi is the machine that executes the task Ti.

2. $Input(T_i)$ is the set of T_i inputs.

3. $Output(T_i)$ is the set of T_i outputs.

4. If $Output(T_i) \subset Input(T_j)$, then T_i is the predecessor of T_j and T_j is the successor of T_i.

5. $Pred(T_i)$ is the set of T_i predecessors.

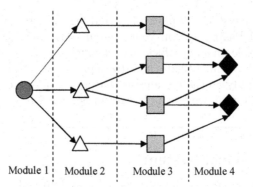

FIGURE 7.1: Distributed application with four modules and ten tasks.

6. $Succ(T_i)$ is the set of T_i successors.

7. If $Cardinality(Pred(T_i)) = 1$, then T_i is called an *Isolated* task.

8. Let $E = Succ(T_i) \cap Succ(T_j)$ with $i \neq j$, if $E \neq \emptyset$, then:
   - T_i and T_j are called *Friend* tasks
   - and $\forall\, T_s \in E$, T_s is called a *Shared* task

9. $Friend(T_i)=\{T_j,\ i \neq j,\ T_i$ and T_j are *Friend* tasks $\}$

10. If T_i and T_j are friends, then M_i and M_j are friends too.

11. A Shared task is a non-isolated task and vice versa.

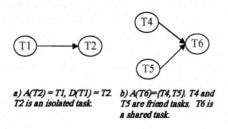

*a) A(T2) = T1, D(T1) = T2.   b) A(T6)={T4,T5}. T4 and
T2 is an isolated task.   T5 are friend tasks. T6 is
                              a shared task.*

FIGURE 7.2: Two cases to illustrate the concept of *friend*, *shared*, and *isolated* tasks.

7.2.1   How a User Can Describe His Distributed Application

For our system, a user provides a compressed package containing all the necessary data, binary files, and the dataflow graph of the application (where nodes are tasks and edges are communication between tasks). Indeed, he represents the data flow graph in an XML file, called *Application.xml*. The user should specify also, in the same file, the execution requirements of each application task.

Now, using *Application.xml*, *PastryGrid* generates, for each task T_i of the data flow graph, a file $Task_i.xml$ that contains the following information: the application name, the module name, the task name, $Input(T_i)$, $Output(T_i)$, $Friend(T_i)$, $Succ(T_i)$, and the execution requirements for running $Succ(T_i)$. $Friend(T_i)$ and $Succ(T_i)$ will be useful for the M_i machine (carrying out T_i) to participate in resources discovery and tasks assignment in order to ensure a decentralized coordination (see Section 7.2.4 for a detailed description).

PastryGrid provides a tool, called SDAD (System for Distributed Applications Description), to help the user in preparing the compressed file of his application. We conceived this tool with a graphical mode for simple applications and an advanced mode for complex ones. So, a user has just to draw the data flow graph of his application or follow a special wizard for complex applications. Thus, SDAD generates the *Application.xml*, puts data and binary files in a directory tree, and compresses all these data in a zip file.

7.2.2 Design of PastryGrid

PastryGrid is composed of four components: the addressing scheme we use to identify machines and applications, the protocol of resource discovery, the RDV point, and the coordination approach between machines carrying out a given application.

7.2.2.1 Addressing Scheme

In centralized DG systems, a dedicated server (master) has a global view of all resources (i.e., machines and user applications). This architecture does not need to have a special addressing scheme to reference resources. A simple identifier mechanism can be powerful. However, when we want to switch to a decentralized style, we have to withdraw this mechanism. Modern P2P networks, like Pastry [RD01], CAN [RFH$^+$01] and CHORD [SMK$^+$01] propose another addressing scheme style based on DHT (Distributed Hash Table). In this kind of addressing scheme, nodes can join and leave the network; the hash table entries are automatically mapped to new physical nodes. References to these entries (i.e., keys) do not need to be updated. They use a fully decentralized routing algorithm that guarantees the delivery of lookup messages to the appropriate physical node in at most $O(\log_{2^b}(N))$ hops, where N is the number of physical nodes currently participating in the network.

Then, we choose to use Pastry as the overlay network for the PastryGrid system, but we could equally have chosen to follow other modern DHT implementations such as CHORD and CAN, which achieve the same effect using slightly different approaches. PastryGrid assigns a 128-bit *nodeId* to each physical node when it joins the network. In addition, an application is assigned and referenced via a 128-bit *ApplicationId* identifier. This identifier is hashed using the application and the user name. When a machine joins the network, an XML description file of its characteristics (CPU, RAM, OS) is generated and stored in the install directory of PastryGrid (this file will be useful in resource discovery).

7.2.2.2   Resource Discovery

Resource discovery consists of looking for a set of available machines to run applications. The protocol we propose for this task is collaborative and fully decentralized, as detailed later. Before that, let's justify why we have proposed our own protocol. Although there are several works done in this field of research, we preferred to conceive a new protocol for two reasons: 1) simplicity of grafting and controlling the discovery module; and 2) in most systems, a machine sends its local information (CPU, RAM, OS, Cost) to one or many servers and information on resources cannot be updated in real time, so we believe that it is better if we can access the machine and check its real characteristics directly. Certainly, this can be expensive in terms of communication, but we show, in the experimental evaluation, that for an institutional desktop grid size, this discovery protocol performs well. Moreover, we do not focus, at this level, on the performance evaluation of our protocol of discovery, we just need a functional one to establish a fully decentralized execution of a distributed application with precedence between tasks.

Before detailing the resource discovery algorithm, let's note that when a machine executes a task, it is locked during runtime. We have opted for this choice because we consider that a single task could require lot of resources.

When PastryGrid assigns a task T_i to a machine M_i, the machine executes T_i and, then, may possibly search necessary machines for tasks of $Succ(T_i)$ (This is not compulsory; Section 7.2.4 gives more details of research assignment strategy). Suppose that M_i looks for N machines (with the same configuration for the sake of simplicity in the current implementation).

1. M_i checks its leaf set. Indeed, it constructs a vector $V_a(H, W, R)$ where a is the application ID, $H = M_i$ *node handle* $\cup$ *nodes handles of its leaf set*, W contains names of tasks to be assigned $(Cardinality(W) = N)$, and R describes the execution requirements. M_i sends $V_a(H, W, R)$ to the head of H $(head(H))$.

2. $M = head(H)$ checks if it is free and fits R: if yes, M takes $T = head(W)$ to execute it and deletes it from W $(W = W - T)$.

3. If $W = \emptyset$, then the discovery process is accomplished. GOTO 6). Else M deletes its node handle from H $(H = H - M)$.

4. If $H = \emptyset$, then M updates with a new leaf set: $H = leaf\ set\ of\ M$. This case, certainly, occurs if $N > size(leaf\ set)$. It may also occur even though $N < size(leaf\ set)$: when no sufficient free machines fit R.

5. M forwards the new vector $V_a(H, W, R)$ to the head of H, which is

another machine in the old leaf set, if 4) is not done, or the first machine in a new one. GOTO 2)

6. End of tasks assignment.

Now, PastryGrid has assigned the N tasks to N machines, but how a machine can start executing its task will be detailed in Section 7.2.4.

7.2.3 Concept of RDV

In centralized DG systems, there is a coordinator, a dedicated server, which accepts execution requests coming from clients, assigns tasks to the workers according to a scheduling policy, transfers binary and data files to workers and supervises tasks executions. Moving all this work in a decentralized manner is not trivial, especially when there are dependencies between application tasks. PastryGrid focuses on these kinds of applications and executes them in a fully decentralized way. The main idea is to create, dynamically, a new RDV point for each new application submission. Each application has a single *ApplicationId* identifier. The RDV is the machine that *nodeId* is numerically closest to *ApplicationId*. It represents a communication mean between tasks of a given application. A machine needs to know only the *ApplicationId* to reach the RDV machine. Each one can send/receive data/request to/from the RDV.

7.2.3.1 Execution Initialization

As mentioned in Section 7.2.1, a user submits his application providing a compressed file containing the necessary data for execution. The submission machine (i.e., the machine receiving the compressed file) generates the *ApplicationId* identifier hashing the application and the user names. Thereafter, it initiates a new RDV, sending an *InitApplication* message with an attached zip file (the compressed package of the application) to the node that *nodeId* is numerically closest to *ApplicationId*. When receiving an *InitApplication* message, the RDV extracts the package to rebuild the initial tree structure of the application. Then it generates the $Task_i.xml$ file for each task (see Section 7.2.1 for information about $Task_i.xml$). After that, the submission machine must initialize the execution phase by assigning initial tasks (i.e., the tasks of the first module) to the adequate machines according to the discovery protocol described in Section 7.2.2.2.

7.2.3.2 Communication with RDV

In PastryGrid, there are two kinds of communication between nodes. The first one consists of sending, directly, a message to the node using its node handle. The second type consists of routing a message, among participating nodes, to the node that *nodeId* is closest to the key indicated in the message. Then, knowing the *ApplicationId* name only, a machine can send a message to the RDV hosting the application *ApplicationId*.

7.2.4   Coordination and Data Transfer

During the discovery process, when characteristics of a machine M match the execution requirements of a given task, M gets the task name and the application identifier *ApplicationId*. The identifier is useful to reach the RDV of the application and the task name is useful to select necessary data for the execution of the task. Indeed, the machine contacts the RDV sending a *DataRequest* message containing the task name, to incorporate necessary data. Now, the RDV generates a package with the data, the binary, and the file $Task_i.xml$; then, it compresses this package and sends it directly to the concerned machine in a message called *YourData*.

As mentioned before, in addition to the execution of the task T_i, the machine M_i can participate in the discovery of machines needed for the execution of T_i successors. So M_i looks for and selects free machines according to criteria imposed by tasks to be assigned among $Succ(T_i)$. It is important, here, to note that M_i can start this work only if tasks to be assigned can really start (i.e., they no longer wait for any input). When M_i would like to affect a task T of $Succ(T_i)$ to a node, it should discover an adequate machine M that fits T requirements. When M is found, T is assigned to it. Now, to start the execution, M needs $Input(T)$. Thus, M asks M_i for its results and asks the RDV to send it the other inputs (data, the binary file, and the $Task_i.xml$).

Until now, we have said that M_i could search machines for tasks of $Succ(T_i)$. When does this occur and how? Let us note T_s a task of $Succ(T_i)$. We distinguish two cases:

1st case: If T_s is an isolated task, the machine M_i directly launches the search of a free machine without asking for RDV authorization. Once found, the machine M_s receives, directly, the outputs of M_i without passing by a RDV point.

2nd case: If T_s is a shared task, M_i will not systematically search a machine for it. In fact, a competition on the step for machine discovery for T_s will take place between M_i and its friend machines executing $Pred(T_s)$. Indeed, when M_i finishes its execution, it sends a *SearchRequest* message to the RDV point to request an authorization to look for a machine to execute the shared task T_s. The RDV checks if $Friend(M_i)$, executing $Pred(T_s)$, have accomplished friend tasks of T_i:

- If so, the RDV returns a fulfillment message *SearchRequestAck* to authorize the machine M_i to start the discovery process. Now, M_s, the machine found for T_s, will not take inputs only from M_i but also should take other inputs from RDV. This one sends to M_s the results of execution of the other predecessors of T_s on other machines: outputs of T_i friends included in $Pred(T_s)$. In addition to these outputs, the RDV also sends binary, data file and $Task_s.xml$ of T_s.

- Else, if the friends of M_i executing $Pred(T_s)$ are not finished yet, the RDV returns a *SearchRequestReject* message to prevent re-search. Thus, the machine M_i transmits $Output(T_i)$ to the RDV with a *WorkDone* message.

In this way, we succeed in avoiding that each machine executing $Pred(T_s)$ launches a discovery of a machine for T_s, then notifies its friends. We optimize also the use of the machines; it is not necessary to hold a machine for a task when it must wait for a result to start its execution. It is also important to mention that we have implemented a synchronization technique to avoid concurrent access on the same resource in the RDV.

7.3 Installing PastryGrid

PastryGrid is entirely developed in Java. We have used *FreePastry* API to create an overlay network of nodes and implemented hash tables manipulation functionalities.

To install PastryGrid, you need just to install a Java development kit (version > 1.5) and the FreePastry library[2]. After that, you can download the PastryGrid middleware from sourceforge at http://pastrygrid.sourceforge.net.

7.4 Deploying an Application

We illustrate the different steps for the traveling salesman application. We remind you that the traveling salesman problem is to find a shortest possible tour that visits each city exactly once from a list of cities and their pairwise distances.

The traveling salesman application is composed of five modules. We describe each module as follows (see Figure 7.3):

GenerateAllChannels: Generating all Hamiltonian edges of the graph given as input (a graph described in a text file) and returns as output a text file (AllPaths.txt) containing N paths.

SplitFile: This module splits the output files from the previous module into several files. The resulting files are named "AllPaths0.txt; AllPaths1.txt; ... ; AllPathsN.txt."

FindMinPath: This module looks for the path of minimum weight of each of the output files of the previous module. This is done in parallel by the different nodes of the PastryGrid network. It outputs N files. Each file will contain the path of minimum weight.

AssemblerFiles: This module assembles the resulting files from the pre-

---

[2]http://www.freepastry.org/FreePastry/, release 2.1.

vious modules in a single file.

FindFinalMinCircuit: This module will be applied to the resulting file from the previous module to generate the final path of minimum weight (Min-Path.txt).

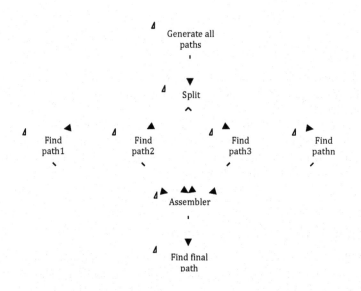

FIGURE 7.3: Modules of traveling salesman application.

First of all, we need to describe the tasks, the dependencies between tasks, and the binary and data files path. All this information can be found in the xml file "table.xml". After that, we prepare the tree containing the binary and data file.

Figure 7.4 describes the differents modules of the application according to PastryGrid specifications.

To run PastryGrid:

1. Type under the command prompt: `java -jar pastrygrid.jar`

   The following menu will be displayed:

   ***** PastryGrid *****
   0) display menu
   1) join ring
   2) run application
   3) get list of nodes executing tasks

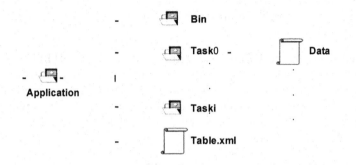

FIGURE 7.4: Application tree.

4) stop working node
5) get application result
6) display neighbors
7) disconnect node
8) restart node
9) usage
q) exit
Type 1 or 2 or 3 etc... or q to exit.

2. First, you need to connect to the PastryGrid system, so select the first choice in the menu "1) join ring". Now your machine is able to work.

To run an application in PastryGrid system:
- Type a letter or "0" (zero) to re-display the menu.
- Select the second choice "2) run application" and then give the path of the zipped folder that contains the application and its data.

The application folder must be zipped (compressed) to ".zip"; it contains:
- a file "table.xml" that describes tasks to execute.
- a directory "data" that contains files data.
- a directory "bin" that contains the binary files to execute tasks.

7.5 Glossary

Overlay Network: An overlay network is a computer network that is built on top of another network. Nodes in the overlay can be thought of as being connected by virtual or logical links, each of which corresponds to a path, perhaps through many physical links, in the underlying network. An overlay network is an abstraction of a physical network and implements a graph topology.

Distributed Hash Table: A distributed hash table (DHT) is a class of distributed system that provides a lookup service similar to a hash table where (key, value) pairs are stored in a DHT organized according to a network topology. In general, scientists consider that retreiving the node containing the information by traversing $\log n$ nodes for an overlay of n nodes is a good compromise.

Peer-to-peer: Peer-to-peer (P2P) computing is a distributed application architecture that partitions tasks or workloads between peers. Peers are equally privileged in the application. They are said to form a peer-to-peer network of nodes because they can play any role: master, slave, dedicated for accounting. Based on a grid model a peer-to-peer system, or P2P system, is a collection of applications run on several local computers, which connect remotely to each other to complete a function or a task.

Grid: *Grid computing* is a term referring to the combination and the coordination of computer resources from multiple administrative domains to reach a common goal. What distinguishes grid computing from conventional high-performance computing systems such as cluster computing is that grids tend to be more loosely coupled, heterogeneous, and geographically dispersed.

Bibliography

[ACJ08]   Heithem Abbes, Christophe Cérin, and Mohamed Jemni. *Pastry-Grid*: decentralisation of the execution of distributed applications in desktop grid. In Bruno Schulze and Geoffrey Fox, editors, *MGC*, page 4. ACM, 2008.

[ACJM10] Heithem Abbes, Christophe Cérin, Mohamed Jemni, and Yazid Missaoui. Fault-tolerance for pastrygrid middleware. In *IPDPS Workshops*, pages 1–8. IEEE, 2010.

[RBR+04] Mema Roussopoulos, Mary Baker, David S. H. Rosenthal, Thomas J. Giuli, Petros Maniatis, and Jeffrey C. Mogul. 2 p2p or not 2 p2p? In Geoffrey M. Voelker and Scott Shenker, editors, *IPTPS*, volume 3279 of *Lecture Notes in Computer Science*, pages 33–43. Springer, 2004.

[RD01]   Antony I. T. Rowstron and Peter Druschel. Pastry: Scalable, decentralized object location, and routing for large-scale peer-to-peer systems. In Rachid Guerraoui, editor, *Middleware*, volume 2218 of *Lecture Notes in Computer Science*, pages 329–350. Springer, 2001.

[RFH+01] Sylvia Ratnasamy, Paul Francis, Mark Handley, Richard M. Karp, and Scott Shenker. A scalable content-addressable network. In *SIGCOMM*, pages 161–172, 2001.

[SMK+01] Ion Stoica, Robert Morris, David R. Karger, M. Frans Kaashoek, and Hari Balakrishnan. Chord: A scalable peer-to-peer lookup service for internet applications. In *SIGCOMM*, pages 149–160, 2001.

Part II

Maturity and Beyond

Chapter 8

Challenges in Designing Scheduling Policies in Volunteer Computing

Trilce Estrada

University of Delaware, Newark, USA

Michela Taufer

University of Delaware, Newark, USA

8.1 Introduction

Volunteer Computing (VC) is a paradigm that allows the use of heterogeneous computing resources (e.g., desktops, notebooks) connected through the Internet and owned by volunteers to provide computing power needed by computationally expensive, loosely coupled applications. For such applications, VC systems represent an effective alternative to traditional High Performance Computing (HPC) systems because they can provide higher throughput at a lower cost.

Many VC projects involve simulating phenomena in nature. Examples of VC projects are Docking@Home [TAC+09], which targets drug design for breast cancer and HIV through protein-ligand docking simulations; Folding@Home [ea03], which explores the physical processes of protein folding; and climatepredicton.net[1], which predicts climate phenomena such as El Niño. The main challenge in VC projects is that the computers, also called *workers* or *hosts*, are non-dedicated, volatile, error-prone, and unreliable. Indeed, workers donate idle cycles and therefore are not fully dedicated to the execution of VC

---

[1]http://climateprediction.net.

applications. Their volatility is due to the fact that they may suddenly leave the project without returning any results (or returning partial results) for the assigned computation. Network and execution errors can take place at any time and cannot be predicted. Finally, the results returned may be affected by malicious attacks, hardware malfunctions, or software modifications and therefore can be invalid [TACI05]. When any of these conditions happens, a VC system should be ready to discard and redistribute the affected computation at the cost of decreasing the overall throughput (i.e., number of jobs completed) and increasing latency (i.e., job lifespan).

To lessen the impact of these conditions, VC projects rely on a scheduler. The scheduler takes decisions about the type and amount of computation or jobs that should be assigned to a worker. Because of the heterogeneity of the resources, in terms of performance and availability, the scheduler cannot apply the same distribution criteria to every machine. Moreover, both performance and availability of a worker can suddenly change between two requests for computation. Ideally, schedulers should be able to intelligently adapt to the characteristics of the VC environment as the simulation evolves.

Such an adaptation has to take into account the two conflicting perspectives of the key players in a VC project: the scientist's and volunteer's perspective. The scientists consume VC resources for their project while the volunteers provide these resources. More in detail:

> **Scientist perspective** defines needs of scientists for which VC provides computational resources. The scientists will always demand that their jobs are returned as fast as possible and as reliable as possible. In order to guarantee the reliability of computed jobs, VC uses a validation step. Replication and comparison of independently computed jobs is frequently used to verify the reliability of the computation. Replication can also be used as a way to guarantee faster turnaround of jobs; several jobs are distributed to different hosts, but only the first one that is returned will be used. Replication implies computation redundancy and therefore wasted resource. The bottom line of this perspective is: the more resources provided by the VC system, the better.

> **Volunteer perspective** determines the volunteers' willingness to donate computational resources. This perspective is crucial in the survival of the VC paradigm and is directly associated with the efficient use of donated resources. Volunteers expect to be acknowledged in terms of *credits* or points that are assigned proportionally to the amount of computation provided. Many volunteers place great value on credits and expect a prompt credit assignment. The path leading from the distribution of a job to a donated resource, through eventual job execution and completion, to the final assignment of credits is not always straightforward. Credits are only awarded after the validation of job results. Awarding credits with some delay has a negative impact on the volunteers' willingness to donate their resources. Also, volunteers expect that

their resources will be constantly used in non-redundant, meaningful computation.

These two perspectives are mutually exclusive and play a key role in scheduling policies. Scheduling policies driven by the scientist's perspective tends to (1) maximize the availability of resources, (2) minimize total latency (faster turnaround of jobs), (3) maximize the throughput (higher number of jobs completed), and (4) maximize reliability (higher replication of jobs and subsequent validation). Ideally, to provide some quality of service to the scientist, a scheduler must supply practically unlimited access to resources, use aggressive replication, and prefer strict validation policies. On the other hand, scheduling policies driven by the volunteer's perspective tends to (1) minimize validation latency (by using fast validation of jobs and credit assignments), (2) maximize utilization (by using constant supply of computation to his or her own resources), and (3) minimize resource wasting (by reducing redundant computation). In order to provide some quality of service to the volunteers, a scheduler should be able to constantly generate a diverse workload, which is distributed across different types of participating hosts. It should favor validation mechanisms that reduce replication and validate jobs as soon as possible, even if in some cases it means missing aspects such as reliability of results. The volunteers' perspective in VC can be relevant to other domains, such as Grid and Cloud computing, as the need for efficiently using resources to minimize costs is applicable to all three domains. In the same way that volunteers relative "happiness" drives the success of VC, the efficient use of resources, and therefore minimized costs, drives the success of non-donated distributed paradigms.

With these two perspectives in mind, the creation of an effective scheduling policy is usually a time-demanding and subjective tuning process in which the project developer has to use his or her knowledge of the specific application and the workers participating in the VC project. Project administrators often spend months observing the behavior of a VC project before changing any parameter setting. Measuring the effectiveness of a scheduling policy and tuning its parameters may also take several months. Finally, extensive tuning does not guarantee that the optimal scheduling policy will be found, as the volunteer community is constantly changing. Automatically generated scheduling policies can address the challenges listed above although their implementation is not trivial.

This chapter gives an overview of possible solutions by presenting tools and environments for the study of new scheduling policies (Section 8.2) as well as giving an overview of different policies that incrementally become more complex and sophisticated in taking automatic decisions (Section 8.3). The chapter concludes with the new challenge that the VC community is called to address, that is, going beyond traditional scheduling policies to build autonomic VC environments driven by application QoS (Section 8.4).

8.2 Simulation and Emulation Tools to Design and Test VC Scheduling Policies

In general, distributed systems incorporate different types of policies that drive their functionality, and VC is no exception. A VC server embodies a number of scheduling policies and parameters that have a large impact on the project throughput and other performance metrics. Unfortunately, it is difficult (if not impossible) to test new policies, under controlled conditions, in the context of a large volunteer computing project. There are many factors that cannot always be reproduced, and poorly performing mechanisms can waste potential computational power, driving away volunteers donating resources. In general, several challenges arise when scheduling policies within real VC environments:

- The time required to measure project throughput, total number of results delivered to the scientists in a given amount of time, under different parameter settings can be significant.

- Problems due to testing might upset volunteers, even to the degree that they leave a project.

- Every experiment is unique and unrepeatable, making the generalization of conclusions difficult.

Consequently, it is preferable to conduct performance studies in simulation environments, where it is possible to test a wide range of hypotheses in a short period of time without affecting the VC community. More generally, exploring new policies and tuning setting parameters for high performance can be done in simulated environments, where it is possible to test a wider range of hypotheses in a shorter period of time without affecting the BOINC community. A simulated environment can be based on a simulator implementing an abstract model of a system, an emulator using all or part of a real system, or a combination of both (hybrid approaches).

In the past there have been several attempts to model VC systems in simulators. Relevant VC simulators include SimBOINC[2], and SimBA [TKE+07]. SimBOINC simulates desktop grids and volunteer computing systems. This simulator is based on the SimGrid toolkit and uses traces obtained from real BOINC clients. This project is no longer being supported and no further changes have been made to the simulator. Thus, this project has been left behind in the state of the art in BOINC development.

SimBA (Simulator of BOINC Applications) is a discrete event simulator that models the main functions of VC environments. SimBA simulates the generation and distribution of jobs executed in highly volatile, heterogeneous, and

---

[2]http://simboinc.gforge.inria.fr/.

distributed environments, as well as the collection and validation of completed jobs. SimBA also simulates the creation, characterization, and termination of workers by using trace files obtained from real BOINC projects, for example, Predictor@Home [TAKC06]. A trace file contains information that characterizes workers, for example, creation time, processor architecture, operating system (OS), and life span. In general, SimBA

- Generates jobs and creates for each job a number of instances or replicas according to the project replication policy

- Distributes instances according to worker requests and the selected scheduling policy

- Models worker volatility and heterogeneity using the worker's characterization obtained from the trace file

- Determines the status of a worker's returned results (successful or unsuccessful, that is, erroneous) using the worker's error rate characterization obtained from the trace file

- Determines the validity of successfully completed results using a quorum, i.e., the required number of results in agreement as set by the project's validation policy

- Computes the performance of the simulated VC project in terms of throughput

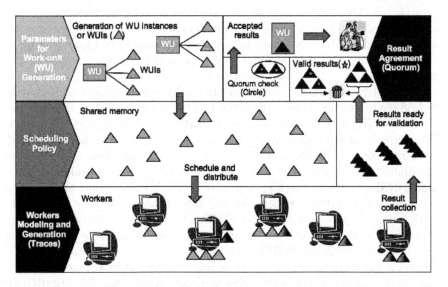

FIGURE 8.1: Overview of SimBA components.

Figure 8.1 gives a general overview of SimBA and its components. SimBA is built upon a modular framework that can easily be extended to include new features and to model a wide range of VC projects: distinct SimBA modules implement various flavors of events. This makes SimBA a general-purpose, discrete event simulator for BOINC environments. Like many other discrete event simulators, SimBA consists of entities (e.g., worker generator, worker node, job generator, jobs, and job instances), events (e.g., generate worker, generate jobs, generate job instances, assign replica to worker, check status), and simulator resources (e.g., BOINC shared memory).

SimBA allows project designers and administrators to extensively study improvements in VC projects, such as new scheduling strategies and validation techniques, without affecting the VC community. SimBA outputs include total number of replicas generated and distributed, workers that connected and disconnected, and percentage of successfully completed jobs. Simulations are set up and results visualized with a GUI. In [TKE+07], validation studies showed that SimBA's predictions of Predictor@Home scheduling policies are within 5% of the performance reported by this BOINC project. Table 8.1 summarizes the average percentage of valid, invalid, error, and timed-out results as well as those still in progress at the end of the simulations as compared to the number of distributed job instances for both applications used by Predictor@Home [TAKC06]: (1) conformational search using a Monte Carlo simulated-annealing approach using MFold [KS98] and (2) protein refinement, scoring, and clustering using the CHARMM Molecular Dynamics simulation package [Bea83].

A general drawback in simulators is that the VC server logic is abstractly represented. Thus, the simulators may not be able to model a broad variety of server logics accurately without each time manually requiring the user to tune the DES logic. A more abstract model of VC is limited by system complexity. Complexity can be kept under control in hybrid emulated environments, making emulators the valuable approach to study performance for a broader sets of scheduling policies. Very few simulation environments are based on hybrid environments, among them the BOINC client emulator [KAM07] and EmBOINC [ETA09]. The BOINC client emulator uses the BOINC client code. Its goal is to study and replicate the way in which job instances are scheduled and executed inside a single BOINC client. As its name also indicates, the BOINC client emulator is not able to study server scheduling policies. EmBOINC is a trace-driven emulator of BOINC projects that allows developers to tune scheduling policies for VC systems and accurately predicts the performance of new policies. As for SIMBA, it embodies a discrete event simulator (DES) that models the volunteer hosts, including their heterogeneity and volatility. While EmBOINC simulates the VC hosts, it emulates the VC server; it interacts directly with the actual BOINC server daemons, triggering the generation, distribution, collection, and validation of jobs. EmBOINC thereby minimizes the maintenance burden caused by the rapid development of BOINC. Further, by plugging directly into a BOINC server, EmBOINC

TABLE 8.1: Comparison of SimBA Predictions with Predictor@Home Results

	MFold P@H			CHARMM P@H		
	Real Project (%)	SimBA Simulation (%)	Difference and Conf. Interval (%)	Real Project (%)	SimBA Simulation (%)	Difference and Conf. Interval (%)
Distributed	100.00	100.00	0 ± 0.02	100.00	100.00	0 ± 0.01
Valid	71.89	67.94	3.9 ± 2.58	54.30	54.08	0.2 ± 0.32
Invalid	1.76	2.68	0.9 ± 1.42	5.78	8.52	2.7 ± 1.21
Error	1.88	6.53	4.6 ± 0.88	15.09	11.92	3.1 ± 0.54
Timed-out	13.83	14.81	0.9 ± 3.26	12.44	9.74	2.7 ± 2.16
In Progress	10.64	8.04	2.6 ± 1.72	12.39	15.74	3.3 ± 2.03
Execution Time	15 days	107 min		8 days	44 min	

allows testers to directly and accurately tune diverse policies for different host populations. Figure 8.2 gives an overview of the EmBOINC framework from [ETA09], with its three components that can be set up by the user:

- Host modeling: Model a dynamic set of hosts and their (static and dynamic) attributes. This model can be based on real BOINC traces or on synthetic ones generated by EmBOINC (which can generate traces). Dynamic attributes can be modeled with continuous or discrete probability distributions.

- Task modeling: Model heterogeneous or homogeneous workloads for one or more applications. Jobs can have different lengths, numbers of replicas, quorums (number of results needed for validation), sensitivities (impact of OS and architecture on the results of two instances of the same job), and platform requirements (e.g., 32 or 64 bits, GPU or CPU).

- VC setting: This is currently synonymous with BOINC setting. EmBOINC employs BOINC settings such as size of job cache, homogeneous redundancy level, maximum number of job instances assigned per request, and selection of scheduling policy.

Currently, EmBOINC can monitor (1) throughput-based metrics: project throughput (rate of valid jobs for the entire project) and application throughput (rate of valid jobs per application); (2) latency-based metrics, each measuring an average time: distribution latency (from job generation to job instance distribution), in-progress latency (from job distribution to the reporting of the completed job), execution latency (a job's completion time), and validation latency (from a results collection to its validation); (3) starvation-based metrics: capability of the BOINC server to keep the hosts busy. EmBOINC collects the performance metrics throughout the simulation. Final evaluations are stored in log files accessible via the EmBOINC interface. EmBOINC's accuracy has been validated via diverse VC projects: Predictor@Home [TAKC06], Docking@Home [TAC+09], World Community Grid (WCG) projects [ETR09, ETA09]. Figure 8.3 shows a validation study from [ETR09] that compares the number of jobs collected daily with EmBOINC versus. three real BOINC WCG projects.

There are several reasons for using a hybrid environment like EmBOINC rather than a pure simulator. First, EmBOINC is easy to install: as easy as creating a new BOINC project. It is also easy to use: the selection of host population and parameters can be done through the BOINC interface. The analysis of results can also be done through the BOINCWeb interface. EmBOINC is easy to maintain: EmBOINC is integrated directly into BOINC code and can be used with the very last version of the BOINC server (BOINC scheduler and daemons). Last but not least, its integration in BOINC makes EmBOINC convenient for testing and tuning: every policy tested or tuned with EmBOINC works unchanged with BOINC.

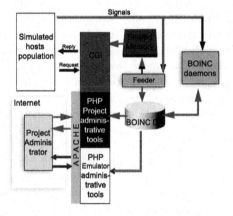

FIGURE 8.2: Overview of the *EmBOINC* framework integrated in BOINC.

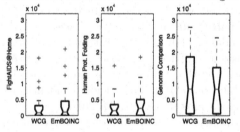

FIGURE 8.3: Statistics of job collected per day with *EmBOINC* and three real BOINC projects.

8.3   Scheduling Policies

8.3.1   Naive and Knowledge-Based Scheduling Policies

Most existing policies that schedule jobs in VC projects are based on heuristics and can be classified in two classes: naive or knowledge-based. *Naive scheduling policies* assign computation without taking into account the history of the workers. Examples of naive scheduling policies are (1) First-Come-First-Serve ($FCFS$), for which job instances are sent to any host that applies for computation [ACW05]; (2) Locality scheduling policy, for which job instances are preferentially sent to hosts that already have the necessary data to accomplish the work [Bel96]; and (3) Random assignment, for which job instances are selected randomly. *Knowledge-based scheduling policies* look at the history of the worker applying for computation and the whole community. Examples of knowledge-based scheduling policies are as follows: (1) Fixed thresholds (FIX_T), which check the availability and reliability values of the requesting host; if they are above a certain predefined threshold, the scheduler assigns the requested work to that host [EFT+06]; (2) Variable thresholds (VAR_T), which is similar to the fixed thresholds, but the scheduler varies the thresholds at runtime; if the number of job instances waiting for distribution is greater than the number of requests generated by the hosts, then the thresholds decrease, otherwise they increase; and (3) World Community Grid scheduling policy (WCG), which assigns computation based on the average turnaround time of the worker, which is the average time a worker needs to return a result[3]. Both naive and knowledge-based scheduling policies can use Homogeneous Redundancy (HR) for the distribution of multiple job instances. HR distributes instances of the same jobs to workers that are computationally equivalent, meaning that they have the same operating system and processor vendor (e.g., Intel or AMD). This yields bit-identical successful results, even for chaotic applications [BUA+97, TACI05].

Scheduling policies may have different levels of implementation complexity. Policies such as FCFS are very easy to implement in VC projects. FCFS, also called bag of tasks [ACW05, ea04], consists of assigning the first available job instance to the first requesting worker. FCFS can be extended to have the worker match the HR criteria, if HR is used and at least another instance of the job has already been distributed. In other words, the instances are assigned to the workers as if they were in a queue; when a worker requests a new task, it receives the oldest instances in the queue. This policy has the disadvantage of assigning tasks even to workers that in the past have been shown to lack reliability in their results or in availability for the project.

Knowledge-based scheduling policies can, among others, include resource knowledge and can rely on thresholds to take decisions on what job to dis-

---

[3]http://worldcommunitygrid.org.

tribute to what resources. Threshold-based policies take into account specific features of the hosts (resources) they are distributing work to, for example, availability and reliability of the host, error rate, matchmaking, and other characterization of the hosts. For instance, fixed or variable threshold policies can use availability and reliability as driving features for the distribution of job instances. In VC environments in general and in BOINC environments in particular, an available worker is not on all the time, as is the case in cluster or Grid computing. Instead, an available worker is one that is productive, that is, it returns a number of results over a certain interval of time. The worker can be on but because of volunteer-imposed restrictions, such as exclusive CPU use by other applications, it may not be dedicated to BOINC projects. Consequently, the availability of a $worker_i$ at time t is defined as

$$availability_i(t) = \frac{WUIcompleted_i(t)}{WUIdistributed_i(t)} \qquad (8.1)$$

where $WUI_{completed}$ is the number of job instances completed successfully (and therefore sent to the VC server), and $WUI_{distributed}$ is the number of job instances that were assigned to $worker_i$. A worker may not return all its assigned instances successfully because of errors or timeouts.

At the same time, an available worker is not necessarily a reliable worker: returned results may be affected by hardware malfunctions, incorrect software modifications, or malicious attacks. For example, participants in BOINC projects might increase their CPU frequency to gain more credits. This overclocking can cause hardware bit errors (hardware malfunctions), and these errors can affect floating-point calculations but do not crash the computer and might occur sporadically, for example because of fluctuations in ambient temperature. Such errors can be detected by comparing the results of multiple instances of the same jobs. In particular, in BOINC these checks are performed on multiple instances of the same jobs and for projects such as P@H are based on strict equality comparison combined with Homogeneous Redundancy (HR). The reliability of a $worker_i$ at time t expresses the level of trust in terms of returned results and is defined as

$$reliability_i(t) = \frac{WUIvalidated_i(t)}{WUIcompleted_i(t)} \qquad (8.2)$$

where $WUI_{validated}$ is the number of valid results (instances that have passed the validity check) over the total number of completed instances.

Availability and reliability are not directly correlated: an available worker is not necessarily reliable. A certain level of availability and reliability characterizes each $worker_i$ at time t:

$$worker_i(t) ==> \{availability_i(t), reliability_i(t)\}. \qquad (8.3)$$

A scheduling policy driven by availability and reliability can associate a rating of both metrics, that is, availability and reliability, to each worker updating its ratings each time it returns results.

The scheduling policy takes into account the worker classification in terms of availability and reliability ratings when assigning a new bundle of job instances to workers. In general, the policy establishes thresholds, one or several prioritized thresholds for worker availability ($threshold_{availability}(i)$ where $i >= 1$) and one or several prioritized thresholds for worker reliability ($threshold_{reliability}(i)$ where $i >= 1$), and uses these thresholds to determine the distribution of instances to workers. The scheduler does not distribute job instances to workers with availability and reliability ratings that are below certain threshold values. Threshold values can range between zero and one and are not necessarily equal. For threshold levels of 0 for both availability and reliability, this scheduling policy degrades to the First-Come-First-Serve policy.

Worker availability and reliability thresholds can be fixed or can change during a project to better meet the performance goals of the project. When a worker requests a new bundle of job instances, the server computes the worker's availability and reliability ratings based on the worker's performance history. If both ratings are above the predefined thresholds, then the worker is assigned a bundle of job instances. Otherwise, the worker is not assigned work. That is, $worker_i(t)$ receives instances, if and only if $availability_i(t) >= threshold_{availability}$ and $reliability_i(t) >= threshold_{reliability}$. This policy has the disadvantage of starving workers that are considered to be unsuitable to perform the work at hand and, thus, virtually removes these workers from a project. To alleviate this situation, the policy shall allow workers to improve their ratings. But this can only be achieved by assigning job instances to these unsuitable workers and, thus, permitting them to demonstrate that they are more available or reliable than indicated by their histories. It has been observed that the availability and reliability ratings of workers change slowly, therefore continuing to assign instance bundles to a worker for a time after which the worker is first deemed unsuitable seems to be a reasonable strategy.

In this chapter, SimBA is used to compare the performance of the two scheduling policies described above, that is, First-Come-First-Serve (FCFS) and a fixed threshold-based policy, and to understand the behavior, in particular, of the latter. For the comparison, three sets of experiments were used for two application traces, that is, an MFold trace and a CHARMM trace, both taken from the Predictor@Home project. For each set of experiments the random-number generator seed at the beginning of the simulation was fixed, making sure that all the simulations are repeatable. A set of experiments consists of twenty-five simulations, each of which is driven by one of the two application traces. Every experiment in a set is unique in terms of the worker availability and reliability threshold tuple used in the experiment. The value of any of the two thresholds is 0, 0.25, 0.5, 0.75, or 0.95. The selection of these values was chosen to uniformly sample the tuple search space. For each experiment, we measured the total number of generated, distributed, successful, and unsuccessful results, as well as the total number of valid and invalid results. The results for the simulations driven by the MFold and CHARMM traces are shown in Figures 8.4 and 8.5, respectively. Each figure has the same x- and y-axes, which represent the availability and reliability thresholds, respectively. The z-axis indicates the simulation results. Note that for clarity the first two pictures in the figures have different orientations.

Values in the two figures are normalized with respect to the total number of generated job instances. This normalization permits us to make a fair comparison of the results across all simulations. This is necessary because the incidence of unsuccessful or invalid results causes the master to generate additional instances of related jobs. When comparing only the values on the main diagonals, the figures show that for both the MFold and CHARMM trace-driven simulation the threshold-based policy with the worker availability and reliability thresholds both set to 0, which is equivalent to the FCFS policy, has the worst behavior. The figures also reveal how for both MFold and CHARMM, we can identify a sweet spot where the number of valid results that we can return to the scientists reaches higher values and their differences are within 1%. This spot is delimited by the values of 0.50 and 0.75 for both the availability and reliability thresholds. This shows that when the availability and reliability thresholds are selected properly, the number of trusted results delivered in a given amount of time increases. In contrast, if the selected thresholds are at either end of the spectrum (i.e., 0 or 0.95), the simulation results are poor.

8.3.2 Adaptive Scheduling Policies

Adaptive scheduling policies require intelligent techniques to reset scheduling parameters. Among the possible techniques, a distributed, evolutionary method can efficiently search over a wide space of possible scheduling policies for a small subset of policies that minimizes both errors and invalid results while maximizing project throughput. For the sake of generality, a large, initial set of possible scheduling policies mut be considered. In the attempt to define a rigorous syntax of the scheduling rules, each scheduling policy can be defined as a variable number of IF-THEN-ELSE rules driving the job assignment; for example, a BNF (Backus-Naur Form) grammar can model the structure of the rules. The components of the grammar are defined as follows: A rule is a logical expression or an inequality. A logical expression can be two rules joined by AND, a negated rule, the logical constant TRUE, or the logical constant FALSE. An inequality is two arithmetic expressions joined by a comparison operator. Note that logical expressions and inequalities can produce only Boolean values as results. Arithmetic expressions may be two arithmetic expressions joined by an arithmetic operator, or an operand. An operand is a number or a factor. An arithmetic operator can be an addition, subtraction, or multiplication. Note that the operator division is omitted because it can be replaced by the inverse of a multiplication. A comparison operator may be greater than, less than, greater than or equal, or less than or equal. A number is a random value given by a uniform distribution between 0 and 1, or a uniform distribution between 0 and 100. A factor is a feature characterizing the

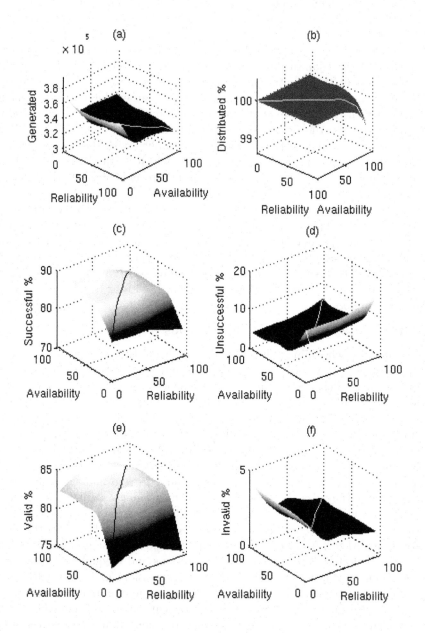

FIGURE 8.4: MFold simulations with FCFS, and threshold-based policies.

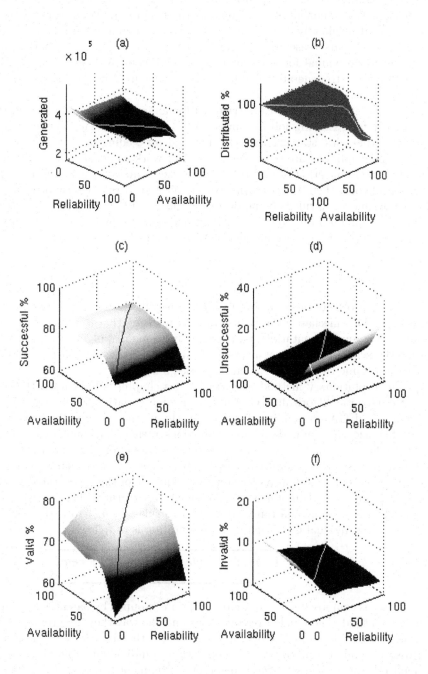

FIGURE 8.5: CHARMM simulations with FCFS, and threshold-based policies.

worker community or the application. The set of factors must be large; the evolutionary component of a proposed method is in charge of discarding those factors that ultimately do not play any relevant role in improving project performance. A parser uses the grammar to generate a scheduling policy that acts as an individual for the evolutionary algorithm. A code generator uses the individual to generate the scheduling policies used in the computational environment.

The ordered IF-THEN-ELSE rules within an individual can be used by the scheduler for assigning job instances (WUI) to workers. The scheduler goes through the rules, starting with the first and proceeding until it finds a rule whose preconditions are true. The amount of work assigned to the worker is related to the position of this rule. If a worker requests k job instances and meets rule one, then it gets all the work requested. If rule one is not met, the scheduler moves to rule two and decreases the amount of work assigned based on the formula in Equation 8.4, where, given an individual of m rules and the worker meeting rule s, the number of WUI assigned to the worker is $\tilde{k}$:

$$\tilde{k} = k - \frac{k * (s - 1)}{m} \tag{8.4}$$

If a worker does not meet any of the m rules, it will still receive at least one job instance. This mechanism prevents starving workers, that is, workers that do not receive any computation and therefore continue reapplying for jobs. Note that the scheduler does not necessarily distribute computation to a worker using the same rule during the entire worker's lifespan.

The evolutionary method proposed in this chapter is built upon the concept of Genetic Algorithms (GA) in which some variations of the traditional definition of GA are applied. First of all, the method uses individuals with variable size, and individuals are expressions rather than numerical arrays, that is, floating point or integer. Moreover, the operations have been adapted to deal with expressions rather than numbers. The i_{th} individual $C_{i,j}$ in generation j is represented as a concatenated sequence of rules, generated by the BFN grammar. A delimiter is used to separate two rules. The length of an individual is variable and ranges from 1 to 10 rules. If $m \leq 10$ is the number of rules of individual $C_{i,j}$, then its representation is as follows:

$$C_{i,j} = \quad \overline{\left| \; rule_1 \; , \quad rule_2 \; , \quad rule_3 \; , \quad ... \quad rule_m \; \right|}$$

Each rule is composed of one or more conditions grouped by parenthesis, which may be nested. The number of conditions and the way they are grouped vary from rule to rule. Four GA operations are used, that is, selection, mutation, crossover, and elitism, to evolve a population of individuals from one generation to the next. The *selection* is based on tournament selection of individuals where two individuals are selected randomly to compete in the tournament; the individual with the best fitness wins and is chosen to pass its rules to the individuals in the next generation. Tournament selection, in contrast to roulette wheel selection, helps in keeping the diversity of the selected individuals [BT95]. In the *elitism* operation, the best individual of generation j is passed to generation $j + 1$ without any change. When passed by elitism, individuals do not have their fitness evaluated again. In the first generation, the individual passed by elitism is a manually designed scheduling policy that combines attributes used by the fixed-thresholds scheduling policy. In *crossover*, two individuals previously selected combine their rules to form two new individuals that will be evaluated in the next generation. An algorithm to combine the rules is presented in Figure 8.6. Note that for crossover, the order of the rules in new individuals is the same as in their parents and that the new individuals contain at least the same number of rules as the smallest parent and only one of them can be as long as the longest parent.

```
, , "
Select individuals " C_{x,j}  and  C_{y,j}
with  size_x  and  size_y  number of
rules respectively.
Assuming that  size_x ≥ size_y
Create empty individuals  C_{x,j+1}  and  C_{y,j+1}
For i=1 to  size_y
  Get rule i from  C_{x,j}
  Get a random\index{random} boolean variable. If 1 then:
      Append rule i from  C_{x,j}  to  C_{x,j+1}
      Append rule i from  C_{y,j}  to  C_{y,j+1}
  Otherwise:
      Append rule i from  C_{x,j}  to  C_{y,j+1}
      Append rule i from  C_{y,j}  to  C_{x,j+1}
If  size_x > size_y
  For i=1 to  size_x
      Get rule i from  C_{x,j}
      Get a random\index{random} boolean variable. If 1 then:
        Append rule i from C_{x,j} to  C_{x,j+1}
      Otherwise:
        Append rule i from C_{x,j} to  C_{y,j+1}
```

FIGURE 8.6: Pseudocode of the crossover.

Three levels of *mutation* are used in the version of the method proposed in this chapter: individual-level, rule-level, and condition-level. Individual-level mutation can be applied in three different ways: (1) if the individual has a medium size, the mutation randomly changes the position of one rule in the sequence; (2) if the individual is short, the mutation adds one new rule; and (3) if the individual is long, the mutation deletes a randomly chosen rule. An individual is considered short if it has less than four rules, long if it has more than seven rules, and medium otherwise. The rule-level mutation splits long rules. Because very long rules tend to be too restrictive, the method detects long rules with small usage frequency and splits them, selecting randomly the break point and keeping the order of nested parenthesis. The condition-level mutation replaces either operators or factors using knowledge-based mutation matrices that provide the probability that these changes may occur.

In addition to the four GA operations, *pruning* can be used to reduce the number of rules. The method in this chapter includes counters that monitor how many times a rule is applied; rules whose frequency is below a defined threshold are automatically removed from individuals. Long rules are possible but very unlikely to meet the threshold limits.

A general fitness function, used for driving the GA algorithm, should be given by a multi-objective function that maximizes throughput and minimizes replication of computation due to (a) errors, that is, network and computation failures; (b) timed out results, that is, results too late for the simulation or never returned because of the worker volatility; and (c) invalid results, that is, affected by malicious attacks, hardware malfunctions, or software modifications. A job is successfully completed if it is not affected by errors or timeout. A job is valid if at least *min_valid* job instances have been completed successfully and their results agree; that is, their results either are equal because the Homogeneous Redundancy (HR) policy is applied [TACI05] or are within a certain range if no HR is used. The throughput, $Throughput_{wu}$, is the total number of successfully completed, valid jobs at the end of a simulation. When errors, timeout, or invalid results take place, the system reacts by generating new WUIs for the faulty jobs. *min_valid* and the maximum number of job instances per job that a VC project can generate are both defined by the project designer; usually *min_valid* ranges from two to four and the maximum number of job instances ranges from five to eight. The amount of replication is captured by the average job instances distributed per job, $AvgWUI_{wu}$, over the whole simulation; the fewer faulty jobs, the closer this value is to *min_valid*. The fitness function used in this chapter is modeled as the ratio between throughput, $Throughput_{WU}$, and average WUIs distributed per WU, $AvgWUI_{WU}$. The fitness function is given in Equation (8.5).

$$fitness = \frac{Throughput_{WU}}{AvgWUI_{WU}} \qquad (8.5)$$

The number of generations, the size of the individuals, and the mutation rate are key components of any evolutionary methods.

In the case study presented in this chapter, thirty generations with a population size of seventy individuals (scheduling policies) and a mutation rate of 60% are considered. Such a mutation rate and number of individuals per population were chosen to prevent premature convergence. The number of generations was set to thirty because of time constraints to perform our experiments; however it was observed that thirty generations were sufficient for the convergence of the fitness function. The search for general scheduling policies was distributed across the nodes of a Beowulf cluster with 64 dual-core nodes each with a 2.0 GHz AMD Opteron processor, 256 GB RAM, and 10TB disk space. The method was trained with SimBA using performance traces from two BOINC projects, that is, Predictor@Home with CHARMM and FightAIDS@Home. Best scheduling policies were tested with SimBA using performance traces belonging to other three BOINC projects, that is, Predictor@Home with MFold, Human Proteome Folding and Genome Comparison. Note that SimBA uses traces to emulate the worker community, their error rates, timeout, etc. The features of the training and testing projects are shown in Table 8.2. In this table, *Project* indicates the name of the BOINC project, *Simulated_hrs* is the length of the BOINC project for which the traces have been collected, *Min_valid* is the minimum number of WUIs whose results must agree to consider valid a job, and *Usage* states whether the project was used for training or testing.

TABLE 8.2: Training and Testing VC Projects

Project	*Simulated_ hrs*	*Min_valid*	*Usage*
P@H CHARMM	170	3	TRAINING
FightAIDS	550	3	TRAINING
P@H MFold	340	3	TESTING
Folding	600	11	TESTING
Genome	1200	3	TESTING

Results of the GA approach are presented in Figure 8.7 in which two comparisons are performed. First, the scheduling policies generated by the genetic algorithm (GA-designed) are compared against four manually designed scheduling policies: First-Come-First-Served ($FCFS$), fixed availability and reliability thresholds (FIX_T), variable availability and reliability thresholds (VAR_T), and the World Community Grid scheduling policy (WCG). Note that the World Community Grid policy is the result of several person-years of design effort and includes sophisticated correlated rules. Second, a non-triviality test is performed in which the GA-designed scheduling policies are compared against 2,100 scheduling policies randomly generated. The purpose of this comparison is to estimate whether the quality of the results is accidental or due to the proposed evolutionary approach. To create the random policies, the same set of rules used by the evolutionary method is used, but the evolutionary operator is not used to evolve those policies. The number of randomly generated policies, that is, 2,100, was selected to be equal to the total number of policies generated during the evolutionary process (that is, thirty generations × seventy individuals per generation = 2,100).

Results in Figures 8.7.a, 8.7.b, and 8.7.c compare and contrast the performance of the manually designed scheduling policies ($FCFS$, FIX_T, VAR_T, and WCG), the four GA-designed scheduling policies ($GA1$, $GA2$, $GA3$, and $GA4$), and the randomly generated scheduling policies ($BestRand$ is the best random policy per project, and $MeanRand$ is the average over the 2,100 samples). Note that the best random policy, $BestRand$, refers to a different policy in each of the three projects. In contrast with our method, which was able to find four common scheduling policies in the first 10-best policies, none of the first 50-best randomly generated scheduling policies was common across projects, evidencing their lack in generalization capability. In particular, Figure 8.7.a presents throughput and average WUIs for P@H MFold, Figure 8.7.b presents throughput and average WUIs for Human Proteome Folding, and Figure 8.7.c presents throughput and average WUIs for Genome Comparison (note that the last two projects are part of World Community Grid). The dashed lines in each figure serve as reference to identify areas of the search space in which either GA-designed scheduling policies or randomly generated policies, if falling, perform better than the best manually designed scheduling policies for that project. Policies that fall above the horizontal dashed line outperform the best manually designed scheduling policy in terms of throughput. Policies that fall to the left of the vertical dashed line outperform the best manually designed scheduling policy in terms of error reduction (average WUIs)

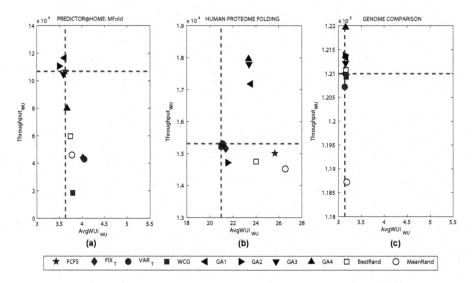

FIGURE 8.7: Throughput and average WUIs per WU for the three testing projects.

In general, the results presented in this chapter show that none of the manually designed scheduling policies works best for all projects. More specifically, if only the throughput is considered, the policy based on variable thresholds is better for P@H with CHARMM and Human Proteome Folding; the FCFS policy is better for P@H with MFold and FightAIDS@Home; and the Fixed thresholds policy is better for Genome Comparison. In terms of average WUI per WU, the World-Community-Grid policy is more effective in reducing the number of errors and timeout for FightAIDS@Home and Human Proteome Folding. Note that the goal of our work is to find scheduling policies that are general enough to improve performance across different projects or within the same project at different times of the project lifespan. The dynamic behavior of VC projects might indeed result in scenarios in which the same project has different resource requirements at different times. On the other hand, the GA-designed policies improve throughput across projects, in particular *GA*1 increases throughput across all the three testing projects (+9.1% for P@H with MFold, +12.2% for Human Proteome Folding, and +0.3% for Genome Comparison) while the other three GA-designed scheduling policies increase throughput for at least two of them. In terms of average WUI per WU, *GA*2 performs similarly to the best manually designed scheduling policy across the three projects with an average of +0.3%. *GA*1, *GA*3 and *GA*4 have a performance similar to the best manually designed scheduling policies for P@H with MFold and Genome Comparison but perform poorly for Human Proteome Folding. This loss in performance can be associated with a possible imbalance of the fitness function. Work in progress is addressing this issue.

Finally, Figure 8.7 shows that none of the randomly generated scheduling policies outperformed both the GA-designed and manually designed scheduling policies. The results in this chapter show that the simple random generation of scheduling policies is not effective, and that the success of our GA-policies is due to the evolutionary process. By using an *"intelligent"* combination of factors, such a process leads to general, effective scheduling rules for VC projects.

8.4 Beyond Scheduling: Building Autonomic VC Environments

Quality of service (QoS) traditionally means to provide guarantees on the ability of the VC system to deliver a predictable number of results within a predictable interval of time. VC applications often lack mechanisms to provide their users (both scientists and volunteers) with QoS. In large-scale distributed systems, there is also the common belief that the solution to achieve QoS is distributing and replicating as many jobs as possible. It is a fact that VC systems are providing larger numbers of resources for scientific applications than ever before.

One aspect that is often neglected in VC is that applications do not use computational resources just for the sake of using them; applications use resources to meet their scientific goals. Then, the concept of QoS should be redefined to include the application's goals, and QoS should ultimately refer to the ability of a VC system to meet application goals within time constraints. With this new definition of QoS in mind, to ultimately provide QoS in VC, the scheduler should act as a resource scavenger and provider for loosely coupled, high-throughput applications with specific scientific goals; in other words, the scheduler should address QoS from specific application perspective.

Application-driven QoS can be expressed in one general objective: mini-

mum number of computations required to achieve meaningful scientific conclusions. Therefore, applications have to define their scientific goals in terms
of concrete metrics such as accuracy of the scientific result, time to converge to
a scientific solution, or coverage on a search space of possible scientific values,
just to mention a few possibilities. By approaching QoS from the application
point of view, the scheduler should focus on actual scientific goals of the application (e.g., accuracy) when distributing jobs, rather than only on aspects like
throughput and latency that may or may not lead to the accomplishment of
these goals. At the same time, the scheduler can relax the stringent resource
requirements imposed by the scientist's ever-growing demand for resources,
aiming at using scheduling policies that select and deploy available resources
in a more intelligent way. It is important to recall that distributed paradigms
incur higher monetary costs when using resources indiscriminately. Whether
the cost is paid by the users of the project or by resource providers (volunteers)
is irrelevant; higher costs result in unsustainable paradigms.

In order to provide application QoS, the scheduler must have, to a certain
extent, knowledge of the application behavior and its resource requirements.
This is not trivial. To simplify the representation of this knowledge, one can
look at a scientific problem as a problem that can be solved using different
methods, for example, Monte Carlo or Molecular Dynamics, and methods can
be expressed by different algorithms. Input data and modeling constraints
also play a key role in scientific simulations and can be assigned different values. An application-driven scheduler should gather all these different aspects
under a common umbrella called *parameters*. The combination of these parameters define the parametric nature of VC applications. The selection of one
parameter rather than another may ultimately impact not only the resource
requirements, but also the final scientific achievements. Normally the selection
of parameters is done manually by the scientists and not by the scheduler due
to the dynamic nature of the scientific problem and the VC environment. In
this chapter we claim that the parametric nature of VC applications should be
a key component of scheduling policies while schedulers should autonomously
tune the VC parameters at runtime with the ultimate goal of providing application QoS. We also claim that this is the next challenge the VC community
is called to address: building autonomic VC environments driven by both the
specific application goals and the resource performance (throughput and latency).

Bibliography

[ACW05] D. Anderson, E. Corpela, and R. Walton.   High-performance
Task Distribution for Volunteer Computing. In *Proc. of the First
IEEE International Conference on e-Science and Grid Technolo-
gies*, pages 196–203, March 2005.

[Bea83]   B. R. Brooks and et al. CHARMM: a Program for Macromolecular Energy Minimization, and Dynamics Calculations. *J Comp Chem*, 4:187–217, 1983.

[Bel96]   F. Bellosa. Locality-information-based Scheduling in Shared-memory Multiprocessors. In *Job Scheduling Strategies for Parallel Processing*, volume 1162 of *Lecture Notes in Computer Science*, pages 271–289. D. G. Feitelson and L. Rudolph, 1996.

[BT95]   T. Blickle and L. Thiele. A Mathematical Analysis of Tournament Selection. In *Proc. of the 6th International Conference on Genetic Algorithms*, pages 9–16, 1995.

[BUA⁺97]   M. Braxenthaler, R. Unger, D. Auerbach, J.A. Given, and J. Moult. Chaos in protein dynamics proteins. *PubMed*, 9(4):417–425, 1997.

[ea03]   Vijay S. [Pande et al.]. Atomistic protein folding simulations on the submillisecond time scale using worldwide distributed computing. *PubMed*, 68(91), 2003.

[ea04]   L.B. Costa et al. MyGrid: A Complete Solution for Running Bag-of-tasks Applications. In *Proc. of the Simposio Brasileiro de Redes de Computadores (SBRC04)*, 2004.

[EFT⁺06]   T. Estrada, D. Flores, M. Taufer, P. Teller, A. Kerstens, and D. Anderson. The Effectiveness of Threshold-Based Scheduling Policies in BOINC Projects. In *Proc. of the 2st IEEE International Conference on e-Science and Grid Technologies (eScience)*, December 2006.

[ETA09]   T. Estrada, M. Taufer, and D. P. Anderson. Performance Prediction and Analysis of BOINC Projects: An Empirical Study with EmBOINC. *J. Grid Computing*, 7:537–554, 2009.

[ETR09]   T. Estrada, M. Taufer, and K. Reed. Modeling Job Lifespan Delays in Volunteer Computing Projects. In *Proc. of the 9th IEEE/ACM Int'l Symp. on Cluster Computing and the Grid (CCGRID'09)*, 2009.

[KAM07]   D. Kondo, D.P. Anderson, and J. McLeod VII. Performance Evaluation of Scheduling Policies for Volunteer Computing. In *Proc. of the 3rd IEEE International Conference on e-Science and Grid Computing*, 2007.

[KS98]   A. Kolinski and J. Skolnick. Assembly of Protein Structure from Sparse Experimental Data: An Efficient Monte Carlo Model. *Proteins: Structure, Function, and Genetics*, 32:475–494, 1998.

[TAC+09] M. Taufer, R.S. Armen, J. Chen, P.J. Teller, and C.L. Brooks III. Computational Multi-Scale Modeling in Protein-Ligand Docking. *IEEE Engineering in Medicine and Biology Magazine*, 28:58–69, 2009.

[TACI05] M. Taufer, D. P. Anderson, P. Cicotti, and C.L. Brooks III. Homogeneous Redundancy: A Technique to Ensure Integrity of Molecular Simulation Results Using Public Computing. In *IEEE International Parallel and Distributed Processing Symposium (IPDPS'05)*, 2005. Heterogeneous Computing Workshop.

[TAKC06] M. Taufer, C. An, A. Kerstens, and C.L. Brooks III. Predictor@Home: A "Protein Structure Prediction Supercomputer" Based on Public-Resource Computing. *IEEE Transactions on Parallel and Distributed Systems*, 17(8):786–796, 2006.

[TKE+07] M. Taufer, A. Kerstens, T. Estrada, D. A. Flores, and P. J. Teller. SimBA: A Discrete Event Simulator for Performance Prediction of Volunteer Computing Projects. In *International Workshop on Principles of Advanced and Distributed Simulation (PADS'07)*, 2007.

Chapter 9

Modeling and Optimizing Availability of Non-Dedicated Resources

Artur Andrzejak

Heidelberg University, Heidelberg, Germany

Derrick Kondo

INRIA, Grenoble, France

9.1 Introduction

An inherent drawback of non-dedicated computing resources is low and uncontrollable availability of individual hosts. This phenomenon limits severely the range of scenarios in which such resources can be used and adds a significant deployment overhead. Approaches to compensate for these difficulties use redundancy-based fault tolerance techniques supported by modeling and prediction of availability. In the first part of this chapter we discuss a variety of modeling techniques ranging from probability distributions to machine learning-based prediction techniques. Subsequently we focus on methods to provide resource-efficient and cost-minimizing fault-tolerance. Here redun-

TABLE 9.1: Lower Bounds on Probability of a Failure of *at Least One* Host in a Group of n Hosts (with individual failure probabilities of $p_1, \ldots, p_n$)

		Number of Hosts n							
		2	**3**	**4**	**5**	**6**	**8**	**10**	**16**	**32**
	0.1	0.19	0.27	0.34	0.41	0.47	0.57	0.65	0.81	0.97
$p_i \geq$	**0.2**	0.36	0.49	0.59	0.67	0.74	0.83	0.89	0.97	1.00
	0.4	0.64	0.78	0.87	0.92	0.95	0.98	0.99	1.00	1.00

dancy is mandatory to mask the outages of individual machines, yet on the other hand it might increase overhead and resource cost. We describe how availability models help here to obtain statistical guarantees of (collective) availability and how total costs of the resources can be balanced against reliability properties. We also consider the issue of adjusting application architectures in order to tolerate partial resource failures. This promises to broaden the type of applications deployed on voluntarily computing resources from embarrassingly parallel jobs to Map-Reduce-type applications or even Web services.

9.1.1 Problem of Individual Failures

Outages of individual resources occur in all types of distributed systems. In dedicated environments, such an outage is usually caused by a hardware or software *failure*. To simplify our considerations, we will use the term *failure* when referring to outages due to such a malfunction or because the resource owner has interrupted a "non-dedicated" execution (e.g., by shutting down his computer). The essential difference between dedicated and non-dedicated scenarios is that in the latter case, the probability of a failure (within a time unit) is typically much higher than of dedicated hosts.

Consider a collection of n hosts where each can fail (independently of others) with a probability p_i within a time interval T. Then the probability that *at least one* host fails within T is

$$1 - \prod_{i=1}^{n}(1 - p_i) \tag{9.1}$$

Assuming $p_i \geq 0.2$ for T of e.g. two hours (a realistic assumption e.g. in case of SETI@home hosts [SWB+97]) Table 9.1 shows that already for four hosts a chance that at least one host fails within T surpasses 50% (it is at least 0.59). This high failure probability has severe consequences for the type of applications that can be executed on non-dedicated hosts. Most importantly, it shows that without additional measures (such as redundancy and checkpointing), it is virtually impossible in such environments to execute any parallel applications, that is, applications that assume *synchronized availability* of several (distributed) resources.

Consequently, current applications running on non-dedicated resources are predominantly embarrassingly parallel computations where each host processes a chunk of data (or a parameter configuration) independently and does not communicate with other workers. But even in the latter scenario, a failure might have several negative effects, including:

- Lost data - This is a major problem in systems such as Wuala that use non-dedicated (along with dedicated) resources for data storage

- Lost computation - In case of computational jobs, this includes lost work since last checkpoint; also there is the case where all work since last upload of the results is lost (if hosts drops out permanently).

- Delayed job completion - As the work performed by a host in a time unit is a product of its computing capacity and its average availability, each failure implies less work in a time unit and thus a possible delay of the completion.

- Degradation of overall service availability - If hosts are used in a "service" scenario (as in the case of Wuala), individual outages can lead to an overall failure of a (replicated) service.

- Need for replacement/migration actions - Temporarily or permanently failed hosts trigger need for data migration and replication actions, as well as a fresh scheduling actions.

Obviously, even non-parallel computations require a high degree of redundancy, checkpointing, and other mechanisms if hosted on non-dedicated resources. This significantly increases the overhead and reduces by a considerable factor the overall deployment efficiency along with the computational capacity of such infrastructures.

As mentioned above, cases where unexpected failures of individual resources are common and need to be treated are not limited to non-dedicated resources. An example would be low-cost computational nodes deployed in large-scale data centers as in case of Google infrastructures. In Equation (9.1), large n implies that the probability of at least one failure becomes high even if probability of a single node failure is low. Another example would be recently introduced Spot Instances in the Amazon Elastic Compute Cloud (EC2) [Bez]. These instances (typically spare capacity of EC2) can be revoked *abruptly* due to price and demand fluctuations. While offering lower resource costs their availability behavior introduces the same problems as in the case of non-dedicated hosts.

We see that guaranteeing stable and efficient operation of a system composed of either unreliable or many resources is a problem transcending the domain of voluntary computing. The concepts and methods presented in this chapter attempt to be sufficiently generic to cover this extended set of scenarios.

9.1.2   Chapter Contents

The focus of this chapter is modeling the availability of individual resources and their collections. For availability optimization we turn our attention solely to *collections* of resources. There are several reasons for this:

- While it is possible and beneficial to model availability of individual resources (see Section 9.2.2), we have little power (except for host filtering) to enforce the availability of *specific* hosts.

- Tightly coupled distributed, parallel algorithms and many services can be only provisioned by resource collections.

- Moreover, even in the case of embarrassingly parallel computations, we need to provide redundancy and understand its amount for reasons of efficiency. Here again, it is necessary to consider availability optimization in collections of resources.

In Section 9.2 we present methods for modeling availability of non-dedicated resources on short-term and long-term time scales. Section 9.3 considers approaches for enhancing the availability of resource collections. Several aspects are considered: trade-offs between redundancy and availability guarantees and monetary costs of such guarantees if non-dedicated resources are "enriched" by dedicated (cloud computing-type) resources. Section 9.4 illustrates these methods on empirical studies performed on a large number of SETI@home hosts. The final Section 9.5 contains the conclusions.

9.2   Modeling Availability of Non-Dedicated Resources

Modeling of host availability behavior serves two goals:

- *Analyzing past availability behavior* of individual hosts and their groups. The results can be applied in a multitude of ways, including estimation of cumulative computational capacity; understanding availability and failure rate distributions (or, equivalently knowing share of hosts with average availability / failure rate in a given interval); discovering periodic temporal patterns of availability and long-term trends; filtering of hosts by specific criteria like "stable behavior"; detection of large-scale anomalies or failures such as partial network outages.

- *Predicting short-term availability* of individual hosts or their groups, where short-term means one to several hours. This can be directly applied for availability enhancement discussed in Section 9.3. As detailed below, predictive methods frequently use models of *past* behavior. Also note that long-term availability prediction (beyond one day) is usually not feasible except for a small class of hosts with very "regular" behavior.

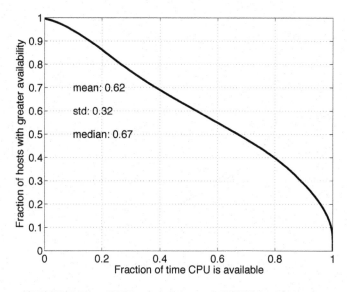

FIGURE 9.1: CPU availability of SETI@home hosts.

Obviously considering more input data and using more sophisticated availability models is likely to yield more precise results (under the assumption that correct, i.e. non-overfitting [WF05] approaches are used). In general, we consider the following classes of availability models with increasing level of sophistication:

- Cumulative models such as distributions of average availability and time between failures (Section 9.2.1)

- Individual models that capture availability behavior of a single host (Section 9.2.2)

- Models that consider individual behavior as well as their dependencies (e.g., short-term correlations between host behavior)

The last case still awaits an empirical study and is a challenge in terms of computational complexity for the amount of hosts considered in Section 9.4.

9.2.1   Availability Distributions

One of the fundamental forms of availability modeling is to consider the fraction of time when each host is available (*aveAva*) and to compute the (cumulative) distribution function of this metric. The distribution is essentially computed by sorting the hosts according to this metric. Figure 9.1 (taken from [KAA08]) shows a plot of such a distribution for a subset of 112,268 hosts participating in the SETI@home project between April 1, 2007, and February 12, 2008 (see [KAA08]); here CPU availability is understood as host availability.

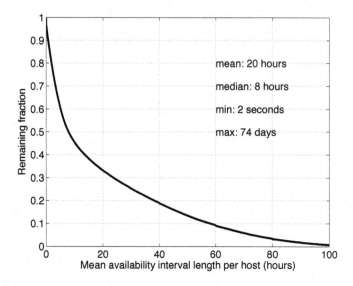

FIGURE 9.2: Average availability run per host in hours.

Obviously 40% of hosts are available at least 80% of the time while almost 70% of hosts are available at least 40% of time.

Even if each host is characterized by only a single metric, their distribution allows us to answer several important questions: What is the cumulative processing capacity of the host population (if all could be used)? Are there many hosts with high average availability? What is the median fraction of time a randomly selected host is available? Answering these questions is helpful, for example, for capacity planning and deciding on which hosts a job is most likely to finish on time.

A related metric is the average duration of host's availability (aka *average availability run*, *aveAvaRun*). Note that for a fixed value of *aveAva*, the metric *aveAvaRun* can vary widely. If a host changes availability state very frequently, *aveAvaRun* might be small (on the order of few minutes); if a host behaves "stable," the value of *aveAvaRun* can reach months. The knowledge of this metric is particularly important to optimize the intervals between result checkpointing. It also helps to select the appropriate hosts for jobs where a non-availability incurs a high cost (e.g., due to data migration overhead). Figure 9.2 (from [KAA08]) shows the cumulative distribution of this metric for the same study as in Figure 9.1. It shows that the median time a host is available is 8 hours, however there are at least 36% of hosts with average uninterrupted availability of more than 20 hours.

Characterizing hosts by such fundamental metrics is simple and beneficial. However, such metrics do not capture more complex temporal patterns (e.g., weekly periodicity) and have limited value for short-term availability predictions of specific hosts.

9.2.2   Individual Availability Models

This section discusses several methods for modeling and forecasting the availability of *individual* hosts. After looking at methods for short-term prediction (partially based on models), we turn our attention to characterization of long-term availability behavior.

9.2.2.1   Short-Term Prediction

Short-term availability prediction attempts to forecast the availability of a specific host in a time interval called *prediction interval* $[t, t + pil]$. Here t is the prediction time (usually the "current time" in the deployment scenario) and *pil* the *prediction interval length*. To simplify, we assume that *pil* is an integer indicating hours and consider a host as available in $[t, t + pil]$ iff it is completely available (i.e., without interruptions) in this time interval. In the following we describe several types of predictors for this scenario.

Last value predictor. The last value predictor (abbreviated `LastVal`) is a simplistic predictor that uses the availability value in the last hourly interval before prediction (i.e., $[t - 1, t]$, where t is in hours) as the prediction of availability for the interval $[t, t + pil]$. Its advantage is virtually non-existing computational and memory cost due to the lack of a past availability model.

Gaussian models of availability runs. This algorithm (abbreviated `Gauss`) models the lengths of both availability and non-availability *runs* (i.e., uninterrupted periods of availability or non-availability). To train a model we compute the average and standard deviation of the length of all availability runs in the training interval data (same for non-availability runs). Subsequently the run lengths are modeled by the Normal (Gaussian) distribution. To predict, we first compute the expected remaining length of the current run. This is given by the number k of hours since last availability change and expected (total) run length (derived from the distribution). If the last observed state is availability, we compute the probability p that the end of prediction interval (i.e., $k + pil$) is still in the current availability run. If the last observed state is non-availability, we use the Gaussian model for non-availability runs to find out whether this run is likely to extend to the first hour of the prediction interval (which gives "non-available" as the prediction value). Finally, if the standard deviation of a run length is above a specified threshold (separately for availability and non-availability runs), we revert to the `LastVal` predictor. This ensures that even if the run lengths are non-stationary, predictions of a reasonable quality can be achieved.

Classification-based predictors. Classifiers are well-studied algorithms with broad applicability in data mining. They are typically used when inputs and outputs are discrete [VAH+02], as in our case. Formally, a *classifier* is a function $f : V \to W$ which assigns to a vector $v \in V$ a label $w \in W$ where W is a discrete set [HTF09]. The scalars in v correspond to *attributes* essentially, properties of an object that we would like to label. In the *learning* or *training phase*, a classification algorithm is presented a set of examples (v, w) (attribute vectors with correct labels) from which it attempts to build a classifier f that captures the relationship between attribute values and labels. Subsequently, *classification* is performed: a vector v with an unknown label is given, and the most likely label $f(v)$ is computed.

In our setting the label w is 1 or 0 depending on whether a host is available in the complete interval $[t, t + pil]$ or not. Attributes are (in the basic version) past availability values of the considered host starting at times $t - k$ (for multiple $k > 0$). Thus, we combine in a training example (v, w) measurements v of past availability with a label w indicating "current"/"future" availability. This effectively turns classification into (availability) prediction. The training examples are obtained from a specific time interval called the *training interval*; essentially, each hour of this interval gives rise to one training example.

Because the above-described "raw" attributes might not be a good data representation to extract availability *information* from the past data we have included additional functions of historical data, (*derived attributes* or *features*) to extend the attribute vector. These features are:

- *Time*: Time and calendar information of the last hour prior to t

- *SwitchAge*: Number of hours since last change of availability until t

- *Averages*: Averages of the host availability over the last 2, 4, 8, 16, ..., 128 hours before t.

We have studied several classifiers, including Naïve Bayes (abbreviated NB) [JL95], Support Vector Machine SMO (SMO) [Pla99], K^* an instance based classifier (K^*) [CT95], and multinomial logistic regression model with a ridge estimator (Logistic) [lCvH92]. Other methods such as a C4.5 decision tree [Qui93] are also possible.

Estimating prediction accuracy. Independent of the method, the prediction results are evaluated on a *test interval*, a data segment following directly a training interval (the training data is ignored for LastVal). To estimate the accuracy, we use a *prediction error* defined as a ratio of mispredictions to all predictions made on the test interval (in case of our 0/1 data, this corresponds to the popular *Mean Squared Error, MSE*). As availability patterns of hosts are likely to change over time, we segment data (for a specific hosts) into a series of partially overlapping training intervals and create a new model for each one. The complete prediction error of model-based predictors is computed over a succession of the corresponding test intervals.

9.2.2.2 Long-Term Modeling and Host Ranking

As noted in Section 9.2, analyzing past availability behavior can be beneficial in a multitude of ways. Our focus here is on exploiting the phenomenon that long-term properties of availability of (individual) hosts can influence short-term prediction accuracy. For example, if a host is switched on and off frequently and randomly, the individual prediction models will be significantly

TABLE 9.2: Long-Term Availability Metrics and Their Sort Order ($\uparrow$ = ascending, $\downarrow$ = descending) for Regularity Rank Computation

aveSwitches	*aveAva*	*aveAvaRun*	*aveNavaRun*	*zipPred*	*MSE*
$\downarrow$	$\uparrow$	$\uparrow$	$\downarrow$	$\downarrow$	$\downarrow$

worse than for hosts that are rarely switched on/off or these events follow a regular pattern.

To this purpose we assign each host a *regularity rank* $r = 1, \ldots, k$, where 1 corresponds to most "unpredictable" ("irregular") hosts and k to most "predictable" ("regular") ones. This rank is assigned according to one of several long-term availability metrics M of hosts ([AKA08, AKA10]) as follows. For a specific M we compute its value for each host and sort them by these values (ascending or descending, depending on M, see Table 9.2). Subsequently, hosts are subdivided into k equally sized groups $r(1), \ldots, r(k)$, each corresponding to a single rank value.

This first metric, called *aveSwitches*, is defined as the number of changes between availability and non-availability (or vice versa) per week. Metric *aveAva* is the average host availability in the training data. The *aveAvaRun* (*aveNavaRun*) is the average duration of uninterrupted availability (non-availability). The metric *zipPred* is more involved and computationally costly. The former is inspired by [KLR04] and is essentially the reciprocal value of the length of a file with the training data compressed by the Lempel-Ziv-Welch algorithm (raw, without the *time* and *hist* features). The rationale is that a random bit string is hardly compressible while a bit string with a lot of regularities is. To compute the final metric, *MSE*, we train a cheap predictor on a part P_1 of the training data (P_2 is not used later) and classify the remaining training data P_2. The Mean Squared Error (Section 9.2.2.1) of the classification on P_2 is used as the error score. The latter metric has turned out to be most useful for finding the regularity ranks on SETI@home hosts and is used in the case studies in Section 9.4.

9.3   Collective Availability

As outlined in Section 9.1.2, generally it is not possible to influence the short-term or average availability of a *specific* non-dedicated host[1]. Consequently, the traditional notion of (individual) availability needs to be relaxed and subsumed by *collective availability* introduced in [AKA08]. The latter is *achieved* if in a pool of N hosts at least n remain available over a specified

---

[1]The exceptions are Amazon EC2 Spot Instances, where availability can be controlled to a large degree by setting different instance bid prices. This mechanism is used in [AKY10] for cost-efficient provisioning of divisible workloads.

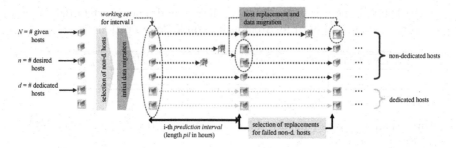

FIGURE 9.3: Illustration of the resource provisioning process across many prediction intervals.

period of time. If an application to be provisioned requests a set of n *desired hosts*, the "resource owner" assigns to it a *working set* of $N \geq n$ hosts with a specific level of *redundancy* $R = (N-n)/n \geq 1$. Obviously, higher redundancy increases the chances that the collective availability is achieved. In addition to adding redundant hosts (i.e., increasing N), it is possible to increase the success rate of achieving collective availability by adding *dedicated* resources (e.g., EC2 instances). The working set is then composed of d dedicated hosts and N minus d non-dedicated hosts. This approach and the availability versus cost trade-offs has been studied in [AKA10].

To ensure uninterrupted operation over a long time period the selection of the working set should be repeated periodically, as illustrated in Figure 9.3. The time interval for which a working set is selected and operated corresponds to the prediction interval *pil* introduced in Section 9.2.2.1. At the end of each prediction interval, hosts that dropped out need to be replaced by other (non-dedicated) hosts (selected via short-term predictions as shown in Section 9.2.2.1). Obviously each replacement of a failed host might incur some costs due to transfer of working data and host setup. To quantify these costs, we introduce the *migration rate* $mr = m/N$, where m is the average number of (non-dedicated) hosts that fail during a prediction interval (and N the working set size). Note that if collective availability is always achieved, mr is never larger than $(N-n)/N$. The selection and assignment of working sets is assumed to be done by a central instance, but there are no inherent constraints to use a fully distributed (Peer-to-Peer) or a hierarchical architecture.

9.3.1 Statistical Guarantees and Cost versus Migration Rate

The notion of collective availability allows us to give (probabilistic) availability assurances. In detail, in a single prediction interval we strive to fulfill an *availability guarantee* y defined as the probability that at least n of the hosts in the (possibly larger) working set remain available over time *pil*. In other words, y is a probability that the collective availability is achieved in a single prediction interval. If there are only non-dedicated resources (i.e., $d = 0$), the only way to keep availability guarantees (for desired n and y) is to adjust the redundancy R (effectively, to change the total working set size N). The minimum sufficient value of R depending on n and y can be derived empirically by simulations on historical data (see Section 9.4); this relation depends on the availability characteristics of considered hosts. Note that by only changing R we have no control on the migration rate.

The possibility to include d dedicated hosts in our working set allows adjusting the migration rate and gives more options how to provide availability guarantees. On the other side, deployment of dedicated hosts incurs additional (monetary) costs. This leads to a cost versus migration rate trade-off: larger d implies lower migration rate yet higher resource costs. We introduce the following schema to offer the "application operator" a spectrum of options with different cost and migration rate (or reliability) characteristics.

For given n and y we compute via simulation a representative set Z of Pareto-optimal combinations of N and d. Each pair (N, d) fulfills the availability guarantee y and is *Pareto-optimal* in the sense that neither N nor d can be decreased (leaving the other parameter constant) without violating y. Each such pair (N, d) has an associated *total cost tc* (sum of the cost of the dedicated hosts and the optional monetary costs of migration) as well as the migration rate mr, which are straightforward to calculate. Subsequently, we sort the pairs (N, d) according to the (increasing) total cost and again according to the (increasing) migration rate. If an application operator has an upper bound on the total costs, to provision the application she chooses a pair (N^*, d^*) with a total cost just below this threshold that minimizes the migration rate. An analogous approach is used if an upper bound on the migration rate is specified. The selected combination (N^*, d^*) is then used in the resource provisioning process over many prediction intervals, as illustrated in Figure 9.3.

9.4 Case Studies for SETI@home-Hosts

9.4.1 Availability Prediction

To illustrate the techniques from Section 9.2 we show selected results of the availability prediction published in [AKA08] and [AKA10]. They were computed on random subsets of more than 112,000 hosts that were actively participating in the SETI@home project [And04] between December 1, 2007, and end of February 12, 2008 (the subset sizes range from 10,000 to about 48,000). About 32,000 hosts had specified host types. Of these hosts, about 81% are at home, 17% are at work, and 2% are at school. We assume that the CPU is either 100% available or 0%, which is a good approximation of availability on real platforms. The availability data per host is a string of bits, each representing non-availability (0) or availability (1) in a particular hour.

We first investigate the influence of the host type and prediction interval length (*pil*) on the error (MSE) of the short-term prediction. Here NB is

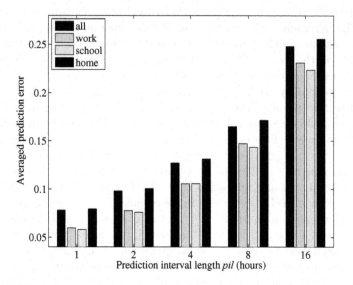

FIGURE 9.4: Prediction error depending on the host type and *pil*.

used as a representative classification algorithm. Figure 9.4 illustrates that the host type influences consistently the prediction error, with work and school hosts being more predictable. The figure shows also a strong influence of the prediction interval length, *pil*, on the prediction error. This is a consequence of increased uncertainty over longer prediction periods. We have also studied impact of the length of the training data interval on accuracy (not shown). It turns out that for training data sizes of more than 20 days, accuracy improvements are marginal. Therefore, a training data interval of 30 days is used in all other experiments.

Next we compare the introduced prediction techniques. Figure 9.5 illustrates for *pil* = 4 the MSE of each prediction algorithms averaged over 500 hosts selected randomly from our data set. The boxplot shows the lower quartile, median, and upper quartile values in the box; whiskers extend 1.5 times the interquartile range from the ends of the box while "crosses" outside them are considered outliers. Obviously NB and Gauss have the smallest median prediction errors but all algorithms except for K* perform similarly well. For *pil* = 1, the highest accuracy was attained by simple predictors like LastVal and NB but again the differences are not very pronounced (see [AKA10]).

Subsequently, we studied how derived attributes influence the accuracy of classifier-based predictors. Figure 9.6 shows averages of MSEs (over 500 hosts as above, *pil* = 4) for such predictors and various derived attributes added one at a time to the base input (see Section 9.2.2.1 for explanation of abbreviations). Except for SMO and time-based derived attributes, all errors are larger. Similarly, this holds for the case when *pil* = 1 (not shown).

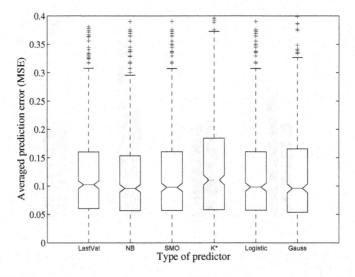

FIGURE 9.5: Comparison of accuracy of the prediction methods ($pil = 4$).

In general, in the case of this data, sophisticated predictors do not have any consistent advantage over an extremely simplistic approach such as LastVal. One explanation is that only feature-scarce data is available for training (essentially as this data is just a bit string representing past availability), which is insufficient to create sophisticated models.

9.4.2   Availability Guarantees

In this section we demonstrate how increasing levels of redundancy R are needed to achieve the desired the availability guarantee y, all this depending on the regularity rank r (Section 9.2.2.2). For this study the same data as in Section 9.4.1 is used; see [AKA08] for all details.

To determine the minimum necessary redundancy level R for the desired availability guarantee y, we change R and in each case observe via trace-based simulations the *success rate*. The latter metric is the fraction of trials (at a fixed $R = R^*$) for which collective availability has been achieved. Interpreting the success rate as the probability of achieving the collective availability for redundancy level R^*, we can find a minimum R that satisfies the desired availability guarantee y by a simple table look-up.

For each trial we randomly choose N number of hosts from the pool predicted to be available for the entire prediction interval (with $pil = 4$ hours). We run trials in this way throughout the test period of 2 weeks. For each data point shown in the figures, we ran about 30,000 trials to ensure the statistical confidence of our results. We also consider two regularity ranks $r = 1, 2$ derived from the metrics *aveSwitches*. Figure 9.7 shows our findings for working set sizes N of 1, 4, 16, 64, 256, and 1,024. As expected, for the lower regu-

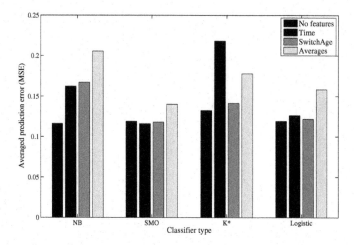

FIGURE 9.6: Influence of the derived attributes for classifier-based prediction and $pil = 4$.

larity rank $r = 1$ much higher redundancy levels are needed to achieve the same success rate as for $r = 2$. For example, with $r = 2$, a redundancy of 0.35 will achieve success rates of 0.95 or higher. With $r = 1$, only the groups with 256 and 1,024 desired hosts can achieve the same level of success rates; at the same time, high redundancy (greater than 0.45) is required. Thus, ranking hosts by regularity levels (here using the *aveSwitches* indicator) significantly improves the accuracy of short-term availability prediction, and consequently the efficiency of achieving collective availability.

9.4.3 Cost versus Migration Rate

As the last showcase of experimental results we consider mixtures of SETI@home hosts and dedicated resources (with costs modeled after Amazon EC2 instances) and study the trade-offs between costs and higher migration rates discussed in Section 9.3.1. To this end we execute trace-driven simulations of the scenario from Figure 9.3 over the entire trace period excluding the training interval.

We consider different pil's values of either 1 or 4 hours and independently also conduct different experiments for four regularity ranks $r(1)$ to $r(4)$ obtained by the MSE metric (Section 9.2.2.2). We also consider different numbers n of requested hosts (50 or 100). For each n, we further vary the working set size N such that the redundancy R varies from 1 (none) to 1.6. For each N, we vary the number d of dedicated hosts allocated from 0 to n. Each such simulation is repeated 4,000 ($pil = 1$) or 2,400 ($pil = 4$) times with varying starting point to ensure statistical confidence.

For each simulation, several metrics are collected. The first one is the success rate defined in Section 9.4.1 (which corresponds to the availability

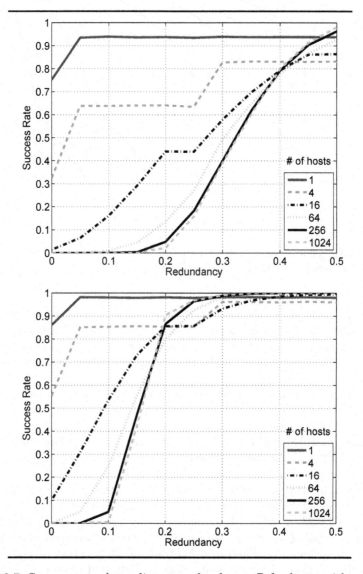

FIGURE 9.7: Success rate depending on redundancy R for hosts with regularity rank $r = 1$ (top: low predictability) and regularity rank $r = 2$ (bottom: high predictability) for $pil = 4$.

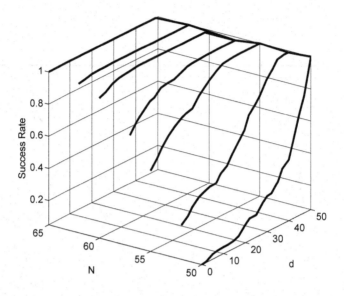

FIGURE 9.8: Success rate depending on the total size N of the working set and number d of dedicated hosts ($n = 50$, $pil = 4$, rank 4 of 4).

guarantee y for each combination (N, d). The second metric is the total cost of achieving this service level. Here as the host cost we assume 10 US cents per hour, which is equivalent to the hourly rate for a small instance on Amazon's EC2. The third metric is the migration rate needed to achieve the service level.

Figure 9.8 shows the first metric (success rate) depending on the total size N of the working set and number d of dedicated hosts for $n = 50$, $pil = 4$, and rank 4 (other cases are shown in [AKA10] with rank order "reversed"). From these result we obtain a set Z of Pareto-optimal combinations of N and d by "intersecting" the surface shown in Figure 9.8 with a plane parallel to x-y axis at a desired value y of the success rate. Each pair (N, d) in Z obviously fulfills the availability guarantee y (and is Pareto-optimal with respect to N or d).

Figure 9.9 shows these sets for three levels of availability guarantees y of 0.90, 0.93, and 0.96. By computing the total cost and the migration rate for each pair (N, d) in Z, we can investigate the trade-off between these two metrics. Figure 9.9 illustrates this trade-off for $y = 0.96$ at the extreme ends of the solution spectrum: while using no dedicated resources eliminates costs, the migration rate is 0.1 (and working set size N becomes 62). On the other hand, using many dedicated resources ($d = 43$) incurs significant costs (4.3 USD per hour) but lowers the migration rate to 0.01. This effect allows us to select an appropriate (usually application-specific) combination of dedicated and shared hosts in order to optimize monetary costs, or the migration rate or both.

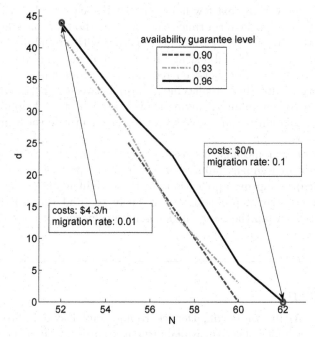

FIGURE 9.9: Pareto-optimal combinations of N and d for $y = 0.9$, 0.93, and 0.96 ($n = 50$, *pil* $= 4$, rank 4 of 4).

9.5 Conclusion

Lack of control over availability on individual hosts is one of the major obstacles for effective usage of non-dedicated resources. In this chapter we have outlined some approaches to overcome this problem. These approaches combine availability modeling and schema for masking outages of individual hosts via redundancy. Availability modeling comprises distribution models as well as short-term and long-term characterization of individual hosts. The major benefit of such models is their ability to identify hosts that are likely to be available within next few hours or that behave "nicely" with respect to availability on a longer-term time scale. As a method for masking outages we have introduced collective availability together with probabilistic guarantees. To achieve collective availability at a minimal monetary and resource, we have investigated the relation between redundancy, probabilistic guarantees and the migration rate (host replacement rate) between "provisioning epochs". In addition to redundancy, mixing-in dedicated hosts is helpful to increase collective availability level at a low host replacement rate yet it implies higher monetary costs.

The proposed mechanisms assume applications that perform efficiently even in face of frequent failures of a significant part of resources (such as MapReduce). Adapting applications to these challenging deployment conditions is a non-trivial problem that is likely to (and should) drive part of the activities within the researchers community devoted to non-dedicated resources.

Bibliography

[AKA08]   Artur Andrzejak, Derrick Kondo, and David P. Anderson. Ensuring collective availability in volatile resource pools via forecasting. In *19th IEEE/IFIP Distributed Systems: Operations and Management (DSOM 2008)*, pages 149–161, Samos Island, Greece, September 22–26 2008.

[AKA10]   Artur Andrzejak, Derrick Kondo, and David P. Anderson. Exploiting non-dedicated resources for cloud computing. In *12th IEEE/IFIP Network Operations & Management Symposium (NOMS 2010)*, Osaka, Japan, Apr 19–23 2010.

[AKY10]   Artur Andrzejak, Derrick Kondo, and Sangho Yi. Decision model for cloud computing under SLA constraints. In *18th Annual Meeting of the IEEE/ACM International Symposium on Modeling, Analysis and Simulation of Computer and Telecommunication*

Systems (MASCOTS 2010), pages 257–266, Miami Beach, FL, August 17–19 2010.

[And04]   David P. Anderson. BOINC: A system for public-resource computing and storage. In Rajkumar Buyya, editor, 5th International Workshop on Grid Computing (GRID 2004), pages 4–10, Pittsburgh, PA, USA, November 4 2004. IEEE Computer Society.

[Bez]   J. Bezos. Amazon.com: Amazon EC2, Amazon Elastic Compute Cloud, Virtual Grid Computing: Amazon Web Services. http://www.amazon.com/gp/browse.html?node=201590011.

[CT95]   John G. Cleary and Leonard E. Trigg. K*: an instance-based learner using an entropic distance measure. In Proc. 12th International Conference on Machine Learning, pages 108–114. Morgan Kaufmann, 1995.

[HTF09]   T. Hastie, R. Tibshirani, and J. Friedman. Elements of Statistical Learning: Data Mining, Inference, and Prediction (2nd Ed.). Springer Verlag, New York, 2009.

[JL95]   George H. John and Pat Langley. Estimating continuous distributions in Bayesian classifiers. In Proc. 11th Conference on Uncertainty in Artificial Intelligence, pages 338–345. Morgan Kaufmann, 1995.

[KAA08]   Derrick Kondo, Artur Andrzejak, and David P. Anderson. On correlated availability in internet-distributed systems. In 9th IEEE/ACM International Conference on Grid Computing (Grid 2008), Tsukuba, Japan, September 29–October 1 2008.

[KLR04]   Eamonn J. Keogh, Stefano Lonardi, and Chotirat (Ann) Ratanamahatana. Towards parameter-free data mining. In Proceedings of the Tenth ACM SIGKDD International Conference on Knowledge Discovery and Data Mining, pages 206–215, August 2004.

[lCvH92]   S. le Cessie and J. C. van Houwelingen. Ridge estimators in logistic regression. Applied Statistics, 41(1):191–201, 1992.

[Pla99]   J. Platt. Fast training of support vector machines using sequential minimal optimization. In B. Schölkopf, C. J. C. Burges, and A. J. Smola, editors, Advances in Kernel Methods — Support Vector Learning, pages 185–208, Cambridge, MA, 1999. MIT Press.

[Qui93]   J. R. Quinlan. C4.5: Programs for Machine Learning. Morgan Kaufmann, San Mateo, CA, 1993.

[SWB+97] W. T. Sullivan, D. Werthimer, S. Bowyer, J. Cobb, G. Gedye, and D. Anderson. A new major SETI project based on Project Serendip data and 100,000 personal computers. In *Proc. of the Fifth Intl. Conf. on Bioastronomy*, 1997.

[VAH+02] R. Vilalta, C. V. Apte, J. L. Hellerstein, S. Ma, and S. M. Weiss. Predictive algorithms in the management of computer systems. *IBM Systems Journal*, 41(3):461–474, 2002.

[WF05] Ian H. Witten and Eibe Frank. *Data Mining: Practical Machine Learning Tools and Techniques*. Morgan Kaufmann, San Francisco, 2nd edition, 2005.

Chapter 10

Security and Result Certification

Filipe Araujo

University of Coimbra, Coimbra, Portugal

Patrício Domingues

Polytechnic Institute of Leiria, Leira, Portugal

10.1 Introduction

This chapter explores the security and dependability issues related to the desktop grid. As stated by Choi and Buyya [CB10], desktop grids are also called *volunteer computing*, *global computing*, *public resource computing*, and *peer-to-peer computing*. In this chapter, we use these terms interchangeably.

As in many other areas of computing, security and dependability are of paramount importance in the desktop grid ecosystem. In such environments, threats exist for both the donated resources and the submitters. On one hand, volunteers need protection against code that inadvertently or maliciously might harm their resources. Indeed, a volunteer project disrupting the regular working of the volunteer machines would have serious public relations consequences, seriously hindering the public perception of volunteer computing. On the other hand, the behavior and results produced by volunteers must be properly scrutinized to detect inappropriate actions and/or erroneous results, and thus ensure the dependability of the performed computations. Therefore, security and dependability are two major goals that need to be carefully considered for a successful desktop grid experience. However, as we shall see in this chapter, security and dependability are complex issues, with many desktop grid frameworks addressing both issues in a less than perfect way. Thus, the volunteering of resources and submission of tasks still require a fair amount of implicit trust between donors and harvesters.

This chapter aims to present the state of the art regarding both security and dependability for desktop grids. The first part presents an overview of the existing issues regarding the security and dependability of desktop grid computing. The second part provides a formal approach on several techniques oriented toward the efficient detection of malicious behavior, namely collusion.

10.2   Architectural Model

The architectural model for a typical desktop grid comprises three types of entities: clients, workers, and a server-side [CB10]. Clients are the ones that submit tasks to the system, having an high interest in a correct and fast execution of their tasks. The workers are the entities that actually carry out the work, providing the computational effort by volunteering their resources for the execution of the clients' tasks. They work in a loop, contacting the server-side to obtain tasks, returning back the computed results, and so forth. Finally, the server-side provides the infrastructure that centralizes the whole management, coordinating the creation, distribution of tasks (workunits in the BOINC lexicon [And04]), the reception and validation of results.

To make available their resources, volunteers need to install the worker-side software infrastructure at their machines and attach to the project(s) they want to contribute to. The worker-side software is comprised of two main components: 1) the management module and 2) the worker software. The management module handles all requests of the worker software and communication to the outside server. For instance, it is the management module that requests tasks from the server and uploads the results when tasks are completed. It also downloads, for each project, the appropriate executable that actually performs the computation. The management module also handles the scheduling of tasks, enforcing the volunteer's defined configuration (e.g., no computation before 8 pm or 30% of CPU time for project A, 45% for project B and 25% for project C). In summary, the management module copes with most functions that interface either with the local machine (file access, checkpoint), etc.) or with each project's server-side. Regarding the worker software, it is the binary that actually computes the tasks of the project, that is, it actually performs the computational work. Similar to the server-side infrastructure, the worker software is under the sole responsibility of the project.

Because computing is prone to hardware, software, and human errors, desktop grid computing needs strong validation mechanisms to ensure the correctness of results. A widely used approach to assess results is to have the work replicated among several workers and compare their results at the end, a technique known as *majority voting*. A result is accepted when a majority of workers returns the same result. BOINC uses the term *quorum* to designate the number of results that need to match in order to validate a result. The quorum is set per project, with a value of 2 being used in several projects. However, validation through replication means that a workunit is executed more than once, and thus computing power is effectively lost. As we shall see later (Section 10.5), other techniques exist for validation of results, although majority voting is the most used, mostly due to its simplicity.

Regarding security, desktop grids need to hedge against two distinct kinds of threats: 1) incorrect results due to either malfunctioning hardware/software or malicious behavior from some of the volunteers and 2) vulnerabilities of the desktop grid software that might expose volunteers to risks that can disrupt their computing activities or data. The desktop grid middleware needs to deal with both cases. We identify the former as *threats to projects* and the latter as *threats to desktop grid resources*. In the next sections, we first analyze the threats to projects and then focus on the threats to desktop grid resources.

10.3 Threats to Projects

Although volunteer resources appear as costless resources from the project point of view, their usage does not go without problems. A computational project that resorts to volunteer resources needs to properly protect itself against the effects of bogus, faulty, and from malicious resources and donors. Otherwise, the project is soon disrupted and its results might be worthless. Additionally, the Web-based infrastructure of the project must also be protected against the common threats that plague such infrastructures.

The main types of problems that volunteer-based projects have to deal with are 1) non-completed tasks, 2) erroneous results, 3) tampered results and 4) attack on the server-side infrastructure. Next we focus on these issues.

10.3.1 Non-Completed Tasks

Several causes can contribute to the existence of non-completed tasks. The most common is a worker leaving, either temporally or permanently, a project without having completed the tasks that were assigned to it by the server-side. This is a frequent situation because volunteer projects are known to have high attrition rates [AF06], especially in projects that have high demands regarding resources, like climateprediction.net [CAS05]. Therefore, non-completed tasks are a common occurrence that any desktop grid middleware must be prepared to deal with. This is usually achieved through a simple timeout mechanism, with the server-side promoting a re-execution of a task whenever the corresponding result has not arrived within the deadline defined for the pair task/volunteer.

In summary, non-completed tasks are often no more than a overhead nuisance that stems from the high attrition rate and volatility of volunteered machines. Through reissuing non-completed tasks, a result can always be obtained at the cost of the CPU lost in non-completed executions.

10.3.2   Incorrect Results

Incorrect results can be due to malfunctioning hardware, software, users, or a combination of all of these. Errors can be classified into two different types of errors: fail-stop and fail-silent [MS94]. The former forces the application to unexpectedly stop (for instance, through a crash), thus being easily detectable. On the contrary and as the name suggests, fail-silent errors occur silently and might go unnoticed. Incorrect results can have devastating effects, as wrong conclusions can be drawn from these results, especially when dealing with fail-silent errors.

A commonly cited cause of hardware malfunctioning is overclocking [And04], when the CPU and/or GPUs have their internal frequency bumped, a practice that might provoke crashes and errors. This is confirmed by Nightingale et al. in their study regarding the crash patterns of nearly one million Windows machines [NDO11], observing that the probability of a crash for machines with an overclocked CPU can increase as much as 20 times, depending on the CPU vendor (the study does not identify the CPU vendors).

Likewise, some laptops have heat issues when running for long periods at 100% CPU. This might provoke miscalculations and, hence, wrong results. This issue can be somewhat circumvented if the desktop grid middleware allows the volunteer to specify a threshold to throttle CPU usage (e.g., 85%), like BOINC does with the *use at most x% of cpu time* option, thus reducing the probability of overheating. Interestingly, and contrary to common belief, Nightingale et al. report that laptops are more robust than desktops, being between 25% and 60% less likely than desktop machines to crash from a hardware fault [NDO11]. In their study regarding five years of hosts volunteer in the SETI@home project (a total of 2.7 millions hosts), Heien et al. report that in 2007, 5.5% of processors were Pentium M, a model mainly used in laptops [HKA11]. This value serves as a minimum, as we believe that the percentage of laptops in volunteer projects is actually much higher.

Some errors are transparently dealt by the infrastructure. For instance, errors that affect the integrity of downloaded or/and uploaded files are caught through simple file integrity verification procedures such as checksums, and desktop grid middleware supports partial transfer for download and upload, thus minimizing the hurdles from network errors.

On the contrary, silent errors, that is, errors that still produce a result, albeit an incorrect one, and go undetected at the worker (the execution appears a regular one), should be dealt with by the server-side, as we see in Section 10.5.

Examples of silent errors include memory-induced ones. According to Schroeder et al. [SPW09], a memory error is an event that leads to the logical state of one or multiple bits being read differently from how they were last written. Memory errors fall into one of two categories: 1) permanent or 2) transient. The former corrupts bits in a repeatable manner because of a physical defect, while the latter randomly corrupts bits without leaving physical damage [SESK09]. In their large-scale study involving 50,000 GPUs of the volunteer project folding@home, Haque and Pande [HP10] report that as many as two-thirds of tested GPUs exhibited a pattern-sensitive susceptibility to soft errors in GPU memory or logic. Yim et al. [YPS+11] studied the behavior of GPUs when faults are injected. They report that, on average, 16 to 33% of injected faults cause silent data corruption (SDC) errors in the HPC programs executing on the GPU. The fragilities of GPU are relevant for desktop grid computing because some volunteer projects have recently embraced the adoption of GPUs in order to harness the massive performance of these devices for certain types of applications. Examples include the volunteer-based projects MilkyWay@home [DWMI+10] and Folding@home [BEJ+09]. For instance, in the folding@home project, as of March 2011, on average an NVIDIA GPU was computing as many as 322 x86 CPUs, while an ATI GPU was worth roughly 144 x86 CPUs. Even if the latest GPU models resort to ECC memory, budget devices might not adopt ECC because graphics (the main reason for GPUs after all) are somewhat fault tolerant to memory errors, and thus the motivation for the extra cost of ECC is not high.

10.3.3 Tampered Results

Several motivations can drive a volunteer to try to tamper with the results. One of these motivations is *credit greed*, that is, the attempt to obtain the highest amount of credits using the minimum amount of computing resources. Indeed, many public computing projects reward their volunteers with virtual credits accordingly to the computational effort performed by the volunteer's machine(s) to successfully compute tasks. These credits serve to organize rankings of performers, a practice that is quite popular among volunteers, especially the most devoted ones [And04]. Therefore, obtaining credits is an important motivation for some volunteers, driving some of them to devise strategies to game the credit system. For instance, not longer after its debut, the well-known SETI@home project was forced to cope with cheaters. Indeed, some malicious workers started to return bogus results like resending the same results over and over [Mol00]. Other workers adhered to schemes where the worker application of the project was replaced by an unofficial optimized one. One example was Akos Fekete's episode. Akos Fekete was an Einstein@home volunteer from Hungary, and used his knowledge of assembly programming to produce an optimized version based on the official binary application [Kni06]. His version more than doubled the performance of the original one, all of this done at the binary level, as he did not have access to the application source code[1]. The patched version rapidly spread through the public forum of the project, with enthusiastic volunteers replacing the official x86 binary with the patched one, in order to have their resources processing more work per time unit, and thus earn more credits. All of this was done without the approval of the project coordinators. Although the outcome of this episode actually benefited the project some of the improvements were later included in the official version of the application it illustrates some of the fragilities of the BOINC client security model. Indeed, a malicious user could follow a similar approach to spread a patched binary that would disrupt the computations, and in fact further patched versions from the well-intentioned Akos Fekete were banned by the project coordinators, fearing that an optimization could disrupt the integrity and soundness of the results.

An example of disruption of a volunteer project due to an optimized version of a worker code occurred with the SETI@home project. Specifically, the codenamed CUDA_V12 optimized application for NVIDIA's CUDA devices behaved erratically when run over the more advanced Fermi class of devices, returning immediately after it had started with a "-9 result_overflow" warning message, along with a (erroneous) result. When paired with a machine running the same problematic version of the software and that returns the same result, the result is validated (SETI@home uses majority voting with a 2-quorum) and thus accepted as valid. Although the number of machines with the problem was relatively small (around twenty machines[2]), the fact that each misconfigured worker only spent some microseconds per task meant that such a worker was able to download tasks until it had exhausted its daily quota, roughly 800 tasks per day. The problem was made worse because GPU have rather large daily task quotas (compared to CPUs), due to their performance.

Another menace to the correctness of results is sabotage, where in a saboteur disguised as a regular user might try to disrupt the project by injecting fake results [Sar02]. A saboteur may not directly benefit from the actions he performs and thus his motivations may simply be to hinder the project. The set of techniques devised to protect projects from sabotage is known as sabotage tolerance [Sar02] and is reviewed in Section 10.5.

10.3.3.1   Attacks to the Server-Side Infrastructure

The server-side is often the weakest element of the desktop grid infrastructure and a single point of failure. Indeed, limited budgets frequently impede a more robust organization of the server-side, leading to systems with no or

---

[1]Although BOINC is open source, public projects rarely release their applications source code, in part to avoid the surge of unwanted versions or for fear that a security issue can be exploited.

[2]http://setiathome.berkeley.edu/forum_thread.php?id=62573.

TABLE 10.1: Downtimes and Interruptions of Einstein@home in 2010.

Date	Event
2010.02.01	Slow database for several days due to large number of results
2010.03.17	1 hour for database maintenance
2010.03.31	The database ran out of space requiring reconfiguration (several hours downtime)
2010.05.31	One of the project server ran out of disk space (12 hours offline)
2010.06.13	Network outage (maintenance)
2010.10.02	Error in the generation of workunits provoked all to fail with validation error
2010.10.20	Few hours (relocation)
2010.11.02	One hour downtime (moving servers)

few redundancies and thus prone to common failure such a broken hard drive, a failed power sources or a corrupted database that disrupts the system, as, usually no spare system exists. Therefore, limited availability and unavailability do occur, even in major desktop grid projects. For instance, Table 10.1 identifies the outages and downtime of the Einstein@home project for the year 2010.

Attacks on the server-side infrastructure can seriously disrupt projects in several ways. According to the *unofficial BOINC Wiki*[3], the menaces to the server-side infrastructure fall within the following categories:

- Denial of service of the server-side infrastructure

- Theft of information

- Distribution of malicious executables

As the name suggests, a denial-of-service attack aims to disrupt the services of the server-side infrastructure, rendering it inaccessible and thus preventing the regular flow of tasks/results. Regarding desktop grid projects, two main types of denial-of-service attacks can occur: 1) network flooding and 2) file flooding. The former consists of traditional flooding of the server with a huge amount of network packets [LRST00]. The latter occurs at a higher level, with the attempt to swamp the server-side with valid operations but executed in a deviant way. For instance, attempts to download a large number of tasks or to repetitively upload large files with bogus results. Desktop grid projects can protect themselves by defining (*quotas*) task requests per worker per day and by imposing a maximum size for result files. This maximum size depends on the project.

---

[3]http://www.boinc-wiki.info/Security_in_BOINC.

Theft of information, whether it is data from the project (e.g., result files) or information relative to volunteers (e.g., e-mail addresses, logins and passwords) can be prevented with traditional security practices for Web infrastructures. Similarly, to prevent the automated creation of (fake) accounts and posts with spam content to forums coupled to the project, robust captcha systems need to be installed and maintained [vABHL03]. It is important to note that the efforts needed to run and maintain a secure Web infrastructure for a desktop grid project might be beyond the financial capabilities of some organizations that precisely seek the use of the desktop grid due to budget constraints. Additionally, volunteers should apply common security measures for accounts on public Web sites, like avoiding the use of a login name and password that are identical or similar to the ones used in more sensitive areas, such as their workplace or in private accounts.

Frequently, hacked websites are used to spread malicious payloads, namely malware applications. Desktop grid projects such as the ones supported by BOINC middleware protect themselves against tampered executables by only releasing digitally signed code and tasks to workers [And04]. Likewise, workers only run executables that have been digitally signed by the project the worker is connected to [DFG+08]. The same goes for workunits, which are digitally signed by the project and whose signature is verified by the worker software.

A hacked project website might leak some relevant data regarding the privacy of a user, like for instance the e-mail address and the IP addresses used by his machines. Nonetheless, a user can register for a volunteer project using an e-mail address created for the purpose on one of the many free webmail sites, thus acquiring an identifier possibly unrelated to its identity[4]. Additionally, because practically no relevant private data is stored by a desktop grid project regarding their volunteers, the value of attacking a project website seems to be relatively low. Ironically, the SETI@home project has been credited with having enabled the localization of a stolen laptop. Indeed, the victim found out that his stolen machine was still contributing to SETI@home by consulting his account, alerting law enforcement authorities that obtained the laptop IP address from the SETI@home project[5]. Other similar cases have been reported and a section of BOINC wiki is devoted to locating stolen computers[6].

10.4 Threats to Desktop Grid Resources

An important issue regarding desktop grids is that volunteering a resource should not, in any way, expose the resource to any kind of danger besides the ones that might affect Internet-connected resources. Indeed, exposing volun-

---

[4]As we shall later on, this practice can be performed by malicious users to conceal their real identities.

[5]http://www.techdirt.com/articles/20070222/080431.shtml.

[6]http://boinc.berkeley.edu/wiki/Locating_stolen_computers.

teered resources to a given vulnerability would probably have serious consequences, and would certainly end up driving volunteers away.

Threats to desktop grid resources can be broadly classified as 1) accidental
or 2) intentional. Accidental threats can be caused by a buggy application
or desktop grid middleware. They can disrupt the regular usage of the local
machine, and cause crashes, high resource usage, and breaches in data privacy
and data integrity. As the name suggests, intentional threats are intentionally
devised to cause prejudice and possibly benefits for the attacker. For instance,
a malicious desktop grid application to be run at volunteered resources can act
abusively, stealing and/or tampering with sensitive data [Dus11] or attaching
the local machine to a malicious *botnet* [McC03].

10.4.1 Protecting Desktop Grid Resources

Sandboxing is a possible approach to protect volunteered resources from
misbehaved code. Sandboxing is defined as the process of running arbitrary code in a safe way by containing it in a closed environment, that is,
where arbitrary code can run without compromising the system it is running
on [FC08, Fer10].

Sandboxing has been used in XtremWeb desktop grid middleware since
its inception, using operating system support where available, such as BSD
jails or OSXs sandboxing platform [FGN01, Fer10]. BOINC implements a
mild form of sandboxing, by running under a specially created unprivileged
account with no access to user data. This is done in the various platforms that
it supports, namely Linux, Windows, and Mac OS X.

A more in-depth path to sandboxing can be achieved through virtual machines that provide isolation from the host [Dus11]. Therefore, by running
workers inside virtual machines, full protection of worker resources can be
achieved. Moreover, if a fixed guest OS is used along with the virtual machine, for example Linux (or any open source OS), it becomes possible for
project owners to only develop the worker application for the selected guest
OS, instead of having to support the major platforms. This way, application development is simplified. The drawback of the virtual machine-based
approach lies in the overheads, namely the need to have (or install) virtual
machine software plus the guest OS image. Additionally, virtual machines consume resources and impose a performance overhead. Resource consumption
stems from the disk space and RAM memory used by the virtual machine
when it runs the guest OS. Performance overhead is due to CPU usage by
the virtual machine environment in executing the guest OS and applications,
although for CPU-bound applications, this overhead can be as low as 10%,
and thus almost meaningless [DAS09] when weighed against the benefits that
such a solution provides.

Several solutions for running BOINC workers over virtual machines
are proposed by Marosi et al. [CMKFL08], Lombrana-Gonzalez et
al. [LGdVT+07], and Ferreira et al. [FAD11].

10.5   Result Certification

Validating the results that worker nodes submit to the central supervisor is one of the fundamental problems in Volunteer Computing Systems [UJP05].

In his seminal work [Sar02], Sarmenta formalized a number of mechanisms to certify results, which we present here: majority voting, spot checking and credibility-based systems. In **majority voting**, worker nodes participate in a voting pool, where votes are the results of the computation. For $2m-1$ nodes, the central supervisor accepts the result that shows up m or more times. The obvious shortcoming of majority voting is the high level of redundancy. To avoid computing the same result multiple times, in **spot-checking**, the central supervisor sends some special jobs to test workers. The assumption here is that malicious nodes cannot differentiate these special jobs from the remaining. Then, as workers pass these spotter jobs, the central supervisor can have higher confidence in their results (and conversely, it trashes results from workers that fail these jobs). Unfortunately, spot-checking does not solve every problem. In particular, consider the whitewashing attack [FPCS04], where nodes can easily get new identifiers and submit different results with these new identifiers. Spot-checking is clearly powerless against these forms of attacks. Despite having a central entity, we cannot rely on complex authentication mechanisms, or volunteers will simply give up the project.

To overcome this problem, **credibility-based systems** use the contrary approach: the worker is not trusted until it proves to be trustable, and confidence in results must grow beyond some point before the central supervisor accepts the computation. In this section, we review these three basic mechanisms.

10.5.1   Majority Voting

Maybe the most obvious way of overcoming unreliable workers is to replicate the computation with a configured number of volunteers. Instead of sending a workunit only once to a worker, the central supervisor sends replicas of this workunit to other worker nodes. This replication raises the problem of possibly different results among the workers. The minimum number of results that must match before the project accepts the workunit as complete is the "Quorum of Results." This number can be quite divergent from project to project, but it is often two or three, meaning that volunteers must submit two or three matching results. There is actually no limit to the quorum size. When the project achieves a quorum for a given result, this becomes a "Canonical Result." The project keeps distributing results to volunteers either until it gets a quorum or until it gives up the workunit, if volunteers achieve the threshold for the number of errors.

With respect to certification of results, BOINC has one (of the many) daemon(s) dedicated to check whether results are valid: this is the "validator". BOINC offers a framework that allows project developers to create a validation function. This function tells BOINC when it should consider two results the same. Floating-point implementations may be a typical source for differences in results [TACBI05]. Interestingly, the validation framework also offers two special kinds of validators: `sample_bitwise_validator()`, and `sample_trivial_validator()`. The former validates results on a bit-by-bit basis, whereas the latter considers two results to match if their CPU time exceeds some minimum threshold. Project owners can accept this option if they trust the workers (as in a university laboratory, for instance).

10.5.2 Spot-Checking

A major and quite obvious shortcoming of majority voting is the amount of redundant work. To overcome this problem, Sarmenta [Sar02] suggested a scheme, called "spot-checking," where the central supervisor distributes special tasks, "spotters," which serve to challenge the workers. The central supervisor provides spotter jobs in Bernoulli trials with some probability, say q. This implies a redundancy level of $1/(1-q)$, as only $1-q$ fraction of the jobs are really useful. One should notice that spotter jobs must be indistinguishable from the others, or, otherwise, saboteurs could just compute the spotter jobs and send random results to the remaining.

To be effective, spot-checking requires blacklisting, to let the central supervisor exclude workers that fail the tests. Excluding a worker implies going back to all the results of this worker and discarding them. With spot-checking, if a fraction f of saboteurs behave as Bernoulli processes, outputting an incorrect result with probability s, the *average final error rate with spot-checking and blacklisting*, ϵ_{scbl} is [Sar02]:

$$\epsilon_{scbl} = \frac{sf(1-qs)^n}{(1-f)+f(1-qs)^n} \tag{10.1}$$

In this formula, $(1-qs)^n$ is the probability that a saboteur remains undetected after n turns. This tops at $\epsilon_{scbl} < \frac{f}{1-f}\frac{1}{qne}$. Based on this result, the author concluded that the error rate depends linearly on n, thus suggesting that project owners should make longer series of jobs (batches).

The central supervisor can only perform spot-checking with blacklisting, if it is able to determine the identity of the workers as opposed to their possibly many identifiers (see the very interesting work of Douceur [Dou02] regarding this point). Otherwise, if workers can freely leave and rejoin the system with new identifiers, the effectiveness of spot-checking severely degrades. In the worst case, if saboteurs compute a single result before joining with a new identifier, spot-checking is powerless. Sarmenta deemed this case "spot-checking without blacklisting." In practice, building a blacklist is very difficult, unless project owners are willing to enforce high barriers for project entrance, usually requiring some manual intervention, for example, through a CAPTCHA mechanism [vABHL03]. On the other hand, such a barrier would work both ways, discouraging donors from entering as well.

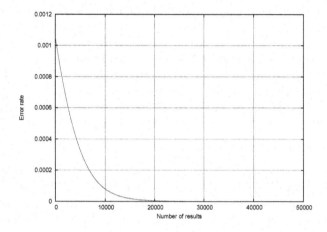

FIGURE 10.1: Error rate as a function of the number of results.

Derrick et al. [KdADS07] point out another practical shortcoming, this one resulting directly from Equation (10.1). The number of actual turns of computation n that each node needs to perform to achieve some error rate ϵ might be prohibitively high. For instance, from experiments in [KAM$^+$07], we can set the values of $s = 0.003$, $q = 0.1$, and $f = 0.35$. For these settings, consider the plot in Figure 10.1. Although we can observe a fast reduction in the error rate, the problem is that we are quite dependent on the number of results that each worker needs to compute. An alternative way of reducing the error rate is to increase the spot-check rate q. We show this plot in Figure 10.2 for $5,000$ results. Fewer results might not even allow us to achieve the error rates we desire. We can clearly see in the plot that to get to the low-end error rates, we need to raise q, with the corresponding impact on redundancy. Unlike this, majority voting allows us to have strict control over the error rates. Because the error rate decreases approximately exponentially [Sar02] with the majority size, m, the owner of a computing project can set m as he is pleased to keep saboteurs out.

10.5.3   Credibility-Based Systems

As we have seen, spot-checking depends on the ability of the central supervisor to figure out the real identity of a worker, something that is often impossible in the wide open Internet. Sarmenta [Sar02] also proposed the inverse strategy: distrust workers by default and make them demonstrate their credibility. The most promising option proposed by Sarmenta was to com-

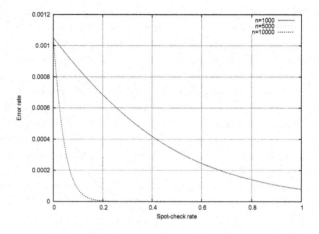

FIGURE 10.2: Error rate as a function of the spot-check rate q.

bine a credibility-based mechanism with spot-checking: workers should pass spot-checking workunits to raise their credibility in the system.

Sarmenta considers four objects of interest that receive a credibility rate: the worker, the results, the result groups, and the work entries. A *result group* is a set of matching results. A *work entry* is the set of all results computed for that workunit. The credibility of the worker determines the credibility of all the results it produces. The credibility of a result group depends on the credibility of the workers outputting those results and on the credibility of other results groups for the same work entry. To determine the credibility of the worker P, Cr_P, Sarmenta used a formulation similar to Equation (10.1). This credibility is computed as the probability that the result from worker P is bad *given* that P survived k spot-checks:

$$P(\text{result from } P \text{ is bad}|P \text{ survived } k \text{ spot-checks}) = \tag{10.2}$$

$$\frac{sf(1-s)^k}{(1-f) + f(1-s)^k} < \frac{f}{1-f}\frac{1}{ke} \tag{10.3}$$

Based on this formula, the credibility of P for spot-checking with black-listing is

$$Cr_P = 1 - \frac{f}{1-f}\frac{1}{ke} \tag{10.4}$$

Credibility of workers can change under a number of circumstances. In addition to growing when they pass spot-checking workunits (if they do not pass, they get blacklisted, if blacklisting is available), the credibility of workers also grows when their results match results from other workers (thereby forming result groups), but can decrease if other workers *sharing result groups* get caught in the spot-checking. The point of the credibility-based scheme is to ensure that the credibility of a work entry reaches some threshold ν. This credibility is a conditional probability that the result is correct, given the results of workers in other work entries and in spot-checking workunits. If the central supervisor accepts a result when it is correct with an average probability of at least ν, then the error rate is at most $\epsilon_{acc} = 1 - \nu$. Putting this the other way around, if the project owner wants the error rate not to exceed ϵ_{acc}, it must configure the central supervisor not to accept any work entry before this reaches a credibility of $\nu = 1 - \epsilon_{acc}$.

From the volunteer's perspective, there is a strong incentive for staying in the system and a weak incentive for cheaters to use multiple identifiers, as new identifiers will always have low credibility. From the project owner's point of view, when they set an error rate of ϵ_{acc}, they are actually setting a point in a trade-off between error rate and replication. Lowering the former involves raising the latter. The quorum of results is rather flexible in this way: instead of waiting for a fixed number of results, the central supervisor waits until the credibility of the result increases.

Credibility-based mechanisms may also work without spot-checking. However, in this case, the central supervisor must assume some bound f for the fraction of malicious workers and assign the same credibility $Cr_P = 1 - f$ to all of them, which seems to be a much less interesting possibility.

10.6   Collusion of Volunteer Nodes

We also review here one form of attack that is especially difficult to tackle. When the colluder nodes are aware of the size of the voting pools (or if they are able to guess the size), they may collectively take some decision to submit wrong results. For instance, a set of malicious workers, after observing that they form a majority in the voting pool, may decide to submit the same incorrect result. This behavior departs significantly from a more naive form of attacking. This latter behavior usually results from nodes that may simply have hardware problems, for example, resulting from overclocked machines, or from some other accidental source of faults, as reported earlier. Colluding nodes pose an interesting challenge for project owners: although they can cause serious harm to the project results, their action leaves some trace, as they submit results that differ from those of correct nodes. Hence, we must find ways to make use of these traces, not only to detect nodes colluding against the project, but also to force colluder nodes to contribute with valid results (or ban them). In this section we overview some mechanisms that can cope with this kind of behavior.

10.6.1   EigenTrust Algorithm

Peer-to-peer networks usually serve the purpose of sharing files between users. However, not all peers have a cooperative behavior and some nodes intentionally introduce damaged content on the network to hinder file spreading [LKXR05]. To control this behavior, Kamavar et al. [KSGM03] proposed the EigenTrust algorithm, which creates a global notion of trust that emerges from experiences of local interactions between peers.

Based on the experience nodes have with their neighbors while exchanging files, they rate each transaction as *satisfactory* or *unsatisfactory*. Each peer node i keeps a *local trust value*, s_{ij}, defined in Equation (10.5) as the difference between satisfactory and unsatisfactory transactions with j:

$$s_{ij} = sat(i, j) - unsat(i, j) \tag{10.5}$$

The key aspect of the work of Kamavar et al. is the transformation of this local notion of trust into a global one common to all peers. First, they resort to a normalization of this metric, to keep it between 0 and 1 Equation (10.6).

$$c_{ij} = \frac{max(s_{ij}, 0)}{\sum_j max(s_{ij}, 0)} \tag{10.6}$$

The definition of c_{ij} eliminates negative interactions by making them similar to cases where no interactions take place. On the other hand, authors use this metric to build a probabilistic model. If one would make a walk on the peer-to-peer network, randomly picking the next hop according to the probabilities c_{ij}, the probability of ending in a reputable peer is higher than ending in a disreputable one. Based on the matrix $[c_{ij}]$, C, we can compute a vector $\vec{t_i} = C^T \vec{c_i}$, as in Equation (10.7), which is the trust that node i has in node k. This weights the opinion that each neighbor j of node i has of other nodes, with the opinion that i has of neighbor j.

$$t_{ik} = \sum_j c_{ij} c_{jk} \tag{10.7}$$

With this equation, trust will be worth two hops. Enlarging the scope of trust is rather easy, as we can count with the acquaintances of acquaintances of nodes. We just use $\vec{t_i} = (C^T)^2 \vec{c_i}$. To go further, we may simply use larger values of n to get the general relation $\vec{t_i} = (C^T)^n \vec{c_i}$. Eventually, for large enough n, we will reach the end of the network (assuming C is irreducible and aperiodic), and $\vec{t}$ will converge to be the same for every peer. Authors have a centralized and a distributed algorithm to compute $\vec{t}$. In our case, we only care about the centralized version, which we show in Algorithm 1. $\vec{e}$ is an m-vector, where all its elements are $1/m$.

Algorithm 1 Centralized EigenTrust algorithm

$\vec{t}^{(0)} = \vec{e}$
repeat
$\quad \vec{t}(k+1) = C^T \vec{t}^{(k)}$
$\quad \delta = ||t^{(k+1)} - t^k||$
until $\delta < \epsilon$

10.6.2 EigenTrust for Volunteer Computing

Intuitively, one can easily agree that the node's "opinion" about a specific neighbor can be more accurate by collecting the opinion that other peers have of the same neighbor.

However, the original scope of EigenTrust is not exactly the same as we consider in this text. In our case, we would rather care about the number of times nodes provide the same result (vote together), versus the number of times they provide different results (i.e., they oppose). And we can take one factor as our advantage: one distinctive feature of volunteer computing grids, when compared to peer-to-peer networks, is the existence of a centralized server. This makes it feasible to collect long-term data on every worker to make an evaluation of the similarity of results they submit. Silaghi et al. [SAS+09] collect historic data about nodes before applying the EigenTrust algorithm to detect colluders. Their idea relies on counting the number of times that each node submits results that differ from other nodes (termed *votes against*). The central supervisor registers each one of the votes against collected by nodes in their voting pools. After some time, we expect honest nodes to have fewer votes against than malicious nodes. Then, using these numbers and the EigenTrust method, we should be able to make a clear separation of nodes into groups in particular, we should be able to tell which nodes are not honest.

Silaghi et al. define the cumulative distribution function, $F_v(i)$ as $F_v(i) = Prob(Y_{v,N} \leq i)$, where $Y_{v,N}$ is a random variable that defines the number of votes against that worker v collects over a total of N voting pools. Assuming three nodes per voting pool, the node can collect between 0 and $2N$ votes against. We have three possible outcomes in each voting pool, which are 0, 1, or 2 votes against. Additionally, authors assume that voting pools constitute independent trials. Hence, the multinomial distribution gives the number of votes against in N trials and we can make use of the coefficients of Equation (10.8), where $p_{v,i}$ is the probability that the node v collects i votes against.

$$(p_{v,0} + p_{v,1}X + p_{v,2}X^2)^N \qquad (10.8)$$

For instance, if $(p_{v,0}, p_{v,1}, p_{v,2}) = (0.6, 0.2, 0.2)$, meaning that the probability of a node having 0/1/2 votes against is 0.6/0.2/0.2, the expansion for five voting pools is $0.0778 + 0.1296X + 0.216X^2 + 0.2016X^3 + 0.1776X^4 + 0.1059X^5 + 0.0592X^6 + 0.0224X^7 + 0.008X^8 + 0.0016X^9 + 0.0003X^{10}$. The

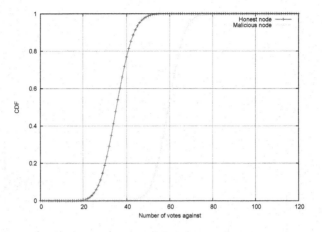

FIGURE 10.3: Cumulative distribution function of $Y_{v,N}$.

coefficient of degree k gives the probability that the node collects k votes against. For example, the node may get four votes against with a probability of 17.76%. If we sum all the coefficients up to degree i, we get the probability that the node has i or fewer votes against, $F_v(i)$.

Based on the observation of the votes against that nodes do in all their voting pools, we can compute the sampled coefficients of $p_{v,0}$, $p_{v,1}$, $p_{v,2}$ as in Equation (10.8). This immediately makes a distinction between honest and malicious workers, as the former should have more voting pools without votes against. As we expand the coefficients according to the previous formula, we obtain important differences between the worker nodes. For example, we may consider one node where $(p_{v,0}, p_{v,1}, p_{v,2}) = (0.6, 0.2, 0.2)$ and another node that collected more votes against (probably a malicious one): $(p_{v,0}, p_{v,1}, p_{v,2}) = (0.3, 0.4, 0.3)$. Based on these coefficients, we expand them to $N = 30$ voting pools, and depict the corresponding cumulative distribution functions in Figure 10.3. The difference between nodes with more and less votes against is clear from the figure, as the curve grows much faster for nodes with fewer votes against, which, in general, should correspond to honest nodes. Unlike this, curves for dishonest nodes should grow more to the right side, with intermediate behaviors in the middle.

Next, based on these curves, authors define a distance between two nodes i and j using the expression of Equation (10.9). These distances define a matrix $D = [d_{ij}]$, with statistical differences between workers. If we normalize this matrix, to get a sum of 1 in each row (to do this we divide the elements of the row by their own sum), we get a matrix C. The C matrix (or a variation

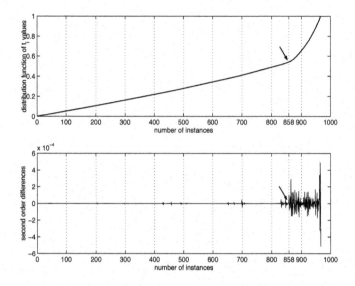

FIGURE 10.4: Sorted t values and second-order differences.

thereof) is used as the input to the EigenTrust algorithm to give a global score to each node. Then, if we sort the scores of the nodes, we can get a curve like the one of Figure 10.4. This curve has an arrow where the inflection approximately separates honest from dishonest nodes. Here, we skip the details on how to identify the inflection point, but authors use the second-order differences plot to identify this point.

$$d\left(v_i, v_j\right) = \sum_k \left(F_{v_i}(k) - F_{v_j}(k)\right) \tag{10.9}$$

A key aspect of the work of Silaghi et al. [SAS+09] is that *after* their algorithm suspects workers, it asks for additional (considered to be) honest workers to confirm results. When this step finishes, authors identify malicious nodes and discard all voting pools where they have voted alone.

10.6.3 Identification of Malicious Results through Correlation

Similar to the work of Silaghi et al., Staab and Engel [SE09] also consider the existence of workers that produce malicious results, under the condition that they win a voting pool. When in the minority, this form of colluders will simply output the correct result, to avoid identification. These nodes are the "Conditional Colluders" (CC). Staab and Engel consider another kind of workers, termed the "Unconditional Colluders" (UC). They assume that there is a fraction p_{mal} of malicious workers in the set of volunteers (either UC or

CC, but not both). Then, authors define the notion of correlation p_c, as the probability that two workers in the same voting pool produce the same result. In other words, if we pick two nodes at random in the same voting pool, we get a high correlation (p_c close to 1) if they probably produce matching results. On the contrary, we will get a low correlation (p_c close to 0) if they probably produce different results.

Authors simulated two volunteer computing experiments: one with honest and UC nodes, the other with honest and CC nodes. The output of these experiments is indeed interesting. They observed that in the case of UC nodes, correlation is very high for nodes of the same kind (UC or honest), but very low if the pair contains one UC and one honest node. Project owners may explore this fact to isolate colluding nodes, because nodes form clusters based on their correlations. Then, under the assumption that honest nodes define the largest group, the smaller group contains colluders.

The situation is dimmer if the volunteer set includes CC nodes and depending on how many of them do exist. Again, correlation between nodes of the same type is similar. However, CC nodes and honest nodes have a higher correlation than UC and honest nodes. Nevertheless, as the fraction of CC nodes, p_{mal}, grows in the community, these will perform an increasing number of attacks and, as a consequence, the correlation between honest and CC nodes will decrease. Consequently, clustering will, once again, enable their identification. In fact, colluder nodes face an important dilemma: their repeated attacks make them clearly identifiable, while their coverage makes them cooperant.

10.6.4   Online Identification of Colluders

While the collusion-detection methods shown so far collect results for posterior evaluation, Canon et al. [CJW10] created two online algorithms to identify groups of nodes that collude. The output of their algorithms is a set of groups, such that workers inside these groups have similar collusion characteristics; that is, they send the same results, either correct or incorrect. One of their methods is based on collusion representation, while the other is based on agreement representation. For the former case, they keep a matrix $\hat{C}$ with the observed collusion probability between groups, whereas for the latter, they keep a matrix $\hat{A}$ with the observed agreement between groups. $\hat{C} = \hat{c_{ij}}$ is the observed probability that one node from group i and another node from group j collude to produce an erroneous result, while $\hat{A} = \hat{a_{ij}}$ is the observed probability of agreement between those two groups, when two of their nodes submit a result from the same voting pool. In the case of collusion, authors need some external mechanism to certify the results (in order to know if collusion really occurred). On the contrary, in agreement representation, simple comparison of results suffices.

Both the collusion and agreement representations split and join groups according to the behavior of their nodes. In the former case, if two nodes u and w submit different results, either they are split from the same group or the collusion probability between their groups (if different) is decreased. When workers finish some job (comprised of several replicas or results) and submit the same incorrect result, the collusion probability estimation of their groups is increased. Other conditions exist, but the main point is: submission of the same incorrect result increases $\hat{c}_{ij}$, if nodes belong to groups i and j (possibly $i = j$); submission of the same correct result or submission of different results decreases $\hat{c}_{ij}$. Based on the value of c_{ij}, the central supervisor may merge or split groups i and j.

In the agreement representation, we only care about observing whether workers produce the same result or not, without needing any certification for this result. When workers from different groups produce the same result, they increase the agreement probability, which may allow merging their groups; if workers produce different results, either we split them from their group (if it is the same), or decrease the agreement probability estimation.

To compute the quality of the algorithms, authors use a formula for the Root Mean Square Deviation (RMSD), which we repeat in Equation (10.10). g and g' represent two groups of workers. $e_{g \cup g'}$ is the absolute difference between an estimation that all the workers in the same voting pool, from the set $g \cup g'$ return incorrect results, and the actual value of the probability. In practice, authors use bounds to compute $e_{g \cup g'}$. This metric reflects the fact that both methods, collusion and agreement, keep statistics for the interactions between groups. The lower the RMSD, the better the method. In their experiments [CJW10], the authors concluded that both methods identified the appropriate sets of colluders and honest nodes, with the value of RMSD being lower in the collusion-based approach.

$$\frac{1}{|G|} \sqrt{\sum_{(g,g') \in G^2} e_{g \cup g'}^2} \tag{10.10}$$

10.7   Conclusion

Since its inception, volunteer desktop grid middleware has progressed immensely, and reached the ability to harness large-scale systems, such as SETI@home and Einstein@home, which congregate thousands of volunteers. However, this success does not come without problems and many dependability issues exist, not only for the owners of the project, but also for the volunteers.

For volunteers, security is still a weak element in the public volunteer desktop grid chain. Indeed, whenever a volunteer offers his resources to a given project, he is implicitly trusting that the project(s) will not cause any harm, nor engage in malicious behavior, such as disclosing private elements or jeopardizing the integrity of the data. Progress has steadily improved the security of desktop grids, with research focused on virtual machines and sandboxing. These can represent a sound path for making desktop grids simpler for projects and safer for volunteers. It is thus expected that in the near future, more secure desktop grid middleware, possibly based on virtualization/sandboxing will allow for a safer volunteering and harnessing of resources.

For project owners, the main dependability problem is the certification of results computed over volunteer platforms. This is still an active research field because current solutions are either computationally expensive (replication) or dependent on a strong identification scheme that might drive volunteers away. This is the case with methods like majority voting, spot-checking, or credibility-based schemes, although the latter offers some flexibility to find a reasonable balance between error rates and replication. In this chapter we also discussed other fronts of research, namely we described a number of different methods that tackle possible colluding actions from malicious volunteers.

Bibliography

[AF06]   David P. Anderson and Gilles Fedak. The Computational and Storage Potential of Volunteer Computing. In *Proceedings of the IEEE International Symposium on Cluster Computing and the Grid (CCGRID'06)*, pages 73–80, Los Alamitos, CA, USA, 2006. IEEE Computer Society.

[And04]   David Anderson. BOINC: A System for Public-Resource Computing and Storage. In *Proceedings of the 5th IEEE/ACM International Workshop on Grid Computing, 2004*, PA, USA, pages 4–10, 2004.

[BEJ+09]   Adam L. Beberg, Daniel L. Ensign, Guha Jayachandran, Siraj Khaliq, and Vijay S. Pande. Folding@home: Lessons from eight years of volunteer distributed computing. *Parallel and Distributed Processing Symposium, International*, 0:1–8, 2009.

[CAS05]   Carl Christensen, Tolu Aina, and David Stainforth. The Challenge of Volunteer Computing with Lengthy Climate Model Simulations. In *Proceedings of 1st International Conference on e-Science and Grid Computing. December 2005, Melbourne, Australia*, pages 8–15, Los Alamitos, CA, USA, 2005. IEEE Computer Society.

[CB10]   SungJin Choi and Rajkumar Buyya. Group-Based Adaptive Result Certification Mechanism in Desktop Grids. *Future Generation Computer Systems*, 26(5):776 – 786, 2010.

[CJW10]   L.C. Canon, E. Jeannot, and J. Weissman. A Dynamic Approach for Characterizing Collusion in Desktop Grids. In *Parallel & Distributed Processing (IPDPS), 2010 IEEE International Symposium on*, pages 1–12. IEEE, 2010.

[CMKFL08] Attila Csaba Marosi, Peter Kacsuk, Gilles Fedak, and Oleg Lodygensky. Using Virtual Machines in Desktop Grid Clients for Application Sandboxing. *Technical Report* TR-0140, Institute on Architectural Issues: Scalability, Dependability, Adaptability, CoreGRID - Network of Excellence, August 2008.

[DAS09]   Patricio Domingues, Filipe Araujo, and L.M. Silva. Evaluating the Performance and Intrusiveness of Virtual Machines for Desktop Grid Computing. In *3rd Workshop on Desktop Grids and Volunteer Computing Systems (PCGrid 2009)*, Rome, Italy, May 2009. IEEE Computer Society.

[DFG+08]   Marco Danelutto, Paraskevi Fragopoulou, Vladimir Getov, Attila Marosi, Gabor Gombas, Zoltan Balaton, Peter Kacsuk, and Tamas Kiss. Sztaki Desktop Grid: Building a Scalable, Secure Platform for Desktop Grid Computing. In *Making Grids Work*, pages 365–376. Springer US, 2008.

[Dou02]   John R. Douceur. The Sybil Attack. In *IPTPS '01: Revised Papers from the First International Workshop on Peer-to-Peer Systems*, pages 251–260, London, UK, 2002. Springer-Verlag.

[Dus11]   Tobias Dussa. DGVCS Security from a Different Perspective Challenges and Hurdles. In *5th Workshop on Desktop Grids and Volunteer Computing Systems (PCGrid 2011)*, pages 1878–1882, Anchorage, Alaska, USA, May 2011.

[DWMI+10] Travis Desell, Anthony Waters, Malik Magdon-Ismail, Boleslaw K. Szymanski, Carlos A. Varela, Matthew Newby, Heidi Newberg, Andreas Przystawik, and David Anderson. Accelerating the milkyway@home volunteer computing project with gpus. In *Proceedings of the 8th International Conference on Parallel Processing and Applied Mathematics: Part I*, PPAM'09, pages 276–288, Berlin, Heidelberg, 2010. Springer-Verlag.

[FAD11]   Diogo Ferreira, Filipe Araujo, and Patricio Domingues. libboincexec: A Generic Virtualization Approach for the Boinc Middleware. In *5th Workshop on Desktop Grids and Volunteer Computing Systems (PCGrid 2011)*, Anchorage, AK, USA, May 2011.

[FC08]   Bryan Ford and Russ Cox. Vx32: Lightweight Userlevel Sandboxing on the x86. In *Proceedings of the USENIX Annual Technical Conference*, pages 293–306, Berkeley, CA, USA, 2008. USENIX Association.

[Fer10]   Diogo Ferreira. Sandboxes in Desktop Grid Projects. Master's thesis, University of Coimbra, July 2010.

[FGN01]   Gilles Fedak, Cecile Germain, and Vincent Neri. Xtremweb : A generic global computing system. In *In Proceedings of the IEEE International Symposium on Cluster Computing and the Grid (CCGRID 01)*, pages 582–587, 2001.

[FPCS04]   Michal Feldman, Christos Papadimitriou, John Chuang, and Ion Stoica. Free-Riding and Whitewashing in Peer-to-Peer Systems. In *PINS '04: Proceedings of the ACM SIGCOMM Workshop on Practice and Theory of Incentives in Networked Systems*, pages 228–236, New York, NY, USA, 2004. ACM.

[HKA11]   Eric Martin Heien, Derrick Kondo, and David P. Anderson. Correlated Resource Models of Internet End Hosts. In *31st International Conference on Distributed Computing Systems (ICDCS)*, 2011.

[HP10]   Imran S. Haque and Vijay S. Pande. Hard Data on Soft Errors: A Large-Scale Assessment of Real-World Error Rates in gpgpu. *Cluster Computing and the Grid, IEEE International Symposium on*, 0:691–696, 2010.

[KAM+07]   Derrick Kondo, Filipe Araujo, Paul Malecot, Patricio Domingues, Luis Silva, Gilles Fedak, and Franck Cappello. Characterizing Results Errors in Internet Desktop Grids (Best Paper Award). In *Euro-Par*, Rennes, Frances, August 2007.

[KdADS07]   Derrick Kondo, Filipe Araujo, Patricio Domingues, and Luis Silva. Result Error Detection on Heterogeneous and Volatile Resources via Intermediate Checkpointing. In *CoreGRID Workshop on Grid Programming Model Grid and P2P Systems Architecture Grid Systems, Tools and Environments. June 2007, Hellas Heraklion, Crete, Greece*, June 2007.

[Kni06]   Will Knight. Programmer Speeds Search for Gravitational Waves. http://www.newscientisttech.com/article.ns?id= dn9180, May 2006.

[KSGM03]   Sepandar D. Kamvar, Mario T. Schlosser, and Hector Garcia-Molina. The EigenTrust Algorithm for Reputation Management in P2P Networks. In *Proceedings of the 12th International Conference on World Wide Web*, WWW '03, pages 640–651, New York, NY, USA, 2003. ACM.

[LGdVT+07] D. Lombrana-Gonzalez, F.F. de Vega, L. Trujillo, G. Olague, and B. Segal. Customizable Execution Environments with Virtual Desktop Grid Computing. In *Proceedings of the 19 th IASTED International Conference on Parallel and Distributed Computing and Networks*. Acta Press Inc,# 80, 4500-16 Avenue N. W, Calgary, AB, T 3 B 0 M 6, Canada,, 2007.

[LKXR05]   J. Liang, R. Kumar, Y. Xi, and K.W. Ross. Pollution in p2p file Sharing Systems. In *INFOCOM 2005. 24th Annual Joint Conference of the IEEE Computer and Communications Societies. Proceedings IEEE*, volume 2, pages 1174–1185. IEEE, 2005. isbn=0780389689, issn=0743-166X.

[LRST00]   F. Lau, S.H. Rubin, M.H. Smith, and L. Trajkovic. Distributed Denial of Service Attacks. In *Systems, Man, and Cybernetics, 2000 IEEE International Conference on*, volume 3, pages 2275–2280. IEEE, 2000.

[McC03]   B. McCarty. Botnets: Big and Bigger. *Security & Privacy Magazine, IEEE*, 1(4):87–90, 2003.

[Mol00]   D. Molnar. The SETI@Home Problem. *ACM Crossroads Student Magazine*, September 2000.

[MS94]   H. Madeira and J.G. Silva. Experimental Evaluation of the Fail-Silent Behavior in Computers Without Error Masking. In *Fault-Tolerant Computing, 1994. FTCS-24, Twenty-Fourth International Symposium on*, pages 350–359. IEEE, 1994.

[NDO11]   E.B. Nightingale, J.R. Douceur, and V. Orgovan. Cycles, Cells and Platters: An Empirical Analysis of Hardware Failures on a Million Consumer PCs. In *EUROSYS 2011*. ACM, April 2011.

[Sar02]   Luis F. G. Sarmenta. Sabotage-Tolerance Mechanisms for Volunteer Computing Systems. *Future Gener. Comput. Syst.*, 18(4):561–572, 2002.

[SAS+09]   G. Silaghi, F. Araujo, L. M. Silva, P. Domingues, and A. Arenas. Defeating Colluding Nodes in Desktop Grid Computing Platforms. *Journal of Grid Computing*, 7(4):555–573, December 2009.

[SE09]   Eugen Staab and Thomas Engel. Collusion Detection for Grid Computing. In *CCGRID '09: Proceedings of the 2009 9th IEEE/ACM International Symposium on Cluster Computing and the Grid*, pages 412–419, Washington, DC, USA, 2009. IEEE Computer Society.

[SESK09]   G. Shi, J. Enos, M. Showerman, and V. Kindratenko. On testing GPU Memory for Hard and Soft Errors. In *Proc. Symposium on Application Accelerators in High-Performance Computing*, 2009.

[SPW09]   Bianca Schroeder, Eduardo Pinheiro, and Wolf-Dietrich Weber. DRAM Errors in the Wild: A Large-Scale Field Study. In *Proceedings of the Eleventh International Joint Conference on Measurement and Modeling of Computer Systems*, SIGMETRICS '09, pages 193–204, New York, NY, USA, 2009. ACM.

[TACBI05]   M. Taufer, D. Anderson, P. Cicotti, and CL Brooks III. Homogeneous Redundancy: a Technique to Ensure Integrity of Molecular Simulation Results Using Public Computing. In *Proceedings of 19th IEEE International Parallel and Distributed Processing Symposium (IPDPS'05), 2005*, pages 119–127, 2005.

[UJP05]   Christian Ulrik, Sttrup Jakob, and Gregor Pedersen. Developing Distributed Computing Solutions Combining Grid Computing and Public Computing. Master's thesis, University of Copenhagen, Denmark, 2005.

[vABHL03]   Luis von Ahn, Manuel Blum, Nicholas Hopper, and John Langford. CAPTCHA: Using Hard AI Problems for Security. In Eli Biham, editor, *Advances in Cryptology - EUROCRYPT 2003*, volume 2656 of *Lecture Notes in Computer Science*, pages 646–646. Springer Berlin/Heidelberg, 2003.

[YPS+11]   Keun Soo Yim, Cuong Pham, Mushfiq Saleheen, Zbigniew Kalbarczyk, and Ravishankar Iyer. HAUBERK: Lightweight Silent Data Corruption Error Detector for GPGPU. In *International Parallel and Distributed Processing Symposium (IPDPS'2011)*, pages 287–300, Anchorage, AK, USA, May 2011.

Chapter 11

Data-Intensive Computing on Desktop Grids

Heshan Lin

Virginia Tech, Blacksburg, Virginia, USA

Gilles Fedak

INRIA, University of Lyon, France

Wu-Chun Feng

Virginia Tech, Blacksburg, Virginia, USA

11.1 Introduction

Innovations in science and engineering are incrementally driven by intelligently making sense of massive datasets. Advanced simulations and experimental analyses in disciplines such as high-energy physics, climate modeling, astronomy, and life sciences require processing terabytes or even petabytes of data on a routine basis. As such, data-intensive scientific discovery has been identified as *the fourth paradigm*, as an addition to the traditional three scientific paradigms: experimental science, theoretical science, and computational science [HTT09].

Traditionally, Desktop Grids are mainly designed for compute-intensive workloads with data management as a secondary concern. The increasing demand of processing-intensive data from the scientific community has posted a myriad of new challenges to Desktop Grid systems. First, harnessed from donated cycles and storage of volunteer computers, compute resources in Desktop Grids are inherently volatile and heterogeneous. In such environments, even basic data-management tasks such as reliably storing large datasets are very difficult to accomplish. Data-intensive computing techniques in cluster environments, for example, distributed and parallel file systems, assume dedicated, homogeneous resources and hence are not suitable for Desktop Grids. Second, compute resources in Desktop Grids can be geographically distributed and interconnected with low-bandwidth networks, making it hard to design scalable data-intensive solutions on these systems. Third, data privacy and security must be protected on Desktop Grids consisting of untrusted computers. The data protection mechanism may add non-trivial overhead when processing large volumes of data.

Despite research efforts in addressing individual aspects of data-intensive computing on Desktop Grids in past years, production Desktop Grids support only simple program paradigms such as bags-of-tasks. Enabling more sophisticated parallel data processing requires heroic efforts to integrate various non-standard components from different solutions. Recently, emergent data-centric programming models such as MapReduce [DG08] and Dryad [IBY+07] have proven to be effective in simplifying parallel data processing on large-scale commodity clusters. In this chapter we highlight two pioneering studies in implementing MapReduce on Desktop Grids. The experiences and lessons learned from these studies suggest that the MapReduce paradigm, combining an intuitive programming interface and a scalable data-management abstraction, can bring new opportunities to offering unified solutions for data-intensive computing on Desktop Grids.

11.2 State-of-the-Art Data Processing on Desktop Grids

We review several state-of-the-art efforts related to supporting data-intensive computing on Desktop Grids.

11.2.1 Storage Harnessing of Desktop Computers

With the continuing decrease in price and increase in capacity of hard drives, today the storage of a desktop computer within an organization is mostly underutilized [ABC+02, DB99]. Many approaches have been investigated in harnessing the unused disk space on desktop computers to provide cost-efficient storage service. Farsite [ABC+02, DW01] is a shared, secure file system built atop distributed storage from untrusted personal computers (PCs). Farsite achieves data availability and reliability through replication, and it uses data encryption to improve security and ensure user access privacy. To improve global data availability, Farsite continuously monitors machine availability and relocates files accordingly to equalize availability across all files in the system. Glacier [HMD05] is a decentralized storage system designed to tolerate large-scale correlated failures. Assuming no a priori knowledge of machine availability, Glacier relies on massive data redundancy realized with erasure code to ensure high data availability. To achieve a balance between replication cost and high availability, ThriftStore [GKR10, GR09] adopts a hybrid architecture that combines a large number of unreliable nodes with a set of low-throughput durable nodes (e.g., Automated Tape Library). By doing so, ThriftStore enables high-throughput read access with the aggregate bandwidth of unreliable nodes, while guaranteeing user-defined levels of service quality with highly available durable nodes.

There have also been storage scavenge systems tailored to meet the I/O requirements of high-performance computing (HPC). FreeLoader [VMF+05] is a distributed cache/scratch system specially designed for write-once-read-many data access. With an asymmetric data striping design, FreeLoader can store scientific datasets that are much larger than the disk space of a desktop computer and provide high read bandwidth. Stdchk is designed to offer low-cost checkpoint storage for a Desktop Grid [AKRVG08]. Stdchk allows applications to access the scavenged storage through a traditional file-system interface, and it provides a set of features optimized for checkpointing I/O such as high-throughput write I/O, incremental checkpointing and automatic checkpoint image removal. File replication is also adopted to tolerate node failures.

11.2.2 Data-Aware Scheduling and Computing

Traditionally, desktop grids follow a computation-centric design, where data management is implicitly handled to accommodate the requirement of computational tasks. For instance, in BOINC [And04], input data is downloaded by a client as needed by a compute task. Such a computation-centric approach is inadequate for supporting complex scientific workflow in large-scale distributed environments. Although more sophisticated data management can be achieved using manual scripts, such an ad hoc solution is cumbersome and error-prone.

As an effort toward promoting data management as "a first-class citizen" in grid computing, Stork [KL04, KL05] proposed a data-aware scheduler that allows data placement activities to be scheduled as computational jobs in a grid. Stork allows users to specify dependencies between data placement jobs and computational jobs in a Directed Acyclic Graph (DAG). Such a DAG can then be processed by a planner such as Directed Acyclic Graph Manager (DAGMan) [DAG] to ensure the data placement and computational jobs are properly coordinated. In Stork, data recovery from failures is transparently handled by the runtime system, freeing users from the tedious work of writing manual scripts.

BitDew [FHC08, FHC09] is a programmable framework aimed at providing transparent and robust data management for Desktop Grid computing. Leveraging peer-to-peer techniques such as distributed hash tables (DHTs), BitDew provides a set of flexible APIs that allow users to easily manage various data behaviors including replication, placement, failure tolerance, lifetime control, and data transferring protocols. The rich set of data-management APIs makes BitDew ideal for developing higher-level data-aware programming frameworks.

11.2.3 Programing Models

Because of the resource volatility in Desktop Grid environments, parallel programming models for cluster computing, for example, the message passing model [Mes95, Mes97], are ill-suited for such as environments. As such, up to now, production Desktop Grid systems such Condor [TTL04], BOINC and Entropia [CCEB03] support only simple programming models that exploit bags-of-tasks or master-worker [pGKLY00] styles of parallelism. Programming models that support more sophisticated parallel-execution patterns are needed for promoting wider adoption of Desktop Grids in data-intensive computing. In canonical grid computing, Pegasus proposes to use DAGs as an abstraction for mapping complex workflow to distributed computing resources [DShS+05]. Recently, programming abstractions that focus on specific scientific processing patterns such as All-Pairs [MBTF08] have also been proposed for campus grid environments. Although these studies suggest that abstractions such as DAGs and All-Pairs can be powerful tools for non-experts to adapt parallel data processing, enabling these abstractions on Desktop Grids entails many technical challenges.

11.3 MapReduce Overview

MapReduce is a programming model introduced by Google for large-scale data processing on compute clusters [DG08]. In MapReduce, users write two primitives for an application: Map and Reduce, which are inspired by functional languages such as Lisp. Figure 11.1 shows the task-execution flow of a MapReduce program. First, the input data is split into chunks, and a set of Map tasks is launched to process individual chunks in parallel. The intermediate output generated by a Map task is partitioned and transfered to corresponding Reduce tasks, where the reduce function is executed to produce final output. Note that transferring intermediate output to reduce tasks (termed *shuffle phase*) follows an all-to-all communication pattern. The input, intermediate, and output data are all stored as key-value pairs on a scalable distributed file system such as Google File System (GFS) [GGL03].

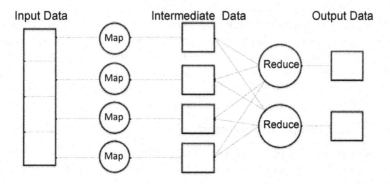

FIGURE 11.1: MapReduce programming model.

In MapReduce, parallel task-execution and fault tolerance are automatically handled by the runtime system, enabling programmers to process large datasets on thousands of distributed computers without intimate knowledge of parallel computing. MapReduce is attractive for Desktop Grid computing for several reasons. First, MapReduce presents a standard, easy-to-use programming interface, which can help users quickly develop portable applications across different Desktop Grid systems or even clusters and clouds. Second, the data flow is well defined on a MapReduce job, thus allowing Desktop Grid designers to transparently optimize data management without involving application developers. Third, MapReduce is friendly for fault tolerance, making it more favorable on volatile resources compared to other parallel programming models such as MPI.

In the remainder of this chapter, we highlight two projects that enable MapReduce on Desktop Grids. The first project, called MOON, short for MapReduce on Opportunistic eNvironments [LMA+10, LMF11], explores the feasibility of extending an existing cluster-based MapReduce framework, that is, Hadoop, to support Desktop Grid environments. The second study aims at designing a new MapReduce system that is tailored for Desktop Grids, using the programmable BitDew data-management framework.

11.4   MOON: MapReduce on Opportunistic eNvironments

MOON is a research effort aimed at extending cluster-based MapReduce frameworks, for example, Hadoop, for institutional Desktop Grids where PCs are connected by local area networks with relatively high bandwidth. Unlike cluster environments, in Desktop Grids, the resources can freely come and go without notifications, and large-scale resource unavailability is not uncommon. For instance, a large percentage of volunteer computers in a student lab can simultaneously become unavailable when a class begins. Relying on volunteer desktop computers alone is difficult to sustain reliable MapReduce computing. Consequently, MOON adopts a hybrid architecture by provisioning a set of dedicated computers to supplement volunteer desktop computers. Such a hybrid architecture allows the system to meet certain levels of service requirement even when volunteer computing resources are highly volatile. For cost efficiency, MOON assumes the number of dedicated computers is much smaller than that of volatile computers.

11.4.1   Hadoop Background

Hadoop is an open-source MapReduce implementation written in Java [Ano]. It consists of a Hadoop Distributed File System (HDFS) and a runtime system for executing MapReduce jobs. HDFS follows a master-slave design similar to the Google File System. The metadata of the file system is managed by a master node termed NameNode, and the actual files are stored as blocks on worker nodes termed DataNodes. In Hadoop, data is replicated, for example, three times by default, to achieve data reliability and availability. To achieve good scalability on very large clusters, HDFS adopts a relaxed consistency model compared to traditional distributed file systems such as NFS.

The MapReduce runtime system in Hadoop also follows a master-slave design. The master node, termed JobTracker, performs bookkeeping and the scheduling of jobs in the system. The slave node, termed TaskTracker, periodically fetches Map or Reduce tasks from the master and executes them on local resources. The MapReduce runtime and HDFS typically run on the same cluster nodes. The scheduling of Map tasks will take into account data locations to reduce network transfer. After all original tasks are scheduled, Hadoop issues backup tasks (also termed speculative tasks) for slow-running tasks to help improve the job response time when the overall job progress is hindered by "stragglers" on some abnormal nodes.

11.4.2 MOON Data Management

Because of the data dependencies within a MapReduce job, in order to provide high-quality MapReduce service, data needs not only to be reliably stored but also to be highly available. For instance, in Hadoop, a running Map task will fail if the corresponding input data split is not available. If a Map task is not able to finish after several reties, the entire job will fail. Thus, a MapReduce job will not be able to complete if a piece of input data is not available during the retry window of time. Similarly, the availability of intermediate data is also critical to the completion of reduce tasks and, in turn, the entire job.

While Hadoop's data management works well on cluster environments, straightforward deployment of Hadoop on Desktop Grids is woefully inadequate. *First*, in Hadoop, the intermediate data generated by a Map task (also called a Map output) is stored on the local disk of a compute node without replication. Consequently, the intermediate data can easily become unavailable, hindering the completion of reduce tasks. Moreover, when the JobTracker finds that multiple reduce tasks report missing a Map output, the corresponding Map task will be re-executed to generate the missing data. This task re-execution caused by missing data can lead to significant waste of system resources. *Second*, even input and output data are replicated on HDFS, the replication cost for achieving high availability is prohibitive when desktop resources are highly volatile, especially with the presence of large-scale correlated node unavailability. For instance, researchers have found that even with erasure coding, achieving six nines of availability under 60% node unavailability requires *eleven* data replicas [HMD05].

To address the above issues, MOON data management adopts a multi-dimensional replication design that leverages the hybrid resource architecture. To replicate data on both dedicated computers and volunteer desktop computers, MOON extends HDFS to differentiate two types of DataNodes: *dedicated* and *volatile*. The replication factor of a file is given as a pair $\{d, v\}$, where d and v specify the number of data replicas on the two types of resources. Because the availability of dedicated nodes is much higher than volatile nodes, hybrid replication can greatly reduce the storage cost compared to replication on volatile nodes alone. Nonetheless, MOON's hybrid replication also raises several design challenges, which are discussed as follows.

11.4.2.1 Non-Uniform Replication and I/O Request Distribution

Although dedicated nodes are highly available, their aggregate I/O bandwidth is scarce because the number of dedicated nodes is much smaller than that of volatile nodes. Consequently, dedicated nodes can easily become a performance bottleneck if replica placement and I/O request serving are not carefully planned. For example, if tasks running on all nodes simultaneously request to place a dedicated replica for their output data, the dedicated nodes will likely be overloaded and in turn cause delay in task completion. As such, one major challenge in designing the hybrid replication is to maximize the utilization of dedicated nodes in improving the MapReduce service quality. MOON tackles this challenge by reserving scarce dedicated I/O resources for data that are more critical to the job completion.

As explained in Section 11.3, there are three types of data in a MapReduce jobs: input data, intermediate data, and output data. The loss of an input data block can force the re-execution of the corresponding Map task and in turn leads to a failure of the job execution. Therefore input data always needs to be highly available. The loss of intermediate data can delay the completion of the corresponding reduce tasks. However, intermediate data can be recovered by rerunning the corresponding Map task, providing that its input data blocks are available. Finally, during the job execution the loss of output data can be recovered by re-executing the corresponding reduce task, although this may trigger re-execution of Map tasks if any intermediate data is missing. However, output data cannot be recovered if the input data of a job has been removed from the system.

One important observation is that different types of data have different tolerance levels of data loss. Accordingly, MOON adopts a non-uniform replication design for different types of data. Specifically, MOON defines two types of files: *reliable files* and *opportunistic files*. Reliable files are used to store critical data that cannot afford any loss, whereas opportunistic files are used to store transient data that can be recovered with additional computation. MOON always maintains one or more dedicated replicas for reliable files in order to maintain high availability for these files. In contrast, dedicated replicas are not guaranteed for opportunistic files.

MOON prioritizes write I/O requests to dedicated nodes according to the distinction between reliable and opportunistic files. The write requests of reliable files to dedicated nodes are always satisfied. However, the write requests of opportunistic files to dedicated nodes can be denied when there is not sufficient I/O bandwidth on these nodes. Realizing this I/O prioritization design requires determining a saturation point of dedicated I/O bandwidth, which will be discussed later in the section.

For read access, MOON gives priority to volatile nodes if a data replica can be found on those nodes. In other words, MOON schedules a read request on dedicated nodes only when the request cannot be satisfied on volatile nodes. Such a design allows users to control read traffic on dedicated nodes by adjusting the number of volatile copies. For instance, suppose the node unavailability rate is p and the replication factor of a file is $\{1, v\}$; then the probability of the read request that will be serviced by dedicated nodes is p^v.

Putting all things together, Figure 11.2 depicts the data-management flow of MOON. The system data and input data are stored as reliable files on the extended HDFS. To improve availability of intermediate data, they are stored on the extended HDFS as well but as opportunistic files. Because output data

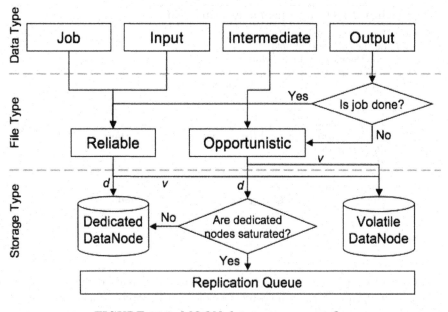

FIGURE 11.2: MOON data management flow.

can be regenerated during the job execution, these data are first stored as opportunistic files and converted to reliable files after the job completion.

11.4.2.2 I/O Load Monitoring

One interesting problem in the I/O prioritization discussed above is to determine the saturation point of a dedicated node. Currently MOON uses a throttling algorithm based on a simple heuristic: when a node closes to saturation, the aggregate I/O bandwidth will plateau when more requests are added. As shown in Algorithm 2, a dedicated DataNode periodically monitors its I/O bandwidth in a window of time. If the increase of I/O bandwidth during the past window on a dedicated DataNode is less than a given threshold, this DataNode will not take new I/O requests.

11.4.2.3 Adaptive Replication of Opportunistic Files

Although data stored as opportunistic files can be regenerated, the availability of these data is critical to the job response time. If the number of volatile copies of an opportunistic file is set too low, its availability may suffer when a dedicated copy cannot be maintained because of the bandwidth constraint on dedicated nodes. On the other hand, if the number of volatile copies is set too high, there will be an unnecessary waste of storage when the dedicated copy is successfully placed. Instead of using a static replication factor, MOON adaptively places replicas on hybrid resources to achieve a balance between replication cost and availability. The basic idea is to use a low

---

Algorithm 2 I/O throttling on dedicated DataNodes

Let W be the throttling window size
Let T_b be the control threshold
Let bw_k be the measured bandwidth at timestep k
Input: current I/O bandwidth bw_i
Output: setting throttling state of the dedicated node

$avg\_bw = (\sum_{j=i-W}^{i-1} bw_j)/W$
if $bw_i > avg\_bw$ **then**
    if $(state == unthrottled)$ **and** $(bw_i < avg\_bw * (1 + T_b))$ **then**
        $state = throttled$
    end if
end if
if $bw_i < avg\_bw$ **then**
    if $(state == throttled)$ **and** $(bw_i < avg\_bw * (1 - T_b))$ **then**
        $state = unthrottled$
    end if
end if

---

number of volatile copies for an opportunistic file when a dedicated copy can be acquired, and to use a high number of volatile copies otherwise.

There are different ways of implementing the adaptive replication discussed above. For instance, users can define a lower bound and an upper bound for the number of volatile copies, which will be chosen by MOON when replicating an opportunistic file depending on whether a dedicated copy can be successfully placed. MOON can also adjust the number of volatile copies according to a user-defined availability level a (e.g., 0.9) assuming the node unavailability rate p can be measured at runtime. More specifically, MOON dynamically maintains v volatile copies such that $1 - p^v > a$ until a dedicated copy is eventually placed on dedicated nodes.

11.4.3 MOON Task Scheduling

One important fault-tolerant feature of Hadoop task scheduling is speculative task execution, which is designed to prevent tasks running on abnormal nodes from elongating the overall job response time. This is implemented by issuing backup tasks to "stragglers" after all original tasks have been scheduled. A straggler is a task with a progress score that is 20% behind the average task progress, where the progress scores is calculated based on the proportion of processed input data.

On Desktop Grids, tasks running on a volunteer computer may be frequently interrupted by the owner's activities, thus adversely impacting the overall job response time. Hadoop's speculative task execution cannot be directly applied here for two reasons. First, Hadoop task scheduling design assumes that a task's progress rate is constant on a compute node. On Desktop Grids, the task progress rate can vary significantly depending on the owner's activities. Second, Hadoop identifies "stragglers" solely by task progress scores, which can be inaccurate in Desktop Grid environments. For instance, a task with a high progress score can suddenly be suspended and may not be able to resume for a long period of time, in which case this high-scoring task may finish much slower than a lower-scoring task that runs without interruptions.

11.4.3.1 Identifying Stragglers on Opportunistic Environments

MOON assumes that its processes are suspended on a desktop when the owner's activities are detected, and the suspended tasks will be resumed later when the desktop becomes idle again. With the original Hadoop task scheduling, a suspended task with a high progress score may be classified as a normal task, in turn causing system resources to be wasted on replicating healthy tasks with a low progress score. Also, Hadoop issues at most one backup copy for each task; if the original and backup copies of a task are both suspended, no progress will be made for this task.

To accommodate frequent task interruptions on Desktop Grids, MOON's task scheduling defines two categories of stragglers: *frozen* tasks and *slow* tasks. A frozen task is the one with all copies being suspended[1], and hence no progress is being made for such a task. A slow task is a straggler according to Hadoop's criteria, that is, a task with a progress score 20% behind the average progress. Accordingly, MOON maintains two separate lists for the frozen and slow tasks, respectively. Higher priority will be given to the frozen tasks when issuing a backup task. To ensure task progress under high resource volatility, MOON will always issue one backup copy to the frozen task, regardless of the number of copies that have been scheduled for this task.

11.4.3.2 Two-Phase Scheduling

When a job gets close to finish, the sudden suspension of the last batch of tasks can considerably elongate the job response time, because the detection of suspension and the execution of the backup task can take a non-trivial amount of time. MOON remedies this by aggressively replicating tasks toward the end of the job. Specifically, the execution of a job is divided into two phases: *normal phase* and *homestretch phase*. The homestretch phase starts when the number of tasks falls below $H\%$ of the execution slots in the system[2], when MOON starts proactively maintaining R *active* running copies of the remaining tasks regardless of their progress scores. Suppose the unavailability rate of the desktop computer is p; in the homestretch phase, the probability of a task being frozen reduces to p^R. To prevent task replicas from consuming too many system resources, MOON limits the number of task replicas under $H\%$ of the available execution slots in the system. Both H and R are parameters configurable by users.

MOON's two-phase scheduling algorithm assumes the availability of a desktop computer cannot be known a priori and trades additional usage of

---

[1]MOON detects a suspended task using the heartbeat mechanism in Hadoop.

[2]In Hadoop, each node has two Map and two Reduce execution slots.

system resources with better job response time. More efficient algorithms can be designed in desktop grids where the machine availability follows predictive models. For instance, a group of desktop computers may be reserved for a class, in which case the scheduler can avoid scheduling tasks to the reserved computers during the class time. Studies have also shown that in some Desktop Grids, there is considerable correlation between the availabilities of different computers [KFC+07]. In this system, placing two copies of a task on machines that do not have correlated availabilities can help prevent both copies from being suspended simultaneously.

11.4.4 Performance Evaluation

This section presents some preliminary results of MOON on an emulated Desktop Grid system. The experiments are executed on the System X cluster at Virginia Tech, comprised of Apple Xserve G5 compute nodes with dual 2.3GHz PowerPC 970FX processors, 4GB of RAM, and 80 GB hard drives. System X uses a 10Gbs InfiniBand network and a 1Gbs Ethernet for interconnection. To closely resemble Desktop Grid systems such as student lab, only the Ethernet network is used in the experiments. The MOON system is developed based on Hadoop 0.17.2. To emulate resource volatility on a Desktop Grid, unavailability traces are generated for each node following a Poisson distribution with a mean node-outage interval (409 seconds) extracted from the aforementioned Entropia volunteer computing node trace [KTI+04]. The unavailability traces are independent between different nodes. At runtime of each experiment, a monitoring process on each node reads in the assigned availability trace, and suspends and resumes all the Hadoop/MOON processes on the node accordingly. The experiments focus on two applications, the `sort` application with 24GB input data and the `word count` application with 20GB input data.

11.4.4.1 Hybrid Replication of Intermediate Data

In a typical Hadoop job, the *shuffle* phase of a reduce task, that is, copying intermediate data from every Map task, makes up a substantial portion of the job execution time even on clusters. Achieving efficient shuffle performance is more challenging on Desktop Grids because the intermediate data could be unavailable due to frequent node outage. This experiment evaluates the efficacy of MOON's hybrid replication of intermediate data with respect to the overall job response time. Specifically, the experiment compares two replications approaches: 1) the *volatile-only* (VO) replication approach that statically replicates intermediate data only on volatile nodes, and 2) the *hybrid-aware* (HA) replication approach described in Section 11.4.2. For the VO approach, the number of volatile copies is increased from 1 (`VO-V1`) to 5 (`VO-V5`). For the HA approach, MOON is configured to store one copy on dedicated nodes when possible, and increase the minimum volatile copies from 1 (`HA-V1`) to 3 (`HA-V3`).

The experiment uses sixty volatile nodes and six dedicated nodes or a

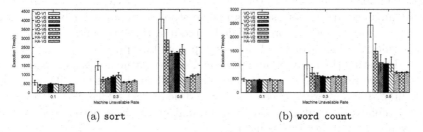

(a) sort   (b) word count

FIGURE 11.3: Impacts of different replication policies for intermediate data on execution time.

10:1 volatile-to-dedicated (V-to-D) ratio. To focus solely on intermediate data, the input/output data use a fixed replication factor of $\{1, 3\}$. The two-phase scheduling algorithm introduced in Section 11.4.3 is used with $H = 20$ and $R = 2$.

Figure 11.3a shows the results of sort. As expected, enhanced intermediate data availability through the VO replication clearly reduces the overall execution time. When the unavailability rate is low, the HA replication does not exhibit much performance gain. However, HA replication significantly outperforms VO replication when the node unavailability level is high. While increasing the number of volatile replicas can help improve data availability on a highly volatile system, this incurs significant overhead caused by the extra I/O. As a result, there is no further execution time improvement from VO-V3 to VO-V4, and from VO-V4 to VO-V5, the performance actually degrades. With HA replication, having at least one copy written to dedicated nodes substantially improves data availability, with a lower overall replication cost. More specifically, HA-V1 outperforms the best VO configuration, that is, VO-V3 by 61% at the 0.5 unavailability rate.

With word count, the gap between the best HA configuration and the best VO configuration is smaller. This is not surprising, because word count generates much smaller intermediate/final output and has many fewer Reduce tasks, in which case the cost of fetching intermediate results can be largely hidden by the execution of Map tasks. Also, increasing the number of replicas does not incur significant overhead. Nonetheless, at 0.5 unavailability rate, the HA replication approach outperforms the best VO replication configuration by about 32.5%.

11.4.4.2   Overall Performance Impacts of MOON

This experiment evaluates the overall performance impact of MOON by comparing an augmented version of Hadoop that replicates intermediate data on HDFS. The vanilla version of Hadoop is not used because it performs poorly on volatile resources without replication of intermediate data. To establish a comparison baseline, the experiment assumes that the availability of a dedicated node is equivalent to that of at least three volatile nodes. Consequently, the augmented Hadoop is configured to store six replicas for both input and output data. MOON stores one dedicated and three volatile replicas, that is, a replication factor of $\{1, 3\}$, for the input and output data.

In testing the native Hadoop system, sixty volatile nodes and six dedicated nodes are used. These nodes, however, are all treated as volatile, as Hadoop cannot differentiate between volatile and dedicated nodes. For each test, the VO replication configuration that can deliver the best performance under a given unavailability rate is used.

The MOON tests are executed on sixty volatile nodes with three, four and six dedicated nodes, corresponding to 20:1, 15:1 and 10:1 V-to-D ratios, respectively. The intermediate data is replicated with the HA approach using $\{1, 1\}$ as the replication factor. As shown in Figure 11.4, MOON clearly outperforms Hadoop-VO for 0.3 and 0.5 unavailable rates and is competitive at a 0.1 unavailability rate, even for a *20:1 V-to-D ratio*. For `sort`, MOON outperforms Hadoop-VO by a factor of 1.8, 2.2, and 3 with three, four and six dedicated nodes, respectively, when the unavailability rate is 0.5. For `word count`, the MOON performance is slightly better than augmented Hadoop, delivering a speedup factor of 1.5 compared to Hadoop-VO. The only case where MOON performs worse than Hadoop-VO is for the `sort` application at the 0.1 unavailability rate and the V-to-D node ratio is 20:1. This is due to the fact that the aggregate I/O bandwidth on dedicated nodes is insufficient to quickly absorb all of the intermediate and output data; a reduce task will not be flagged as complete until its output data reaches the predefined replication factor (including one dedicated copy).

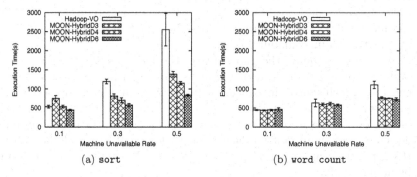

(a) `sort`   (b) `word count`

FIGURE 11.4: Overall performance of MOON versus Hadoop with VO replication.

11.5 Practical MapReduce Computing on Desktop Grids with BitDew

In this section we present an alternative implementation of MapReduce for Desktop Grid. Our focus is on the context of loosely connected Internet Desktop Grid, and in particular the case where resources are provided by volunteer hosts. We believe that applications requiring an important volume of data input storage with frequent data reuse and limited volume of data output could take advantage not only of the vast processing power, but also of the huge storage potential [AF06] offered by volunteer computing systems. Overall, the objective of our work is to broaden the scope of the Desktop Grid beyond bag-of-tasks parallel application to support new classes of scientific applications such as bioinformatics [AGM+90] and large data analysis in general.

However, the context of Internet Desktop Grids raises several specific challenges compared to the Desktop Grid composed of local resources. The first challenge is the support for collective file operation that exists in MapReduce. In particular, the redistribution of intermediate results between the execution of Map and Reduce tasks, called the *Shuffle* phase, presents some similarities with the `MPI_AlltoAll` collective [HLD09]. On the Internet, because of the host's volatility and churn, of extremely varying network conditions and connectivity issues due to NAT and firewalls, collective communication are considerably difficult to achieve efficiently. The second challenge is that the large volume of data to process imposes decentralized mechanisms. For instance, a key security component is the results certification [Sar02] that is needed to check that malicious volunteers do not tamper with the results of a computation. Because intermediate results might be too large to be sent back to the server, results certification mechanism cannot be centralized as it is currently implemented in existing Desktop Grid systems. The third challenge is that dependencies between the Reduce tasks and the Map tasks, combined with host volatility and laggers, can dramatically slow down the execution of MapReduce applications. Thus, we have to propose an aggressive performance-optimisation solution, which combines latency hiding, data and tasks replication, and barriers-free reduction.

11.5.1 Architecture of MapReduce over BitDew

In this section, we first give a brief overview of the BitDew middleware and describe the implementation and optimization of the framework.

BitDew [FHC09] was developed by the INRIA as an open-source solution to manage large datasets on Desktop Grids, Clouds, and Grids. Main BitDew features are fault tolerance, multiple file transfer protocols, and data scheduling. The key concept of BitDew is the *attributes* that are abstractions to drive key data management operations, namely life cycle, distribution, placement, replication, and fault tolerance with a high level of abstraction.

Our implementation of MapReduce on top of BitDew [TMC⁺10] allows us to take advantage of the ability of BitDew to manage large-scale data in a highy volatile environment. We implement MapReduce applications as a dataflow between the map inputs, the mappers, the intermediate results, the reducers, and the results. The BitDew attributes are then used to express the dependencies between data and the fault-tolerance scheme. The BitDew runtime environment takes the responsibilities of scheduling the data to the volunteer nodes. Eventually, when faults occur, BitDew reschedules the data on available nodes to ensure the resilience of the computation.

We now present some of the high-level features and optimizations that have been developed to address the resource characteristics of Desktop Grid systems.

11.5.1.1 Latency Hiding

One of the key features of our implementation is that we target resources that are spread over the Internet. Because it is unlikely that direct communication between hosts is always possible because of firewall settings, the communication latency can be orders of magnitude higher than the network latency found in clusters. One of the mechanisms to hide high latency in parallel systems is overlapping communication with computation with data prefetching. Our implementation features multi-threaded workers, each capable of processing several concurrent file transfers even if the transferring protocol is synchronous (e.g., HTTP). As soon as a file transfer completes, the corresponding Map or Reduce tasks are enqueued, and the ready tasks are processed by the worker concurrently. The maximum number of Map and Reduce tasks running concurrently can be configured, as well as the minimum number of tasks in the queue before computations can start. Data prefetch is natural with MapReduce, as data are staged on the distributed file system before computation tasks are launched.

11.5.1.2 Collective File Operation

MapReduce is composed of several collective file operations, which, in some aspects, look similar to collective communications found in parallel programming such as MPI. These collective operations include 1) *Distribution*, which is the initial distribution of file chunks (similar to MPI_Scatter); 2) *Shuffle*, which is the redistribution of intermediate results between the Mapper and the Reducer nodes (similar to MPI_AlltoAll); and 3) *Combine*, which is the assemblage of the final result on the master node (similar to MPI_Gather). Because collective operations were missing in BitDew, we have implemented the notions of DataCollection and DataChunk, where the former manipulates a set of data as a whole and the latter manipulates an individual part of a DataCollection.

11.5.1.3 Fault Tolerance

Desktop resources can join and leave the network at any time without notice; therefore, the execution runtime must be resilient to a massive number of failures. In the context of MapReduce, failures can happen during the execution of Map or Reduce tasks, or during the communication, that is, file upload and download. If a Map task fails, the Map input chunk must be distributed to another node in order to restart the Map task. To implement this behavior, we simply enable the automatic rescheduling of Map input data that has disappeared. The situation slightly differs if a Reducer node crashes, because all the corresponding intermediate results should be sent to the new Reducer node. To realize this, we set an affinity link between the intermediate results and the Reducer so that all intermediate results are still linked together after the failures. To tolerate file transfer failure, we rely on the reliable file-transfer mechanism featured in BitDew.

11.5.1.4 Barrier-Free Computation

Desktop Grids are not only prone to node crashes, but also to host churn, that is, joining and leaving of nodes because of owner activities or instable network connectivity. For the purpose of fault tolerance, several replicas of the same data can exist in the system. In particular, intermediate results can be duplicated. To handle the replication of intermediate results, we have adopted a de-duplication, which allows Reducer nodes to detect that several versions of the same intermediate file exist in the queue. Thus, only the first version of the intermediate file is reduced. Traditional cluster implementation of MapReduce such as Hadoop introduces several barriers in the computation and in particular between the execution of Map and Reduce tasks. However, there can be a long period of time before nodes reconnect; it is preferable that Reduce tasks start as soon as intermediate files are produced. To enable this feature, in our implementation, the API of the Reduce task allows the programmer to write the reduce function on segments of key intervals. Because the number of file chunks and the number of reducers are known in advance, it is possible to determine when a reduction finishes. The early reduction combined with the replication of intermediate results allowed us to remove the barrier between Map and Reduce tasks.

11.5.1.5 Scheduling

MapReduce computation follows the paradigm of moving the computation to the data. MapReduce on BitDew uses a two-level scheduler. The first-level scheduler is the BitDew scheduler, which ensures the placement of data on the hosts. The placement of data is guided by the attribute properties specified for each piece of data. However, the scheduling at this level would not be enough to efficiently steer the complex execution of MapReduce application. The second-level scheduler allows us to detect *laggers*, that is, nodes that spend an unusually long time to process the data and slow down the whole computation. Workers periodically report to the second-level node, that is, the master node, the state of their ongoing computation. The master node can then determine if there are more nodes available than tasks to execute. In this case, increasing the replication factor of the remaining tasks can avoid the lagger effect.

11.5.1.6 Distributed Result Checking

A key security component of Desktop Grid systems is the result certification, which needs to be checked so that malicious volunteers do not tamper with the results of a computation. Because intermediate results might be too large to be sent back to the server, a result certification mechanism cannot be centralized as currently implemented in existing Desktop Grid systems. As a consequence, we have decentralized the majority voting certification algorithm [MSF11]. We replicate each input map file, which in turn replicates intermediate results. Once a reducer has obtained n out of p intermediate results, the result that appears most often is assumed to be correct. Similarly, we replicate the number of reducers, and the output of the reduce tasks are checked on the master node, which is trustable. Although majority voting involves redundant computation, the heuristic is efficient at detecting erroneous results [KAM+07], which are likely to be significantly higher for I/O bound jobs.

11.5.2 Performance Evaluation

In this section we evaluate the performance of our implementation using the word count benchmark adapted from the Hadoop distribution. Word count counts the number of occurrences of each word in a large collection of documents. The file transfer protocol used to send and receive data is the HTTP protocol.

The first experiment evaluates the scalability of MapReduce/BitDew when the number of nodes increases. Each node has a different 5GB file to process, split into fifty local chunks. Thus, when the number of nodes doubles, the size of the whole document counted doubles as well. For 512 nodes[3], the benchmark processes 2.5TB of data and executes 50,000 Map and Reduce tasks. Figure 11.5 presents the throughput of the word count benchmark in MB/s versus the number of worker nodes. This result shows the scalability of our approach and illustrates the potential for using Desktop Grid resources to process a vast amount of data.

In the second experiment, we vary the number of mappers and reducers and look at the impact on the performances. Table 11.1 presents the time spent in the Map function, the time spent in the Reduce function, and the total makespan of word count for the number of mappers varying from 4 to 32 and the number of reducers varying from 1 to 16. As expected, the

---

[3]GdX has 356 double core nodes, so to measure the performance on 521 nodes we run two workers per node on 256 nodes.

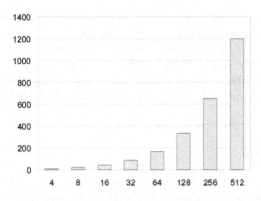

FIGURE 11.5: Scalability evaluation on the `word count` application: the y axis presents the throughput in MB/s and the x axis the number of nodes varying from 1 to 512.

TABLE 11.1: Evolution of the Performance According to the Number of Mappers and Reducers.

#Mappers	4	8	16	32	32	32	32
#Reducers	1	1	1	1	4	8	16
Map (sec.)	892	450	221	121	123	121	125
Reduce (sec.)	4.5	4.5	4.3	4.4	1	0.5	
Makespan (sec.)		473	246	142	146	144	150

Map and Reduce time decreases when the number of mappers and reducers increases. The difference between the makespan and the Map plus Reduce time is explained by the communication and time elapsed in waiting loops.

11.6   Conclusion

Desktop Grids are traditionally designed for compute-intensive workloads. While many studies have been conducted to address individual perspectives of data-intensive computing on Desktop Grids, it still requires heroic efforts to integrate various non-standard components to process large datasets on today's production Desktop Grid systems. In this chapter we highlighted two pioneering studies for data-intensive processing on Desktop Grids inspired by the recent success of data-centric programming models, for example, MapReduce and Dryad, on commodity clusters. The first study introduces MOON, a MapReduce system designed for LAN (local area network) based Desktop Grids within institutions by extending Hadoop, a popular cluster-based MapReduce implementation, on a hybrid resource architecture. The second study focuses on designing and implementing a MapReduce system based on the BitDew data-management framework for Internet Desktop Grids. Both studies reveal that the resource dynamics of Desktop Grids poses great challenges for reliable and efficient MapReduce computing. Yet, the preliminary results of these studies suggest that MapReduce holds great promise for providing standard and unified data-processing solutions on Desktop Grids.

Bibliography

[ABC+02] Atul Adya, William J. Bolosky, Miguel Castro, Gerald Cermak, Ronnie Chaiken, John R. Douceur, Jon Howell, Jacob R. Lorch, Marvin Theimer, and Roger P. Wattenhofer. FARSITE: Federated, Available, and Reliable Storage for an Incompletely Trusted Environment. In *Proceedings of the 5th Symposium on Operating Systems Design and Implementation*, 2002.

[AF06] David Anderson and Gilles Fedak. The Computational and Storage Potential of Volunteer Computing. In *Proceedings of the 6th IEEE International Symposium on Cluster Computing and the Grid (CCGRID'06)*, pages 73–80, Singapore, May 2006.

[AGM+90] Stephen F. Altschul, Warren Gish, Webb Miller, Eugene W. Myers, and David J. Lipman. Basic Local Alignment Search Tool. *Journal of Molecular Biology*, 215(3):403–410, 1990.

[AKRVG08] Samer Al-Kiswany, Matei Ripeanu, Sudharshan S. Vazhkudai, and Abdullah Gharaibeh. stdchk: A Checkpoint Storage System for Desktop Grid Computing. *Distributed Computing Systems, International Conference on*, 0:613–624, 2008.

[And04] D. Anderson. BOINC: A System for Public-Resource Computing and Storage. *Grid Computing, IEEE/ACM International Workshop on*, 2004.

[Ano] Anonymous. Hadoop. http://hadoop.apache.org/core/.

[CCEB03] Andrew Chien, Brad Calder, Stephen Elbert, and Karan Bhatia. Entropia: Architecture and Performance of an Enterprise Desktop Grid System. *Journal of Parallel and Distributed Computing*, 63, 2003.

[DAG] Directed acyclic graph manager. http://www.cs.wisc.edu/condor/dagman/.

[DB99] John R. Douceur and William J. Bolosky. A Large-Scale Study of File-System Contents. In *Proceedings of the 1999 ACM SIGMETRICS international conference on Measurement and mod-*

eling of computer systems, SIGMETRICS '99, pages 59–70, New York, NY, USA, 1999. ACM.

[DG08]   Jeffrey Dean and Sanjay Ghemawat. Mapreduce: Simplified data processing on large clusters. *Commun. ACM*, 51(1), 2008.

[DShS⁺05]   Ewa Deelman, Gurmeet Singh, Mei hui Su, James Blythe, Yolanda Gil, Carl Kesselman, Gaurang Mehta, Karan Vahi, G. Bruce Berriman, John Good, Anastasia C. Laity, Joseph C. Jacob, and Daniel S. Katz. Pegasus: A Framework for Mapping Complex Scientific Workflows onto Distributed Systems. *Scientific Programming*, 13:219–237, 2005.

[DW01]   John R. Douceur and Roger P. Wattenhofer. Optimizing File Availability in a Secure Serverless Distributed File System. In *Reliable Distributed Systems, 2001. Proceedings. 20th IEEE Symposium on*, pages 4 –13, 2001.

[FHC08]   Gilles Fedak, Haiwu He, and Franck Cappello. BitDew: A Programmable Environment for Large-Scale Data Management and Distribution. In *Supercomputing Conference*, 2008.

[FHC09]   Gilles Fedak, Haiwu He, and Franck Cappello. BitDew: A Data Management and Distribution Service with Multi-Protocol File Transfer and Metadata Abstraction. *Journal of Network and Computer Applications*, 32:961–975, 2009.

[GGL03]   Sanjay Ghemawat, Howard Gobioff, and Shun-Tak Leung. The Google File System. In *Proceedings of the 19th Symposium on Operating Systems Principles*, 2003.

[GKR10]   Abdullah Gharaibeh, Samer A. Kiswany, and Matei Ripeanu. ThriftStore: Finessing Reliability Trade-Offs in Replicated Storage Systems. *Parallel and Distributed Systems, IEEE Transactions on*, PP(99):1, 2010.

[GR09]   Abdullah Gharaibeh and Matei Ripeanu. Exploring Data Reliability Tradeoffs in Replicated Storage Systems. In *HPDC '09: Proceedings of the 18th ACM international symposium on High performance distributed computing*, pages 217–226, New York, NY, USA, 2009. ACM.

[HLD09]   Torsten Hoefler, Andrew Lumsdaine, and Jack Dongarra. Towards Efficient MapReduce Using MPI. In *Recent Advances in Parallel Virtual Machine and Message Passing Interface, 16th European PVM/MPI Users' Group Meeting*. Springer, Sep. 2009.

[HMD05]   Andreas Haeberlen, Alan Mislove, and Peter Druschel. Glacier: Highly durable, decentralized storage despite massive correlated failures. In *Proceedings of the 2nd Symposium on Networked Systems Design and Implementation (NSDI'05)*, May 2005.

[HTT09] Anthony J. G. Hey, Stewart Tansley, and Kristin M. Tolle. *The Fourth Paradigm: Data-Intensive Scientific Discovery*. Microsoft Research, 2009.

[IBY+07] Michael Isard, Mihai Budiu, Yuan Yu, Andrew Birrell, and Dennis Fetterly. Dryad: Distributed Data-Parallel Programs from Sequential Building Blocks. In *Proceedings of the 2nd ACM SIGOPS/EuroSys European Conference on Computer Systems 2007*, EuroSys '07, pages 59–72, New York, NY, USA, 2007. ACM.

[KAM+07] Derrick Kondo, Felipe Araujo, Paul Malecot, Patricio Domingues, Luis M. Silva, Gilles Fedak, and Franck Cappello. Characterizing Result Errors in Internet Desktop Grids. In *European Conference on Parallel and Distributed Computing EuroPar'07*, Rennes, France, August 2007.

[KFC+07] Derrick Kondo, Gilles Fedak, Franck Cappello, Andrew A. Chien, and Henri Casanova. Characterizing Resource Availability in Enterprise Desktop Grids. *Future Generation Computer Systems*, 23(7):888 – 903, 2007.

[KL04] Tevfik Kosar and Miron Livny. Stork: Making Data Placement a First Class Citizen in the Grid. In *Proceedings of the 24th International Conference on Distributed Computing Systems (ICDCS'04)*, ICDCS '04, pages 342–349, Washington, DC, USA, 2004. IEEE Computer Society.

[KL05] Tevfik Kosar and Miron Livny. A Framework for Reliable and Efficient Data Placement in Distributed Computing Systems. *J. Parallel Distrib. Comput.*, 65:1146–1157, October 2005.

[KTI+04] Derrick Kondo, Michela Taufer, Charles L. Brooks III, Henri Casanova, Iii Henri, Casanova Andrew, and Andrew A. Chien. Characterizing and Evaluating Desktop Grids: an Empirical Study. In *Proceedings of the 18th International Parallel and Distributed Processing Symposium*, 2004.

[LMA+10] Heshan Lin, Xiaosong Ma, Jeremy S. Archuleta, Wu chun Feng, Mark K. Gardner, and Zhe Zhang. MOON: MapReduce On Opportunistic eNvironments. In *IEEE International Symposium on High Performance Distributed Computing*, pages 95–106, 2010.

[LMF11] Heshan Lin, Xiaosong Ma, and Wu-chun Feng. Reliable MapReduce Computing on Opportunistic Resources. *Cluster Computing*, pages 1–17, 2011. 10.1007/s10586-011-0158-7.

[MBTF08]   Christopher Moretti, Jared Bulosan, Douglas Thain, and Patrick J. Flynn. All-Pairs: An Abstraction for Data-Intensive Cloud Computing. In *Parallel and Distributed Processing, 2008. IPDPS 2008. IEEE International Symposium on*, pages 1 –11, april 2008.

[Mes95]   Message Passing Interface Forum. *MPI: Message-Passing Interface Standard*, June 1995.

[Mes97]   Message Passing Interface Forum. *MPI-2: Extensions to the Message-Passing Standard*, July 1997.

[MSF11]   Mircea Moca, Gheorghe Cosmin Silaghi, and Gilles Fedak. Distributed results checking for mapreduce on volunteer computing. In *Proceedings of IPDPS'2011, 4th Workshop on Desktop Grids and Volunteer Computing Systems (PCGrid 2010)*, Anchorage, AK, May 2011.

[pGKLY00] Jean pierre Goux, Sanjeev Kulkarni, Jeff Linderoth, and Michael Yoder. An Enabling Framework for Master-Worker Applications on the Computational Grid. In *the 9th IEEE International Symposium on High Performance Distributed Computing*, 2000.

[Sar02]   Luis F. G. Sarmenta. Sabotage-Tolerance Mechanisms for Volunteer Computing Systems. *Future Generation Computer Systems*, 18(4):561–572, 2002.

[TMC+10]   Bing Tang, Mircea Moca, Stphane Chevalier, Haiwu He, and Gilles Fedak. Towards MapReduce for Desktop Grid Computing. In *Fifth International Conference on P2P, Parallel, Grid, Cloud and Internet Computing (3PGCIC'10)*, pages 193–200, Fukuoka, Japan, November 2010. IEEE.

[TTL04]   Douglas Thain, Todd Tannenbaum, and Miron Livny. Distributed Computing in Practice: The Condor Experience. *Concurrency and Computation: Practice and Experience*, 2004.

[VMF+05]   Sudharshan S. Vazhkudai, Xiaosong Ma, Vincent W. Freeh, Jonathan W. Strickland, Nandan Tammineedi, and Stephen L. Scott. FreeLoader: Scavenging Desktop Storage Resources for Bulk, Transient Data. In *Proceedings of Supercomputing*, 2005.

Chapter 12

Roles of Desktop Grids in Hybrid Distributed Computing Infrastructures

Simon Delamare

INRIA, University of Lyon, Lyon, France

Gilles Fedak

INRIA, Lyon, France

There is a growing demand for computing power from scientific communities to run large applications and process huge volumes of scientific data. Meanwhile, Distributed Computing Infrastructures (DCIs) for scientific computing continue to diversify. Users can not only select their preferred architectures among Supercomputers, Clusters, Grids, Clouds, Desktop Grids, and more, based on parameters such as performance, reliability, cost, quality of service, or energy efficiency, but can also combine transparently several of these infrastructures together. In this chapter we explore several scenarios where Desktop Grids are combined with Clouds and/or Grids to improve performance and quality of service or decrease the cost usage.

12.1 Hybrid Distributed Computing Infrastructures (H-DCI)

Grand-challenge scientific applications push scientific communities to look for larger Distributed Computing Infrastructures (DCI). European and Chinese Grid infrastructures such as EGI (European Grid Infrastructure) [GJG+05] or ChinaGrid [Jin04], Desktop Grids and Volunteer Computing systems such as XtremWeb [FGNC01, CGC01, CDF+05] and BOINC [And04], and the new emerging Cloud Computing such as Amazon EC2 or Microsoft Azure are representative of this diverse attempts to reach ever higher throughput computing.

Recently, with the success of Cloud Computing platforms, a new paradigm has emerged, pushed by the widespread of virtualization technologies, where the complexity of an IT infrastructure is completely hidden from its users. Cloud Computing provides access through Web services to commercial high-performance computing and storage infrastructure (*IaaS* Infrastructure-as-a-Service), distributed services architecture (*PaaS* Platform-as-a-Service) or even application and software (*SaaS* Software-as-a-Service).

The economical model associated with Cloud Computing, and the ease of use associated with virtualization makes this approach very competitive for scientific communities compared with traditional Grid computing. As a consequence, Cloud technologies are being adopted by the Grid communities. For instance, the StratusLab [MRM+11] project aims at providing Cloud systems based on the OpenNebula technology to the EGI community.

Desktop Grids use computing, network, and storage resources of idle desktop PCs distributed over multiple LANs or the Internet. In the past decade, Desktop Grid computing has been proved an effective solution to provide scientists with tens of TeraFLOPS from hundreds of thousands of hosts at the fraction of the cost of a Grid system. As a result, Desktop Grid technologies are now considered a complement to regular Grid technologies. For instance, the European Union, through several FP7 projects (EDGeS, EDGI, DIGISCO)[BFG+08], supports a large research and development effort to make Desktop Grids easily available to the scientific public and in particular through the EGEE community. Recently, in 2011, several Memorandums of Understanding were signed between EDGI, EMI, the European Middleware Initiative, and EGI, the European Grid Infrastructure to sustain the development of Desktop Grid technologies and bridge middleware.

However, DCIs may have very different characteristics in terms of performance, reliability, power efficiency, costs, QoS etc If we look further at their intrinsic characteristics, we observe that

Reliability, elasticity. Desktop Grids are subject to both volatility and elasticity because they are composed of volunteer PCs that join and leave the network at any time, and the set of nodes involved in the computation can change at any time. Similar to Desktop Grids, Clouds are also elastic. However, these two kinds of elasticity differ in the sense that Cloud elasticity is controlled by the user who can dynamically scale up or scale down the infrastructure according to its needs, while with Desktop Grid, it is the infrastructure itself that changes according to participation of the volunteer resources. At the moment, usage of the Grid for parallel application is mainly rigid and it is not common for Grid resource managers to handle computing resource failures.

Performance and quality of service. Performance depends on many factors, including the number and the power of computing elements. In addition, Cloud performance is also limited by the amount of money a user is willing to spend, but in exchange, the SLA guarantees the availability and the reliability of the machines. Grid users have to share the infrastructure with other users, so the computing power obtained to run their application depends on the usage of the Grid by other users. Ensuring quality of service on the Grids is still considered a research challenge that requires significant advances in scheduling, resource performance modelling and application performance forecasting. Desktop Grids present the most unpredictable performance which depend on many factors: hosts reliability, host heterogeneity as well as application requirements in term of communication and storage.

Economical model. and therefore the usage price radically differs among Volunteer Computing, where volunteers buy the computers, ensure the maintenance; and pay the electricity bill, Grids, where the institutions support all the costs associated with the acquisition and the maintenance of the infrastruceure; and finally the Clouds, which offer the most flexible economical model, where users only pay for resources actually used (storage, bandwith, CPU hours).

Because these DCIs have different characteristics, there are many scenarios that advocate for hybrid infrastructures that assemble Clouds, Desktop Grids, and Grids to mitigate the disadvantages of certain aspects of some infrastructure and enjoy the benefits of others. Actually, it is reasonable to think that such a hybrid infrastructure could become the next trend in distributed computing, as an extension of the concept of *Sky Computing*, introduced by Keahey et al. in [MKF10] to denote an infrastructure composed of multiple Clouds.

Desktop Grid middleware plays a key role in hybrid DCI because it has several desired features to manage BE-DCI resources: resilience to node failures, no reconfiguration when new nodes are added in the system, task replication or task rescheduling in case of node failures, and push/pull protocols that help with firewall issues. Although there are many solutions for BoT execution on cross-infrastructure deployments, we have found that in many cases, Desktop Grid middleware is used to schedule tasks on the computing resources. For instance, early in the development of XtremWeb, we used the prototype to schedule jobs on several Condor pools [LFC+03] using a mechanism known as PilotJobs[1]. The XtremWeb worker is scheduled as a regular Job on the Condor pool. Once the XtremWeb worker is executed on the Condor resource, it retrieves jobs from the server, executes the job and sends the result to the XtremWeb server. The same mechanism can be used to assemble together resources coming from Grid or Cloud infrastructures. In the last case, the worker must be embedded in a Virtual Machine. In [CFH+08], we proposed a security model to create hybrid DCI made of trusted and untrusted resources. In [DHL+04], we demonstrated that it was possible to build a fault-tolerant RPC runtime environment able to gather resources from different Grids.

The first motivation for assembling DCIs is to obtain greater computing power. In this configuration, Desktop Grids can play a supplementary role for Grid users by offering a vast amount of computing power for little additional cost. Not surprisingly, several projects have created bridge technologies that allow Grid users to use resources provided by Desktop Grids. GridBot [SSGS09] puts together Superlink@Technion, Condor pools, and Grid resources to execute both throughput and fast-turnaround oriented BoTs. The Desktop Grid middleware used is BOINC augmented with the matchmaking mechanism of Condor to implement more sophisticated scheduling policies. The European FP7 projects EDGeS (Enabling Desktop Grids for E-science) [UKF+09] and EDGI (European Desktop Grid Infrastructure), have developed bridge technologies to make BOINC [KFF08] and XtremWeb [HFK+10] are transparently available to any EGI Grid users as a regular Computing Element. The cornerstone of the EDGeS and EDGI projects is the 3G bridge software, which implements bi-directional job transmissions between Service Grid and Desktop Grids. New development of the 3G-bridge will allow ARC- and Unicore-based Grid to connect to the European Desktop Grid Infrastructure. Following a similar path, the Latin America EELA-2 Grid has been bridged with the OurGrid infrastructures [BDC+08]. Actually, there exists synergy between OurGrid and EDGI to adopt compatible standards and technologies.

The next scenario, known as *Cloud bursting* [MKF10], is a mechanism that offloads part of the Grid workload to the Cloud when there is peak demand. It is noteworthy that several studies have compared the costs of running large scientific on Clouds. In [PIRG08], the authors investigate the cost and performance of running a Grid workload on Amazon EC2 Cloud. Similarly, in [KJM+09], the authors introduce a cost-benefit analysis to compare Desktop

---

[1]We adopt the following terminology to describe the main components of Desktop Grid middleware: the server, which schedules tasks; the client who submits tasks to the server; and workers, which fetch and execute tasks on the computing resources.

Grids and Amazon EC2. In [BYSS+11], the authors propose a Pareto-efficient strategy to offload the Grid workload, which consists of bags-of-task application with deadlines on the Cloud.

Because Desktop Grids allow us to leverage on the existing infrastructure, it can be used to alleviate access to remote resources. In Freeloader [VMF+05], the available storage of local Desktop PCs is used to create distributed storage. Remote data, usually stored on HPC storage is fetched and locally stored on Desktop PCs. This provides parallel data access and has been shown to significantly improves the performance of data-intense applications. This scenario would be even more beneficial when the remote infrastructure is a Cloud because it allows us to reduce the cost with storage and data transferred. We propose such a solution in Section 12.2. One can imagine other combinations within hybrid DCIs: CATCH [MBV11] uses the Cloud storage service to improve data access between Desktop worker and HPC centers.

However, because Desktop Grids trade reliability against lower prices, they offer poor Quality of Service (QoS) with respect to traditional DCIs [MKF06]. Besides Desktop Grid, other particular usages of an existing infrastructure can also provide unused computing resources without any guarantees that the computing resources will remain available to the user during the complete application execution. For example, the Grid resource managers such as OAR [CDCG+05] manage a best-effort queue to harvest idle nodes of the cluster. Tasks submitted in the best effort queue have the lowest priority; at any moment, a regular task can steal the node and abort the ongoing best-effort task. In Cloud computing, Amazon has recently introduced EC2 Spot instances [Ser09] where users can bid for unused Amazon EC2 instances. If the market Spot price goes below the user's bid, a user gains access to available instances. Conversely, when the Spot price exceeds his bid, the instance is terminated without notice. Similar features exist in other Cloud services [MKF11]. In Section 12.3 of this chapter, we present SpeQuloS, a framework that provides Quality-of-Service for applications executed on Desktop Grids by offloading part of the application on reliable resources provided by Clouds.

12.2 Using Local Desktop Grids as Cache-to-Cloud Resources

In the CloudCash project, initiated by two French research institutes CNRS and INRIA and the Chinese university, HUST plans to investigate one particular setting of hybrid Sky computing where Desktop Grid resources are used to supplement the Cloud infrastructure and/or reduce the usage of Cloud. Figure 12.1 illustrates our approach: let's imagine an enterprise or an institution that uses Cloud resources, such as online file storage, remote database, online services, applications, or VMs running on the bare Cloud infrastructure. The use of these resources costs money, while on the other hand, users might have vast idle local resources available on their local Desktop networks. Our proposition is to design a proxy system that would cache the Cloud resources on the Desktop so that

- It increases the performance of the application by moving the resources closer to the users or by adding additional Desktop resources. For instance, caching the data on the file system built using Desktop PCs storage can reduce data access time.

- It reduces the cost of Cloud utilization by reducing access to the Cloud resources. For instance, when Desktop PCs are idle, EC2 VM instances could be migrated from the Amazon Cloud to continue their execution on the Desktop PCs in order save CPU price utilization.

There exist many different middleware and infrastructures of Grid computing. It is noteworthy that two members of the consortium, IN2P3 and HUST, are involved in the two major production Grid efforts in Europe and China, respectively, EGI and ChinaGrid. Researchers and developers in the first trend create a Grid service, which can be accessed by a large number of users. A resource can become part of the Grid by installing predefined software sets, or middleware. The middleware is, however, so complex that it often requires extensive expert effort to maintain. It is therefore natural that individual people do not often offer their resources in this manner, and Grid is generally restricted to larger institutions, where professional system administrators take care of the hardware/middleware/software environment and ensure the high availability of the Grid.

XtremWeb-HEP [LUD12] is a research project belonging to lightweight Grid systems. It is a free open-source and non-profit software platform to explore scientific issues and applications of Global Desktop Grids, Global Computing, and Peer-to-Peer distributed systems. The aim of the project is to investigate how a large-scale distributed system (LSDS) can be turned into a parallel computer with classical user, administration, and programming interfaces possibly using fully decentralized mechanisms to implement some system functionalities. XtremWeb currently uses a module for LSM (Linux Security Module) called SBLSM to ensure application sandboxing; this method is for Linux only. The principle is to apply a security policy to a set of binary processes. Every time a sandboxed process issues a system call, the module checks a dedicated variable and either grants or denies the request.

Virtual machine (VM) technologies are currently experiencing a resurgence in both industry and research communities. These technologies are widespread, both in the server market (XEN) [BDF+03] and the Desktop market (VMWare, Virtual Box). We think that VMs offer many desirable features, such as security, ease of management, OS customization, performance isolation, checkpointing, and migration, which can offer the technological breakthrough needed to unify the large-scale distributed infrastructure. Amazon's EC2 introduces the notion of Cloud Computing, where virtual machines are

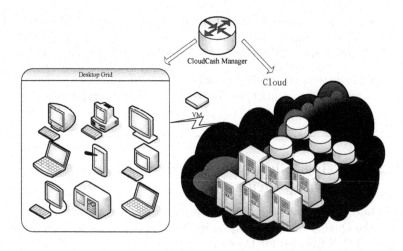

FIGURE 12.1: CloudCash schema–Using desktop grid as cache to cut cloud costs.

dynamically provisioned immediately with customized software, environments and use is charged by the hour. A major advantage of cloud computing is the ability to use a variable number of physical machines and VM instances, depending on the needs of the problem. For example, a task may need only a single CPU during some phases of execution but may be capable of leveraging hundreds of CPUs at other times.

We expect the following innovations in the field of bridging Desktop Grids and Cloud Computing with the use of virtualization technologies:

- Demonstrate that scientific communities can access two computing infrastructures (Desktop Grids and Cloud Computing) as a single and unified system. Thanks to virtualization technologies, the heterogeneity and complexity of the two platforms will be kept hidden from the end user.

- Bring new lightweight VM technologies that are generic enough to be deployed over a large range of high-throughput computing infrastructures. Our VM technology will feature fast deployment VM on-the-fly, VM monitoring, VM migration, and VM scheduling.

- Grid portals offer their scientific communities seamless access to a wide range of applications. With CloudCash, whole scientific communities can cooperate on the construction or collection of scientific applications bundled as VM images, facilitating the portage of complete bundles of application to emerging computing infrastructures such as Clouds or Desktop Grids.

- Enable secure and efficient access to large scientific data sets that cannot be easily moved from one infrastructure to another. However, we have shown that P2P technologies can alleviate this issue [WFC07, FHC08].

- Give users the ability to select the most adequate execution platform according to a price/performance/reliability compromise.

12.3 Using Cloud Resources to Improve the QoS of Desktop Grids

In the section we describe the SpeQuloS framework, which aims at enhancing the Quality of Service (QoS) in Desktop Grids (DGs). SpeQuloS provides several QoS features for Desktop Grids users:

- It reducesthe completion time of Bag-of-Tasks (BoT) executed on DGs.

- It allows stable performances by improving the execution stability of BoT under the same environment conditions.

- It gives a statistical prediction of BoT completion time when executed on DGs.

To support DGs resources and reach this QoS level, SpeQuloS dynamically deploys Cloud resources to assist DG resources in BoT execution. This section describes several strategies to appropriately deploy Cloud resources. Strategies address the questions of when to use the Cloud resources, how many resources to use, and what is the best way to use them.

Performance evaluation shows that SpeQuloS correctly addresses the major cause of QoS degradation in Desktop Grids: The tail effect. Bag-of-Task execution with both XtremWeb-HEP and BOINC middleware is improved, whatever nature of the workload of the DG. BoT completion times are often 2 times lower when SpeQuloS is used. To achieve this, SpeQuloS only requires a small number of tasks to be executed on Cloud resources. In addition, SpeQuloS is able to accurately predict completion times, thus improving QoS delivered to DG users.

This section also describes the implementation and deployment of SpeQuloS inside the production infrastructure of the European Desktop Grid Initiative FP7 project (EDGI).

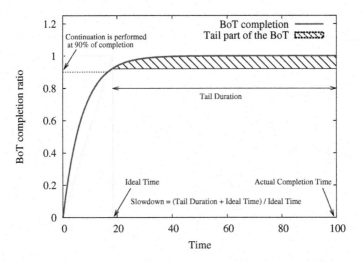

FIGURE 12.2: Example of BoT execution with Tail effect.

12.3.1 Motivation

Desktop Grids are an attractive solution to provide computing resources at low cost. Desktop Grids are made of desktop computers, that pool their computing power to building a computing infrastructure. Resources can be supplied by volunteer individuals or can be collected from existing installed base of computers, like those found on a university campus. In either case, no computing resource deployment is required and the cost of running the Desktop Grid is low.

However, Desktop Grid resource availability can be considered as Best Effort. Usually, computing power is available to the Desktop Grid when the desktop computer is idle. Therefore, a resource becomes unavailable when the computer is used or turned off, sometimes without informing the Desktop Grid server.

The volatile nature of Desktop Grid causes a low level of Quality of Service, especially compared to other distributed computing infrastructures (DCIs), such as regular Grids. This is particularly observable for Bag-of-Tasks executed on DGs, when users wait for all tasks belonging to a BoT to be completed. The execution profile of BoT is often similar to the one denoted in Figure 12.2. It is observed that the execution of the last part of a BoT takes a large part of the total execution time. We call this event **Tail effect**.

Tail effect can be characterized by the "tail slowdown," which is the ratio of an ideal time for BoT completion, defined by assuming a constant completion rate throughout the execution, versus the actual completion time observed.

Desktop Grid middleware, such as BOINC or XWHEP, have features to handle nodes volatility: resilience to node failures, no need for reconfiguration

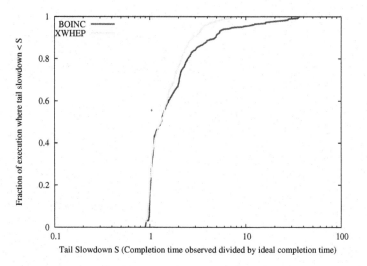

FIGURE 12.3: Execution of BoTs in DGs: Cumulative distribution functions of Tail slowdown. The slowdown denotes the importance of the Tail effect by measuring the BoT completion time divided by the ideal completion time.

TABLE 12.1: Average Fraction of Bag-of-Tasks in Tail (ratio between the number of tasks in tail and number of tasks in BoT) and Average Percentage of Execution Time in Tail (percentage of total BoT execution time spent in the tail)

Avg. % of BoT in tail		Avg. % of time in tail	
BOINC	XWHEP	BOINC	XWHEP
4.65	5.11	51.8	45.2

when new nodes are added, task replication, etc. However, observations made from BoT execution captures show that these feature are not able to remove the Tail effect. Figure 12.3 presents the cumulative distribution functions of "Tail slowdowns" measured, for both BOINC or XWHEP middleware.

Our observations show that the Tail effect affects about one half the executions, where slowdown is higher than 1.33, meaning that the Tail effect slows the execution by more than 33%. Indeed, the Tail effect significantly disturbs BoT executions in some cases: It doubles the completion time from one quarter (XWHEP) to one third (BOINC) of execution. In the worst case, the slowdown can be as high as a penalty of 10.

Tasks being executed during the Tail effect, that is, later than the ideal time, create this slowdown by being longer than others to be completed. Table 12.1 shows Tail characteristics, according to middleware used. One can observe that only a small percentage of tasks are executed in the Tail, but they take an important percentage time of the total execution time.

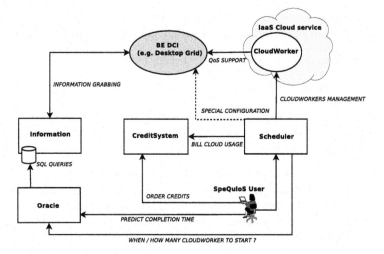

FIGURE 12.4: The SpeQuloS modules and their interactions.

These results show that the Tail effect causes unsatisfactory levels of QoS in Desktop Grids, for both BOINC and XW middleware. This Tail effect strongly slows BoT completion times, and leads to a highly variable and unpredictable execution.

To mitigate the Tail effect and provide a better QoS to Desktop Grid users, we propose to supply additional resources to Desktop Grids, to support execution in particular when the Tail effect occurs. Resources from Cloud computing are the best candidates for this task: They are available on-demand and the resources provided by the public Cloud such as Amazon EC2 are virtually unlimited. This solution is proposed by the SpeQuloS framework.

12.3.2 SpeQuloS Framework Overview

The SpeQuloS framework aims to provide QoS to Desktop Grid users by provisioning Cloud computing resources that are stable and reliable.

SpeQuloS uses Infrastructure as a Service (IaaS) Cloud services to instantiate Cloud instances, called a Cloud worker, to support Desktop Grid computing. A Cloud worker runs the same DG middleware worker software as executed by other workers that connect to the Desktop Grid server.

SpeQuloS implements a whole framework to efficiently use Cloud resources, account for and arbitrate their usage, and provide QoS information to DG users. SpeQuloS is distributed in several modules, as shown in Figure 12.4.

Each module implements a dedicated role in SpeQuloS:

- The Information module collects and stores information from Desktop Grids.

- The Credit System module bills and accounts Cloud resource utilization.

- The Oracle module implements some strategies to efficiently use the Cloud resources, and computes BoT completion time predictions.

- The Scheduler handles Cloud workers management.

Here is a typical use-case scenario of SpeQuloS:

- A user requests QoS support for a BoT execution. SpeQuloS' Scheduler module returns a tag to identify the user BoT. Then, the user can submit its BoT to the DG server.

- A user can request from the Oracle module a prediction on the BoT completion time. If he wants to, he can then order from the Credit System some QoS support for his BoT by allocating some credits. If there are enough credits on the user's account, the CreditSystem creates the QoS order.

- The Scheduler monitors the QoS order created by the Credit System. If some credits have been allocated to a BoT, it asks the Oracle when to start Cloud workers to support the execution of this BoT.

- When appropriate, Cloud workers are launched by the Scheduler and participate in BoT computation.

- Cloud resource utilization is periodically billed by the Scheduler to the Credit System. If credits allocated to a BoT exhaust, Cloud workers are stopped.

- When Cloud usage has ended, the Scheduler requests the Credit System to close the QoS order. If some credits allocated by user remain, they are transferred to the user's account.

12.3.2.1   Monitoring Desktop Grids

SpeQuloS needs to collect information from Desktop Grids related to BoT execution for two reasons:

- Get real-time information on BoT executed in DGs.

- Store previous execution traces to help in future BoT completion time prediction.

The Information module uses a database to store BoT completion history as the evolution of completed tasks against total BoT tasks according to time, for every BoT execution monitored.

The Information module also hides differences between infrastructures, for instance, data collected from BOINC or XWHEP are stored under the same format.

12.3.2.2 Billing Cloud Usage to Users

Using Cloud resources may require us to spend money, and in the case of SpeQuloS, many users may want to use them to support theirs task execution. SpeQuloS' Credit System module implements Cloud usage account and arbitration.

It provides a simple banking system for depositing, billing, and paying with some virtual credits. The CreditSystem uses a fixed exchange rate, where 1 *CPU.hour* of Cloud worker is equivalent to 15 credits. According to the amount of Cloud resources needed, a user has to order the corresponding number of credits, which are withdrawn from his account.

A deposit policy is used to fund users' accounts. According to administrator needs, various policies can be deployed. As an example, this simple policy can be used to limit the usage of 200 Cloud nodes per day and per user: Every 24 hours, deposit $d = max(6000, 6000 - user\_credit\_spent)$ credits to the user's account. Deployment of other policies, much more sophisticated, can be considered.

12.3.2.3 Predicting Completion Time

One major feature to enhance users' perceived QoS is to be able to predict BoT completion time. By analyzing the current state of a BoT execution and history of previous ones from data collected by the Information module, the Oracle is able to compute the predicted completion time of a BoT. If the predicted time is attractive enough for the user, he may decide to order QoS to SpeQuloS.

SpeQuloS uses the following prediction methods: When called by a user for a prediction, the Oracle gets from the Information module the BoT completion ratio (r) and the elapsed time since BoT submission $(t_c(r))$. The predicted completion time t_p is computed as follows: $t_p = \alpha . \frac{t_c(r)}{r}$.

α is an adjustment factor that allows the determination according to BoT executions in a given Desktop Grid. It is set to 1 at initialization, and adjusted after each BoT execution to minimize the average deviation between the predicted completion time and the times really observed. SpeQuloS returns the predicted time associated to a statistical uncertainty as the success rate (with a $\pm$ 20% tolerance) of predictions performed against previous BoT executions stored in the history.

12.3.2.4 Strategies for Cloud Resources Deployment

Several strategies can be considered to decide when to start Cloud workers, and how many of them to use, according to the amount of Credits supplied by an user, and how to use them.

For instance, here are three strategies that could be used to decide when to launch Cloud workers:

- Completion Threshold (9C): Cloud workers are started when 90% of BoT tasks are completed.

- Assignment Threshold (9A): Cloud workers are started when 90% of BoT tasks have been assigned to DG workers.

- Execution Variance (D): Cloud workers are started when task execution time increases: Let $t_c(x)$ be the time at which x percent of BoT tasks are completed and $t_a(x)$ be the time at which x percent of BoT tasks were assigned to workers. The execution variance is $var(x) = t_c(x) - t_a(x)$. A sudden increase in execution variance denotes the beginning of the Tail effect. When execution variance reaches 2 times the maximum execution variance measured during the first half of the BoT execution, the Cloud workers are started.

If users allocate credits equivalent to S cpu.hours of Cloud usage, two approaches on how many Cloud workers to start can be considered:

- Greedy (G): S workers are started. If a Cloud worker does not receive a task to compute, it stops to not waste credit and lets other Cloud workers complete their tasks.

- Conservative (C): At any time t_e, an estimated remaining time to complete a BoT t_r can be computed thanks to the Information module as follows:

$$t_r = t_c(1) - t_e = t_c(1) - t_c(x_e) = \frac{t_c(x_t)}{x_t} - t_c(x_t)$$

where $t_c(x)$ is the elapsed time when x percent of BoT tasks are completed ($\frac{t_c(x)}{x}$ is the BoT completion rate). Conservative strategy starts $\max(\frac{S}{t_r}, S)$ Cloud workers. Therefore, there will be enough credits to run them until the estimated time remaining to complete the BoT.

Several solutions can be investigated to deploy and use Cloud resources:

- Flat (F): Cloud workers behave as any regular workers, and the DG do not make a distinction between them. Cloud and regular workers are in competition to execute tasks in this strategy.

- Reschedule (R): Cloud workers and regular workers are differentiated by the DG server: When Cloud workers make a request for tasks, if all uncompleted tasks are being executed by regular workers, the DG server creates new duplicates of these tasks to assign them to Cloud workers. Therefore, tasks executed on regular workers that may cause the Tail effect are rescheduled on Cloud resources.

- Cloud Duplication (D): Cloud workers and regular workers are totally separate. When Cloud resources are provisioned, a new DG server is created in the Cloud and all uncompleted tasks from the original DG server are copied to the Cloud server. Then Cloud workers process tasks from this server and the results are merged back inside the original server, allowing the task execution by Cloud resources during the Tail effect.

Implementation of these strategies differs in terms of complexity. While Flat does not need any particular DG modification, Reschedule requires us to patch the DG server in order to adapt the scheduling process, and Cloud Duplication needs to implement task duplication and results merging between the original and Cloud DG servers.

12.3.2.5 Cloud Workers Management

The Scheduler module handles the management of Cloud resources. When credits are allocated to support a BoT execution, and according to Oracle strategies on when and how provision Cloud support, the Scheduler module will remotely start, configure, and stop the Cloud workers.

SpeQuloS uses the *libcloud* library, an API to access various IaaS Cloud technologies. Cloud workers are configured using SSH to connect to the appropriate DG.

Scheduler operations to manage QoS support for BoT and Cloud workers are presented in Algorithms 1 and 2.

Algorithm 3 MONITORING BoT

---

for all B in BoTs **do**
  if Oracle.shouldUseCloud(B) **then**
    if CreditSystem.hasCredits(B) **then**
      for all CW in Oracle.cloudWorkersToStart(B) **do**
        CW.start()
        configure(B.getDCI(),CW)
      end for
    end if
  end if
end for

Algorithm 4 MONITORING CLOUD WORKERS

---

for all CW in startedCloudWorkers **do**
  B ← CW.getSupportedBoT()
  if (Info.isCompleted(B)) **or** (**not** CreditSystem.hasCredits(B)) **then**
    CW.stop()
  else
    CreditSystem.bill(B,CW)
  end if
end for

12.3.2.6 Implementation

Each SpeQuloS module is independent and uses Python programming language and MySQL databases for their implementation. Modules communicate using Web services. Several BE-DCIs and Cloud services can be connected at the same time to a single SpeQuloS deployment. The SpeQuloS source code is publicly available[2].

SpeQuloS supports both BOINC and XWHEP middleware, which are used in BE-DCIs. It must be ensured that Cloud workers compute tasks belonging BoTs for which credits have been allocated. In BOINC, this is possible by modifying the matchmaking scheduler. In XWHEP, a configuration option can be used since 7.4.0 version.

SpeQuloS can also take part in larger hybrid infrastructures. For instance, the 3G-Bridge developed by SZTAKI for the EDGI infrastructure is used for Grid and Desktop Grid interconnection. SpeQuloS supports BoTs submitted using the 3G-Bridge, thanks to a dedicated field of 3G-Bridge.

Using libcloud SpeQuloS is able to connect to several IaaS Cloud services including, Amazon EC2, Eucalyptus, Rackspace, OpenNebula, and Nimbus. In addition, we have developed a new driver for libcloud to use Grid5000 [Ba06] as an additional Cloud.

12.3.3 SpeQuloS Performance

This section presents SpeQuloS performance in providing QoS for BoT executed in Desktop Grids. Results presented use Completion threshold, Conservative and Reschedule (9C-C-R) strategies, which have shown to be efficient in removing Tail Effect, while keeping a low credits consumption, measured to be less than 2.5% of the total BoT workload in equivalent Cloud's *cpu.h*.

12.3.3.1 Completion Time Speedup

Figures 12.5a, 12.5b and 12.5c present average BoT completion times observed when SpeQuloS is used or not. Figures show results from BOINC and XWHEP middleware and for SETI@HOME (SETI) and NotreDame (ND) Desktop Grids, for which availability traces have been collected from the Failure Trace Archive[3]. Figures also present results obtained with various BoT workload: "BIG," which is a large BoT of small tasks; "SMALL," which is a small BoT of long tasks; and "RANDOM," which is an heterogeneous BoT with a variable number of tasks of different length.

In any case, SpeQuloS is able to decrease BoT completion time. Depending on the middleware and DG considered, SpeQuloS speeds up the execution from 1.5 to 9 times. Best results are observed with BOINC and SETI, which includes highly volatile resources. As the Tail effect is stronger in this case, SpeQuloS greatly improves the performance.

Results also depend on the BoT workload. BoTs that are made of long tasks (SMALL) or heterogeneous tasks (RANDOM) are more likely to see large improvement. Indeed, without SpeQuloS, it is more difficult to execute this

---

[2]http://graal.ens-lyon.fr/~sdelamar/spequlos/.
[3]http://fta.inria.fr/.

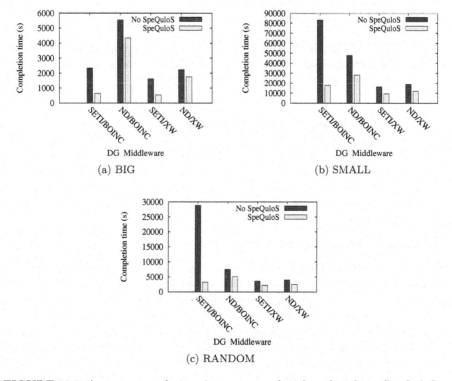

FIGURE 12.5: Average completion time measured with and without SpeQuloS for BOINC and XWHEP middleware in various desktop grids.

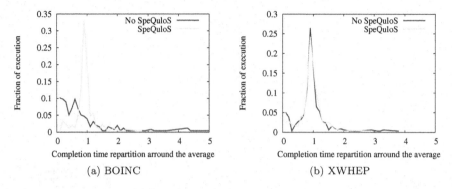

FIGURE 12.6: Repartition functions of measured completion time normalized with the average completion time observed under the same environment. A curve centered around 1 means high stability.

kind of BoT on Desktop Grids, because of the low-power and high-volatility characteristics of their nodes.

BOINC and XWHEP behave differently: BOINC implements tasks replication whereas XWHEP does not, and they also handle node failures in a different fashion. Thus, they cannot be compared, but it can be noted that using SpeQuloS leads to higher performance improvement in BOINC than in XWHEP.

12.3.3.2   Execution Stability

An additional QoS feature provided by SpeQuloS is better execution stability. This metric denotes the ability to reproduce similar performance behaviors under similar environments, such as Desktop Grid considered, middleware used, and BoT workload.

Figures 12.6a and 12.6b present repartition functions of normalized BoT completion times compared to their average. Execution completion times are divided by the average completion time measured in the same execution environment (DG availability traces, middleware used, and BoT workload).

Results show that XWHEP middleware does not benefit from SpeQuloS, as execution stability was already satisfactory. With BOINC, the execution stability is greatly improved by SpeQuloS. When not used, Figure 12.6a shows that many BoT executions have completion times lower than the average, meaning that some lengthy executions exist. SpeQuloS is able to eliminate this critical situation, which leads to completion times that are much closer to the average and a good execution stability.

12.3.3.3   Completion Time Prediction

SpeQuloS prediction results are presented in Table 12.2. Prediction mechanisms described in Section 12.3.2.3, are made at 50% of BoT completion.

TABLE 12.2: SpeQuloS Completion Time Prediction Percentage of Success. (Results obtained for various DGs, middleware and BoT workload are reported. A successful prediction means that the BoT completion time is comprised between ±20% of the predicted completion time).

	BoT Category & Middleware						
	SMALL		BIG		RANDOM		
BE-DCI	BOINC	XWHEP	BOINC	XWHEP	BOINC	XWHEP	Mixed
seti	100	100	100	82.8	100	87.0	94.1
nd	100	100	100	100	100	96.0	99.4
Mixed	97.6	96.1	99.2	93.5	89.6	65.3	**90.2**

A prediction is reported to be successful when the completion time actually observed lies between ±20% of the predicted time.

The α adjustment factor is computed using a "learning phase" of 30 executions using the same DG availability trace, middleware, and BoT workload.

SpeQuloS prediction success rate is high: It is 90% when considering all results mixed. This means that in 90% of cases, SpeQuloS is valid within an uncertainty of ±20%. These results are considered very satisfactory considering Desktop Grids that are composed of volatile resources.

Heterogeneous BoTs (RANDOM) give lower prediction success rates. As task sizes in such BoT can greatly vary, this observation is not surprising as prediction cannot completely rely on past executions.

12.3.3.4 Conclusion on Performance

These results have highlighted the effectiveness of combining Cloud resources with Desktop Grids in order to enhance the QoS of these infrastructures. Using SpeQuloS, BoT execution on DGs can be greatly accelerated by only assigning a small percentage of the workload to the Cloud. In addition, by improving execution stability, SpeQuloS is able to accurately predict the BoT completion time and report this information for users' convenience.

12.3.4 SpeQuloS Deployment in the European Desktop Grid Infrastructure

The deployment of SpeQuloS in the European Desktop Grid Infrastructure (EDGI) is presented in the section. The EDGI project aims at deploying a production DCI interconnecting several private and public Desktop Grids (IberCivis, University of Westminster, SZTAKI, CNRS/University of Paris XI LAL and LRI DGs) to existing Grids (European Grid Infrastructure (EGI), Unicore, ARC), and private Clouds service (StratusLab and local OpenStack, OpenNebula).

EDGI's main goal is to provide the huge computing power available in Desktop Grids to EGI users in a transparent way, so that they can submit their jobs to Desktop Grids in the same way they currently submit jobs to Grids.

SpeQuloS is deployed inside the EDGI infrastructure to enhance the QoS

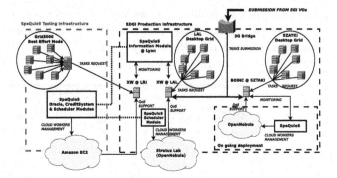

FIGURE 12.7: SpeQuloS deployment as a part of the EDGI infrastructure. SpeQuloS' modules are split and duplicated across the deployment.

of service of jobs executed on the Desktop Grid, which can be much lower than the Grids' QoS that EGI users are used to.

Figure 12.7 presents the current deployment of SpeQuloS in the context of EDGI. This deployment includes a testing infrastructure, used to permanently test SpeQuloS new features and strategies before going into production. To include computing resources that reproduce volatile characteristics of DG nodes, six clusters of Grid'5000 are used in best-effort mode, meaning that they can be preempted at any time by other Grid'5000 users.

The part of the EDGI production infrastructure involving SpeQuloS is currently composed of two DGs, XW@LRI and XW@LAL, which use the XWHEP middleware and are administrated by the University of Paris-XI. Amazon EC2 is used by SpeQuloS to support BoT executed by XW@LRI. The XW@LAL Desktop Grid is composed of laboratory desktop computers and is interconnected to EGI Grids through EDGI's 3G Bridge. The StratusLab infrastructure, which includes an OpenNebula Cloud service, is used connected to SpeQuloS to support BoT executed on XW@LAL. Some interesting behavior can be reported in this kind of hybrid infrastructure. For instance, BoTs can be submitted from XtremWeb-HEP to EGI and eventually be supported by StratusLab through SpeQuloS support.

Others SpeQuloS deployments are in progress to provide QoS support to DGs of the EDGI project infrastructure. For instance, SZTAKI DG is being connected to a dedicated OpenNebula Cloud service through SpeQuloS

Typical applications running in the EDGI infrastructure include DART, a Framework for Distributed Audio Analysis and Music Information Retrieval by Cardiff University; BNB-Grid, which aims to solve hard combinatorial, discrete, and global optimization problems, and ISDEP, which is a fusion plasma application that simulates the Tokamak of ITER. Table 12.3 shows some statistics of infrastructure utilization for the first half of 2011, when SpeQuloS deployment started.

TABLE 12.3: Number of Tasks Executed on XW@LAL and XW@LRI Desktop Grids (University Paris-XI part of the European Desktop Grid Infrastructure) with Number of EGI Tasks and Number of Tasks Assigned by SpeQuloS to StratusLab and Amazon EC2 Cloud Services.

	XW@LAL	XW@LRI	EGI	StratusLab	EC2
#_tasks	557002	129630	10371	3974	119

12.3.5 Related Works

In [VHML11], a framework is presented that would be the closest work to ours. Similarly, Aneka [CVKB11] support the integration between Desktop Grids and Clouds although we went further in terms of implementation and evaluation.

12.3.6 Conclusion

The past years have seen a diversification of Distributed Computing Infrastructures (DCIs) in scientific computing. Users can not only select their preferred architectures among Supercomputers, Clusters, Grids, Clouds, Desktop Grids, and more, based on parameters such as performance, reliability, cost, quality of service, or energy efficiency, but can also combine transparently several of these infrastructures together. The Desktop Grid technologies have a key role to play in this landscape for several reasons: 1) DG middleware allows us to bring together seamlessly and securely resources from various DCIs; 2) DG systems target reuse of existing infrastructures which allows price cuts and energy savings; 3) integration of VM technologies within DG system will allow us to build Cloud-like systems for large-scale distributed computing.

In this chapter, we have explored several scenarios where Desktop Grids are combined with Clouds and/or Grids to improve performance and quality of service or decrease the cost usage. CloudCash is a project that aims to use local Desktop Grids as local cache to remote Cloud system to improve the performance and reduce usage costs.

We presented SpeQuloS, a framework dedicated to increasing the QoS delivered by Desktop Grids by supplying resources available in Cloud computing. SpeQuloS is a relevant example of how infrastructures can be combined to share their benefits and balance their disadvantages.

SpeQuloS supervises BoT executed under Desktop Grids, and supplies Cloud resources to execute the last fraction of BoT to avoid the Tail effect, which harm execution performance in Desktop Grids. Several strategies to use Cloud resources are proposed and performances are analyzed. SpeQuloS is able to significantly improve the QoS of Desktop Grids according to various metrics, completion time, stability, and ability to perform prediction.

SpeQuloS development and deployment have shown benefits and difficulties of hybrid infrastructures. To be flexible for its deployment, the framework is split into several independent and distributed modules that implement a dedicated task, such as information archiving, accounting, prediction, and scheduling.

SpeQuloS deployment inside the European Desktop Grid Infrastructure demonstrates the feasibility and relevance of hybrid infrastructures to address problems of actual high-performance computing users.

Bibliography

[And04]   David Anderson. BOINC: A System for Public-Resource Computing and Storage. In *proceedings of the 5th IEEE/ACM International GRID Workshop*, Pittsburgh, PA, USA, 2004.

[Ba06]   Raphael Bolze and all. Grid5000: A large scale highly reconfigurable experimental grid testbed. *International Journal on High Peerformance Computing and Applications*, 2006.

[BDC+08]   Francisco Brasileiro, Alexandre Duarte, Diego Carvalho, Roberto Barber, and Diego Scardaci. An approach for the co-existence of service and opportunistic grids: The eela-2 case. In *Latin-American Grid Workshop*, 2008.

[BDF+03]   Paul Barham, Boris Dragovic, Keir Fraser, Steven Hand, and Tim Harris. Xen and the art of virtualization. In *Proceedings of the 19th ACM Symposium on Operating Systems Principles*, pages 164–177. ACM, October 2003.

[BFG+08]   Zoltan Balaton, Zoltan Farkas, Gabor Gombas, Peter Kacsuk, Robert Lovas, Attila Csaba Marosi, Gilles Fedak, Ad Emmen, Gabor Terstyanszky, Tamas Kiss, Ian Kelley, Ian Taylor, Oleg Lodygensky, Miguel Cardenas-Montes, and Filipe Araujo. EDGeS: the Common Boundary Between Service and Desktop Grids. *Parallel Processing Letters*, 18(3):433–453, September 2008.

[BYSS+11] Orna Agmon Ben-Yehuda, Assaf Schuster, Artyom Sharov, Mark Silberstein, and Alexandru Iosup. Expert: Pareto-efficient task replication on grids and clouds. *Technical Report CS-2011-03*, Technion, 2011.

[CDCG+05] N. Capit, G. Da Costa, Y. Georgiou, G. Huard, C. Martin, G. Mounie, P. Neyron, and O. Richard. A batch scheduler with high level components. In *CCGRID '05: Proceedings of the Fifth IEEE International Symposium on Cluster Computing and the Grid (CCGrid'05) - Volume 2*, Washington, DC, USA, 2005. IEEE Computer Society.

[CDF+05]   Franck Cappello, Samir Djilali, Gilles Fedak, Thomas Herault, Frédéric Magniette, Vincent Néri, and Oleg Lodygensky. Computing on Large Scale Distributed Systems: XtremWeb Architecture, Programming Models, Security, Tests and Convergence with Grid. *Future Generation Computer Systems*, 21(3):417–437, mar 2005.

[CFH+08]   Gabriel Caillat, Gilles Fedak, Haiwu He, Oleg Lodygensky, and Etienne Urbah. Towards a Security Model to Bridge Internet Desktop Grids and Service Grids. In *Proceedings of the Euro-Par 2008 Workshops (LNCS), Workshop on Secure, Trusted, Manageable and Controllable Grid Services (SGS'08)*, Las Palmas de Gran Canaria, Spain, August 2008. 247–259.

[CGC01]   Vincent Neri Cécile Germain, Gilles Fedak and Franck Cappello. Global Computing Systems. In *Proceedings of the Third International Conference on Large-Scale Scientific Computing (LSCC'01)*, volume 2179 of *Lecture Notes in Computer Science*, pages 218–227, London, UK, 2001. Springer-Verlag.

[CVKB11]   Rodrigo N. Calheiros, Christian Vecchiola, Dileban Karunamoorthy, and Rajkumar Buyya. The aneka platform and qos-driven resource provisioning for elastic applications on hybrid clouds. *Future Generation Computer Systems*, 2011.

[DHL+04]   Samir Djilali, Thomas Herault, Oleg Lodygensky, Tangui Morlier, Gilles Fedak, and Franck Cappello. RPC-V: Toward Fault-Tolerant RPC for Internet Connected Desktop Grids with Volatile Nodes. In *Proceedings of the ACM/IEEE SuperComputing Conference (SC'04)*, pages 39–, Pittsburgh, PA, USA, November 2004.

[FGNC01]   Gilles Fedak, Cecile Germain, Vincent Neri, and Franck Cappello. XtremWeb: A Generic Global Computing Platform. In *CCGRID'2001 Special Session Global Computing on Personal Devices*, 2001.

[FHC08]   Gilles Fedak, Haiwu He, and Franck Cappello. BitDew: A Programmable Environment for Large-Scale Data Management and Distribution. In *Proceedings of the ACM/IEEE SuperComputing Conference (SC'08)*, pages 1–12, Austin, TX, USA, November 2008.

[GJG+05]   Fabrizio Gagliardi, Bob Jones, François Grey, Marc-Elian Bégin, and Matti Heikkurinen. Building an infrastructure for scientific grid computing: Status and goals of the egee project. *Philosophical Transactions: Mathematical, Physical and Engineering Sciences*, 2005.

[HFK+10]   Haiwu He, Gilles Fedak, Peter Kacsuk, Zoltan Farkas, Zoltan Balaton, Oleg Lodygensky, Etienne Urbah, Gabriel Caillat, and Filipe Araujo. Extending the EGEE Grid with XtremWeb-HEP Desktop Grids. In *Proceedings of CCGRID'10, 4th Workshop on Desktop Grids and Volunteer Computing Systems (PCGrid 2010)*, pages 685–690, Melbourne, Australia, May 2010.

[Jin04]   Hai Jin. ChinaGrid: Making Grid Computing a Reality. In *Lecture Notes in Computer Science, Volume 3334*, pages 13–24, Springer-Veralag Berlin Heidelberg, 2004.

[KFF08]   P. Kacsuk, Z. Farkas, and G. Fedak. Towards Making BOINC and EGEE Interoperable. In *Proceedings of 4th IEEE International Conference on e-Science (e-Science 2008), International Grid Interoperability and Interoperation Workshop 2008 (IGIIW 2008)*, pages 478–484, Indianapolis, IN, USA, December 2008.

[KJM+09]   Derrick Kondo, Bahman Javadi, Paul Malecot, Franck Cappello, and David Anderson. Cost-benefit analysis of cloud computing versus desktop grids. In *18th International Heterogeneity in Computing Workshop*, 2009.

[LFC+03]   Oleg Lodygensky, Gilles Fedak, Franck Cappello, Vincent Neri, Miron Livny, and Douglas Thain. XtremWeb & Condor : Sharing Resources Between Internet Connected Condor Pools. In *Proceedings of CCGRID'2003, Third International Workshop on Global and Peer-to-Peer Computing (GP2PC'03)*, pages 382–389, Tokyo, Japan, 2003. IEEE/ACM.

[LUD12]   Oleg Lodygensky, Etienne Urbah, and Simon Dadoun. *Desktop Grid Computing*, chapter XtremWeb-HEP : designing desktop grid for the EGEE infrastructure. CRC Press, 2012.

[MBV11]   H. Monti, A.R. Butt, and S.S. Vazhkudai. Catch: A cloud-based adaptive data transfer service for hpc. In *Proceedings of the 25th IEEE International Parallel & Distributed Processing Symposium (IPDPS 2011)*, 2011.

[MKF06]   Paul Malécot, Derrick Kondo, and Gilles Fedak. XtremLab: A System for Characterizing Internet Desktop Grids. In *Poster in The 15th IEEE International Symposium on High Performance Distributed Computing HPDC'06*, Paris, France, June 2006.

[MKF10]   Paul Marshall, Kate Keahey, and Tim Freeman. Elastic site: Using clouds to elastically extend site resources. In *Proceedings of CCGrid'2010*, Melbourne, Australia, 2010.

[MKF11]   P. Marshall, K. Keahey, and T. Freeman. Improving utilization of infrastructure clouds. In *IEEE/ACM International Symposium on Cluster, Cloud and Grid Computing (CCGrid 2011)*, 2011.

[MRM⁺11]   Alberto Di Meglio, Morris Riedel, Shahbaz Memon, Cal Loomis, and Davide Salomoni. Grids and clouds integration and interoperability: an overview. In *The International Symposium on Grids and Clouds and the Open Grid Forum Academia*, Sinica, Taipei, Taiwan, 2011.

[PIRG08]   Mayur R. Palankar, Adriana Iamnitchi, Matei Ripeanu, and Simson Garfinkel. Amazon s3 for science grids: a viable solution? In *Proceedings of the 2008 international workshop on Data-aware distributed computing*, DADC '08, 2008.

[Ser09]   Amazon Web Services. An introduction to spot instances. Technical report, Amazon Elastic Compute Cloud, 2009.

[SSGS09]   Mark Silberstein, Artyom Sharov, Dan Geiger, and Assaf Schuster. Gridbot: execution of bags of tasks in multiple grids. In *Proceedings of the Conference on High Performance Computing Networking, Storage and Analysis*, SC '09, 2009.

[UKF⁺09]   E. Urbah, P. Kacsuk, Z. Farkas, G. Fedak, G. Kecskemeti, O. Lodygensky, A. Marosi, Z. Balaton, G. Caillat, G. Gombas, A. Kornafeld, J. Kovacs, H. He, and R. Lovas. EDGeS: Bridging EGEE to BOINC and XtremWeb. *Journal of Grid Computing*, 7(3):335–354, September 2009.

[VHML11]   Constantino Vázquez, Eduardo Huedo, Rubén S. Montero, and Ignacio M. Llorente. On the use of clouds for grid resource provisioning. *Future Gener. Comput. Syst.*, 2011.

[VMF⁺05]   S. Vazhkudai, X. Ma, V. Freeh, J. Strickland, N. Tammineedi, and S. L. Scott. FreeLoader: Scavenging Desktop Storage Resources for Scientific Data. In *Proceedings of Supercomputing 2005 (SC'05)*, Seattle, WA, USA, 2005.

[WFC07]   Baohua Wei, Gilles Fedak, and Franck Cappello. Towards Efficient Data Distribution on Computational Desktop Grids with BitTorrent. *Future Generation Computer Systems*, 23(7):983–989, November 2007. Selected from the ISPDC conference.

Chapter 13

Supporting Web 2.0 Communities by Volunteer Desktop Grids

Peter Kacsuk

MTA SZTAKI, Budapest, Hungary

Attila Marosi

MTA SZTAKI, Budapest, Hungary

Lovas Robert

MTA SZTAKI, Budapest, Hungary

Jozsef Kovacs

MTA SZTAKI, Budapest, Hungary

13.1   Introduction

13.1.1   Desktop Grids and Volunteer Computing

Desktop grids (DGs) are grid systems where the resources are collected typically from existing desktop machines via cycle scavenging. The main advantage of DG systems compared to the other forms of grid systems, called service grids (SGs) is that they do not need a large investment and yet they can collect large resource capacity. DGs are also much simpler than SGs concerning the software infrastructure they are relying on. Installation and maintenance of DG resources is extremely simple, requiring no special expertise, so even home computers can easily be provided as computing resources. Typical DG systems are BOINC [And04], Condor [TTL04], OurGrid [CBA+06], XtremWeb [FGNC01], SZDG (SZTAKI Desktop Grid) [KKF+]. These were characterized and compared in detail in [KKF+]. One form of DG system is called volunteer computing when the resources are provided in a volunteer way, typically by home computers. In order to attract home computers, some form of reward mechanism should be introduced in the DG system. BOINC is by far the most advanced volunteer desktop grid middleware, providing credit collecting, administrating, and displaying functionalities for the project participants who donate the spare cycles of their desktop computers. The other important requirement of volunteer computing middleware is to provide some form of protection against erroneous and malicious donated computers. In this respect, again BOINC is the most advanced. It supports redundant computing and result validation. A very important aspect of volunteer computing is its community-based nature. BOINC provides a middleware framework based on which scientific projects like SETI@home [ACK+02], Einstein@Home [A+09], Folding@Home [BEJ+], etc. can be created. However, this framework should be filled in by desktop resources via the activity of a community. This community is the volunteers who want to support the given project due to their personal motivations for supporting science or other reasons. One of the decisive key factors in the success of different volunteer computing projects is their capability of attracting donors to support them. Successful volunteer computing projects are typically able to collect desktop resources in the range of $10,000 - 1,000,000$ that is, much bigger than the typical number of resources collected in SG systems. This shows the power of communities. Of course, these volunteer resources are not as reliable as the managed clusters and supercomputers typically used in SG systems and hence as it was mentioned above this is an important research area of volunteer DG systems: how to handle and manage the unreliable resources.

13.1.2   Web 2.0 Technologies

The Web 2.0 term [O'R05] stands for the second generation internet services based mainly on communities where users create the content as a joint effort or they share information. In Web 2.0 service,s the owner of the server provides only the service framework; the real content is uploaded and maintained, shared, or evaluated by the users. Users communicate with each other and they maintain connections too. Typical Web 2.0 systems are the well-known Facebook [Fac], Youtube [You], and many others like Orkut [Ork] and iWiW [iWi].

Generally, due to the large number of connections and users, Web 2.0 systems should handle heavy data traffic and complex relationships, that need extraordinarily large computational power in many cases. Many Web 2.0 applications deal with large files (e.g., video sharing); conversion or processing of such files demands large amounts of resources. From the Web 2.0 side, numerous application topics with large IT capacity can be solved by Grid technologies like the already-available advisor service system or the media conversion service or new envisioned services like digital signature and watermark service. Further community-specific applications have great potential.

13.1.3   Overview of the Web2Grid Initiative

The main aim of the Hungarian Web2Grid [Web] project is to support Web 2.0 communities in dynamically building the necessary infrastructure that is needed for processing the high-end tasks of their community. The same way as the content is generated as a joint effort of the members, Web2Grid technology enables the community to build the required infrastructure as a joint community effort exploiting the volunteer computing resources of the community members. In order to achieve this goal, Web2Grid exploits the BOINC and SZDG technologies and other earlier initiatives focusing on the integration of Desktop Grid and Web 2.0 technologies. On the one hand, Web 2.0 technologies assist in advertising desktop grid goals and attracting computational resources for Desktop Grid communities. On the other hand, the developed new Web2Grid platform that includes Desktop Grid technology can offer back-end infrastructure for several compute-intensive Web 2.0-related problems:

- For operational requirements of Web 2.0 portals (i.e., large number of users, connections, file processing, and large file conversions)

- For security requirements (watermarking, periodic revision of digital signature, etc.)

- For systems requiring large IT capacity

- For tasks defined by community member end users

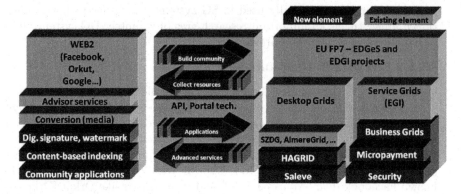

FIGURE 13.1: The main building blocks of the Web2Grid project.

The combination of the Web 2.0 and Desktop Grid technologies provides advantages for both sides (as depicted on Figure 13.1); they can mutually benefit from each other's approach. In such combination, Web 2.0 can extend its capabilities further from the community contents toward shared services with the help of Grid technologies. From the grid point of view (right side of Figure 13.1), the project relies on the results of the EU FP7 EDGeS [UKF+09] infrastructure project that has been further developed by the EDGI [EDG] [KKF+11] project. The main aims of these projects were the investigation of the above-mentioned targets and the development of tools, interfaces, and methodologies to establish the integrated services both for private (local desktop grid) and public (volunteer desktop grid) environments.

In these projects a new bridge was developed that integrates the usual service Grids like EGI [EGI], and the volunteer computing desktop Grids (like SZDG) that are very close to the main philosophy of Web 2.0 technology (right side of Figure 13.1) since Grid resources are provided by a community.

During the Web2Grid project, the aim is to extend these Grids with technologies like micropayment, accounting, business model, and security. Furthermore, the project also integrates other types of Desktop Grid like HA-GRID [Hag] and Saleve [PDS08]. This enhanced Grid system then forms the computational resource needed for the execution of Web 2.0 community services (middle of Figure 13.1) while Web 2.0 community environments help the public Desktop Grid systems collect new resources for the applications. Web 2.0 applications also gain advantages from advanced services like workflow execution provided by the Grid side of the overall system. On the left side of Figure 13.1, the Web 2.0 community using various applications (which may rely on Grid services) and special Web 2.0 applications is supported during the project by showing how these applications can utilize the power of the Grid or Desktop Grid systems. To demonstrate the developed system solution, the Web2Grid project consortium created Web 2.0-based demo applications from different fields, and provided easy-to-use interfaces and recommendations to foster the potential usage of the technologies. The Web2Grid project carefully investigated several aspects of combining Web 2.0 systems and volunteer Desktop Grid systems. Here we summarize the most important aspects of such integration.

13.1.3.1   Aspect 1: Web 2.0, Strength of the Community to Integrate Resources

Like Web 2.0 systems, Desktop Grids are also organized on community bases. For example, the Dutch city initiative of AlmereGrid [Alm] collects resources from the community of inhabitants of a city. In fact, this was the first such city Grid showing a good example for other cities around the world. The initiative of IBM, the World Community Grid [Wcg] gathers donors for several grand-challenge scientific projects via a very attractive Web page that, in fact, plays a role very similar to the attractive Web 2.0 Web pages. The Desktop Grids for eScience Road Map [Roa] mentions communication channels from the Web 2.0 world, too. However, to establish a strong integration of Web 2.0 and volunteer Desktop Grid directions requires more research; how to organize the resources of the Web 2.0 community members to this type of volunteer DGs by the potential groups of enterprises, corporations, academic research institutes, universities, high schools, a city, or on thematic bases.

13.1.3.2   Aspect 2: Tasks Arising from the Web 2.0 Infrastructure

There are several newly launched Web 2.0 services but it is pending what will survive from the new Web. Most of the new services (e.g., graph analysis of connecting people) in the case of a large number of users may critically slow down. In the background, the largest Web 2.0 portals use HPC and HTC technologies (YouTube or Facebook) with thousands of computers in server farms. To provide green Desktop Grid technologies [SE10] for such purposes has not been a well-explored area and there is no well-developed solution so far.

13.1.4   Aspect 3: Critical Tasks with Respect to the Future Perspectives of the Web 2.0 Infrastructure

In the field of Web 2.0 services, the security aspects of the created and available information (text documents, pictures, videos), the security of the transactions and the data, especially identity theft, infringement of copyright etc. causing great harm, are receiving more and more consideration. Digital signatures, watermarking, encrypting of continuously increasing files, and updating and controlling them require extraordinary computing capacity, and this problem does not have an effective enough solution for Web 2.0 service providers. Additionally, using data mining methods, the misuse can be determined and prevented but this also requires large computational capacity. The emphasized security support brings some novel solutions to this field.

13.1.4.1 Aspect 4: Tasks Demanded by the Members of the Web 2.0 Community

The members of each community should be able to use the resources themselves to meet their own objectives (e.g., to process pictures of astronomy, etc.). There is an increasing demand for the Web 2.0 communities to create and develop their own community-specific applications. That is the reason why Facebook has provided the necessary API to develop such applications. In the case of DG systems there are examples of how to run the applications of the members on the community members' computers (XtremWeb, OurGrid, SZDG) but these DG community members are not tightly integrated with advanced Web 2.0 facilities.

13.1.4.2 Aspect 5: Extension of Service/Desktop Grid to Business Grid

Grid systems were not able to penetrate into the commercial world due to the lack of usable business models. Although there have been many attempts to define such business models, they were not taken up by companies for many reasons. In the Web2Grid project, several business models have been investigated on how to attract and pay donors providing resources for the Web 2.0 community and how the community members should pay for the services using the volunteer resources.

13.1.4.3 Aspect 6: Integrating Volunteer Desktop Grids to a Payment System Framework

In the framework of the Web2Grid project, the integration of a micropayment system with volunteer Desktop Grids was formed. The project developed an accounting system that takes over some tasks from BOINC but its role is far more complex in the Web2Grid platform. It motivates donors to join and donate resources by giving virtual credits for their donated CPU time. These credits can later be exchanged for real money and donated to charity. The accounting system maps Desktop Grid tasks to the requests (orders) submitted by the Web 2.0 community members via the Web 2.0 interface of a given application (see details in Section 13.5.2).

13.2 Related Works

A community-driven public service where the service is utilized/used and supported at the same time by the same community is the renderingfarm.fi BOINC project where the members of the community can upload blender files for processing. The real processing is also distributed among the members attached to the BOINC server. As an application, this is a good example of how a community can use its own computational resources for calculating its own requests; however, this project is currently not very strong in Web

2.0 aspects because, they have not integrated their project into an existing Web 2.0 environment like Facebook. An experimental project on linking Facebook and BOINC-based Desktop Grids can be observable in one of Intel's initiative [Int] called Progress Thru Processors in 2009. The aim of this initiative was to promote three different BOINC-based Desktop Grid projects through Facebook communication channels. It was done in collaboration with GridRepublic, which is one of the biggest BOINC user community-building projects. Similarity with our system is that Facebook was used as a front-end for the BOINC projects; contrary to our solution, they did not enable the job submission mechanism for Facebook community members, which is one of the main goals, in our solution. We can say that in this particular case they relied primarily on the power of the community-building feature of Facebook.

The paper by Patoli [PGAB+08] presents an experiment on how to implement a Grid-based high-performance computing solution for a rendering farm utilizing existing institutional Desktop resources by scavenging the idle cycles. To achieve this goal, open source Condor High Throughput Computing software was selected and implemented as a Desktop Grid computing solution. The solution mainly focuses on building up a rendering farm based on Condor. This service then could be accessed by institutional people; however, public availability is not granted. This solution could be a good candidate for a Web 2.0 application, however the infrastructure is based on Condor, which is not really suitable for collecting computational power of PCs owned by people at home. A good example of how a community can provide the needed computational capacity to operate its own services is the CancerGrid Computing System [KKL11] developed by the EU FP6 CancerGrid project. This system aimed to provide a Web-based portal combined with a Desktop Grid infrastructure to perform various chemical-related calculations. Cancer-Grid consists of the gUSE Web-based portal, an integrated molecule database browser, and a BOINC-based Desktop Grid underneath. From an application point of view, several chemical algorithms have been deployed on the BOINC server and these algorithms were combined into workflows handled by the gUSE portal. In this infrastructure the community was formed by chemists, biologists, and physicians, and the computational resources were collected from the institutes these users came from. In this solution, the community was a private one, that is, new members could join only after authentication. The reason behind it is the sensitivity of the data the system was handling, that is, no volunteers were allowed to see the results of the computations. The most interesting example that is closely related to the proposed solution in this chapter is a high-performance online service for genetic linkage analysis, called the Superlink-online [Sil11]. This system enables anyone with Internet access to submit genetic data and analyze it easily and quickly. The running environment is provided by Superlink@Technion - GridBot - Distributed Computing System, which allows the usage of idle cycles of tens of thousands of home Desktop PCs as well as high-performance Grids and Clouds all over the world. This infrastructure receives jobs from the public and processes them on institutional computing resources as well as by a volunteer BOINC project. The two different types of Grid are conducted by a high-level scheduler and using redundancy at job submission. However, this project was intended to support a single application and a single community. Our solution provides a framework for any Web 2.0 community to create and run their own applications.

13.3 BOINC and SZDG

The main goal of the Web2Grid project is to enable Web 2.0 communities to use volunteer computing technology in order to collect computing resources from their own community. This meant that first we had to decide which available volunteer computing technology would be the best to apply in the project. We investigated and compared several existing Desktop Grid technologies, BOINC, Condor, OurGrid, XtremWeb, SZDG, and found that BOINC and its further development SZDG are the two possible candidates because the other three DG systems do not provide any form of attracting volunteer donors.

13.3.1 BOINC and Volunteer Computing

Contrary to the traditional Grid approach, where complex middleware is used to manage the connected dedicated resources, Desktop Grids use a lightweight middleware to attract not just dedicated, but volunteer resources, too. The most well-known volunteer Desktop Grid is SETI@Home which collects about 3 million CPUs worldwide. This Desktop Grid is based on the Berkeley Open Infrastructure for Network Computing (BOINC) platform, which is the most popular volunteer computing and Desktop Grid platform currently. BOINC was developed with volunteer computing in mind; thus its architecture is aims at utilizing the spare cycles of home PCs. It uses a server/client approach where a centralized component is queried for units of work (work units) by its clients. BOINC server deployments are independent of each other and are referred to as projects. Each project aims to achieve a different (scientific) goal, ranging from drug discovery to distributed rendering. Resources can be donated by downloading a lightweight client and connecting it to the public URL of a specific project. After that it will download a binary application and work units that are sets of input files for the application. The client executes the application with one set of input files and when it is completed the client uploads the results to the project. The resource donating persons, referred as donors, are rewarded with virtual credits for the work done by their computers. Credit for a work unit is awarded based on the amount of CPU time spent for computation. The client continuously monitors the CPU time spent and reports it along with the finished result to the project. Using volunteer resources means that the administrator of the BOINC project has no influence over them: 1) they are highly volatile, so they might be turned off for periods of time or might never return the result of the assigned work unit; 2) they are not reliable in the sense that they might be producing erroneous results by purpose or malfunction. The donors might overclock their computers to achieve higher performance, or some malicious donors might try to make the client report more CPU time spent on the work unit than it actually did or harm the project by returning bogus results on purpose.

BOINC uses replication and validation to cope with these problems. Instances of the same work unit are replicated to different resources for processing, and the returned results are compared and validated. Replication can be either static or adaptive. With static replication, new instances of the same work unit are created regardless of whether the computing host of the previous instance(s) can be considered reliable or not (based on the record of previous work done by the host for the project). With adaptive replication, if a work unit is sent to a reliable host and a successful result is returned, no more instances will be created. Result comparison is an application-specific task, as only the developer of the application knows if two results can be considered matching. BOINC provides stock validators for doing binary comparison of output files of results or for comparing text output files, where the line endings may differ depending on the operating system of the computing host (Windows, Linux, and Mac OS X use different characters for marking line endings in text files). If an application produces output files that need more or different types of checking, then the application developer must create a validator component for the application using a framework provided by BOINC. If using redundancy, then usually for the instances of the work unit, either the average of the claimed credit or the lowest credit requested is granted. This filters out the possibility of someone requesting too much credit for a work unit. BOINC also keeps track of which hosts have which work unit instance downloaded and only accepts results of those work units from those hosts. This filters out the possibility of double reporting tasks or reporting the task of some other host.

Usually a BOINC project aims to solve a single scientific problem, meaning it only hosts a single application. Still, the project must support different operating systems and computer architectures. A combination of an operating system and computer architecture (e.g., Linux 64 bit on x86-64) is referred as a platform. Any application the project hosts must at least have binaries deployed for each of the major different platforms (e.g., 32-bit Windows, Linux, and Mac OS X) used widely to be able to attract as many donors as possible.

13.3.2   SZDG (SZTAKI Desktop Grid)

Although BOINC provides very useful components to support volunteer computing (credit system, protection against malicious, and erroneous computers), it has the limitation that it was designed to build individual projects to be run by system administrators without any user interface support. This

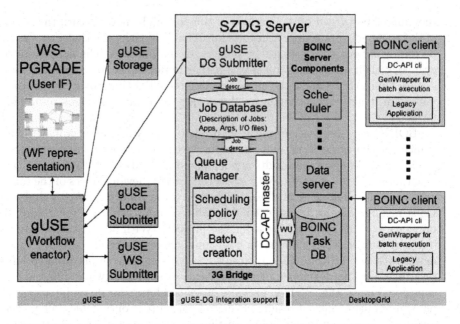

FIGURE 13.2: Architecture of SZDG including WS-PGRADE and gUSE.

feature of BOINC significantly reduces its usability for large communities like the Web 2.0 communities.

SZDG solves this problem by extending BOINC with a high-level workflow-oriented user interface and several other tools that enable the wide usage of SZDG by many different user communities for a wide variety of possible application areas. In fact, the SZDG Desktop Grid technology is basically BOINC extended with the following components:

1. WS-PGRADE Web portal

2. gUSE core services

3. gUSE 3G Submitter

4. 3G Bridge

5. DC-API plug-in

6. GenWrapper

As Figure 13.2 shows, an SZDG server is substantially different from a BOINC server. Beyond the usual BOINC server components, it contains two other important components: 3G Bridge and gUSE DG Submitter. These two components enable the integration of BOINC systems with the WS-PGRADE workflow-oriented portal and the gUSE high-level middleware that contains the WS-PGRADE workflow enactor. With WS-PGRADE and gUSE, users can create workflows in which any node can be a parameter sweep type node that should be executed with a large number of input parameter sets. Every parameter set represents one execution and hence such nodes could be executed $1,000 - 1,000,000$ times. These nodes are ideal to be executed by a Desktop Grid and indeed, WS-PGRADE and gUSE can submit jobs derived from such nodes to the connected Desktop Grid system via the gUSE DG Submitter.

The DC API master component of the 3G Bridge in the SZDG server is responsible automatically generating BOINC work units out of the jobs transferred by the gUSE DG Submitter. These work units are then distributed to the BOINC clients in the usual way. However, in the case of SZDG, these BOINC clients are extended with GenWrapper [MBK09], which can handle the required DC API clients in order to execute those work units that were generated by the DC API master. Furthermore, GenWrapper enables the use of legacy codes in the BOINC client. This means that the typically hard BOINC porting effort can be eliminated by the use of DC API and GenWrapper.

As a summary, we can conclude that the main advantage of using SZDG instead of BOINC is that users can build into their workflows applications that are already registered and supported by DG projects. As a result, large complex workflow applications can exploit various DG systems, including volunteer and institutional ones. This is one of the main ideas of the EDGI project [EDG] that created and maintains an application repository where those applications are stored that were already ported to DG systems and hence can be used in such workflow systems mentioned above. In this way, BOINC systems can be used by large user communities and not only by individual projects.

13.4   Survey on Web2Grid Technologies

In order to measure the potential impact of the described ideas and developments (see Section 13.1.3), a survey has been made among citizens including several groups of questions:

- Use of computers in general

- Use of Web 2.0 applications

- Awareness of Desktop Grid technologies

- Motivations that help offer spare computer capacities

- Obstructive factors

The first three groups of questions allowed us to discover the habits and the background of citizens concerning the related IT technologies. The last

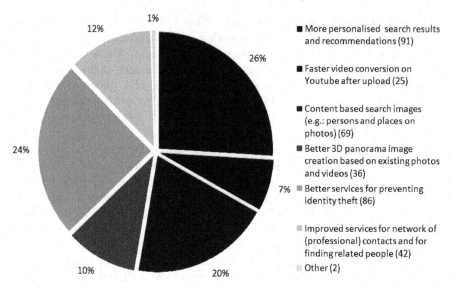

FIGURE 13.3: Survey of what Web 2.0 services users would like to have.

two groups of questions gave more details on the motivations and obstructive factors that would accelerate or slow down the penetration of the developed technologies among the average citizens and the exploitation process based on either volunteer or paid services. Both the motivation and obstructive factors have objective (technology) and subjective (human) aspects that the survey intended to investigate.

The survey has been distributed to 1,000 volunteers in Hungary, and the evaluation is based on 174 replies. Based on the gathered statistics, the survey can be considered not fully representative of the group of Hungarian Internet users over age 14 but covers wide range in age and location.

In general, most of the adult Internet users have Desktop Grid-ready computers; 75% have max. 5-year-old computers, and over 80% have broadband Internet access. They mostly use community portals (only 1 out of 7 had negative, uncertain, or other reply) and would request enhanced Web 2.0-related applications; such as advisory systems, content-based search facilities on photos and videos, and also services that could prevent or at least decrease the number of data and identity thefts (see Figure 13.3).

We experienced low awareness of Desktop Grid solutions. Furthermore, only a low fraction of the surveyed people has ever joined a volunteer project, and the situation is even worse with the volunteers who remain active in the projects. Concerning the motivations, the grand challenge applications with strong impact on the standard of living (e.g., research on climate change) would get the highest support from the volunteers; the majority would help them with spare computer capacities. Two out of five from the sample are ready to support the research pursued by local universities or institutes. Basic

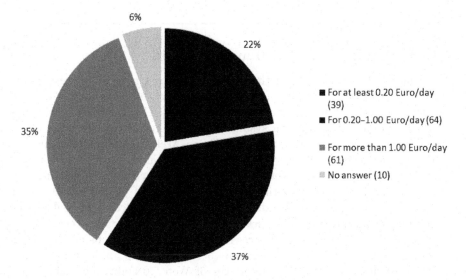

FIGURE 13.4: Survey of financial motivation for donating computing time.

research and the improvement of Web 2.0 and search engines are not popular. In order to attract such volunteers, Desktop Grid operators have to develop advanced methods for motivation in such competitive environments. According to the survey, a strong motivation might be real financial advantage but the collection of virtual credits, obtaining badges' in competitions, and the glory when the volunteers computer finds the breakthrough in the research are not the main factors. One quarter of the sample showed no willingness to join any volunteer project at all. Concerning the financial benefits, one third of the sample would offer a home computer in the range of 0.2 to 1 EUR for at least a 4 hour-long period per day, and one quarter would require even less contribution (0.2 EUR). The financial motivations are shown in Figure 13.4.

The source of major obstructive factors that may keep away the citizens from volunteer projects is the lack of trust. One of the main issues is strongly related to security holes. The possible illegal misuse of home computers by others, and the lack of control over the deployed and executed software packages were also major factors. The fear of slowing down a computer during its everyday usage was reported as well. However, the possible increased electricity cost is not mentioned among the major factors; very likely, the lower awareness of green technologies and environment protection to the generic IT security problems is the reason. The up-to-date and easy-to-understand information transfer about the progress and results of the volunteer applications must be vital. Providing escape routes, for example, easy shutdown, complete uninstall, and incident report functionalities, would help improve the trust between the DG/Web 2.0 providers, together with obtained certificates from well-known sources for the services and applications.

13.5 Bridging Web 2.0 and the Grid

One of the main goals of the Web2Grid project is to solve computation intensive tasks of Web 2.0 communities with the help of volunteer computing by harnessing the power of volunteer Desktop Grids. BOINC is the most popular volunteer computing system; thus the project heavily relies on the BOINC architecture and its SZDG extension. Web 2.0 applications can come from various fields, and also Web 2.0 service providers, external consultants, or community users can develop applications (e.g., using the Facebook API). During the development of the interfaces, Web2Grid project partners were focusing on the protocols needed for commercial services that realize security and accounting functionalities. Figure 13.5 depicts the architecture of the platform developed by the project. It contains four major blocks. First, the platform may contain an arbitrary number of "Web 2.0 applications." These are represented by Web user interfaces of Web 2.0 community sites where users can use the services provided by the applications (e.g., submit images for watermarking). The second block is the Payment system that is used for charging for the Web 2.0 services. Currently, the Abaqoos micro-payment system is supported but the use of the Payment system is optional and is a decision of the application developers. The third block is the Grid that executes the compute-intensive tasks created by the Web 2.0 services: each service is represented as a workflow and computed using volunteer resources. Finally, the Accounting system coordinates the previous three blocks and accounts for the users, work done, and volunteer resources contributed by donors. In the following subsections we detail the main components of the platform.

13.5.1 Web 2.0 Applications

A Web 2.0 application consists of several components. First there is a user interface on one or more Web 2.0 social networking platforms for the application. The goal of the interface is to take orders (requests) from users and deliver the results for them. For examaple, in the case of an image watermarking service, the interface is used for selecting and submitting a batch of images to the service, and viewing the results later. A submission on the Web 2.0 interface is referred as an order because the user makes an order in the application (e.g., to watermark a batch of images) and will get the results (e.g., watermark embedded images) after the order is filled. It is the task of the Order module to track the submissions and to interface with the optional Payment system. If a Web 2.0 application requires payment for its services, the user is redirected to the optional Payment gateway. The order can be filled (e.g., watermarks embedded) only after the payment transaction has finished. The Grid task handler module interfaces with the Grid and with the Accounting system: the Accounting system provides a unique identifier for the order, which is used throughout the system. The Result validator module

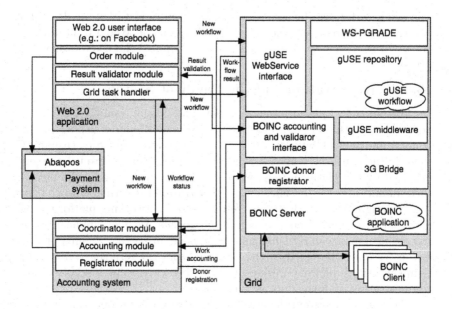

FIGURE 13.5: Architecture of the Web2Grid platform.

is responsible for checking the results returned by the volunteer resources of the Grid. We describe this mechanism in Section 13.5.5. Each Web 2.0 application has two additional components but they are not part of the Web 2.0 application block (in Figure 13.5); rather, they are deployed directly in the Grid. First there is a gUSE workflow representing the Web 2.0 application, and second, there are one or more BOINC applications that are invoked from the workflow.

13.5.2   Accounting System

The project developed its own Accounting system, which takes over some tasks from BOINC but its role is far more complex in the Web2Grid platform. It is the key component as it is responsible (among many other things) for integrating the main components. The Web2Grid platform motivates donors to join and donate resources by giving virtual credits for their donated CPU time. These credits can later be exchanged for real money and donated to charity. Figure 13.5 shows the three major components of the Accounting system. The Coordinator module maps Desktop Grid tasks to one of the orders submitted on the Web 2.0 interface(s) of a given application (one such Web 2.0 orders may result in many Desktop Grid tasks). These tasks need to be accountable, meaning that they must be assigned to a user who sent the order on the Web 2.0 interface and to the donors who should get virtual credit for computing the tasks belonging to the orders. It is achieved by assigning a unique id for each order. This id is used by the Grid system for identification, as a single order may be split to many Desktop Grid tasks. Finally, the Registrator module registers the users signed up on the Web 2.0 interface to be donors in the Desktop Grid server and synchronizes the user database of the two systems.

13.5.3   Registration and the Payment System

The Web2Grid platform recruits users and donors from the Web 2.0 social networking platforms (e.g., Facebook or Orkut). Resource donors (or simply donors) are persons who donate their idle resource capacities (usually CPU time) to the platform, while users are the entities who want to use the services provided by the platform (e.g., want to embed watermarks in the pictures of their albums uploaded to Facebook). Registration for both donors and users is done through Web interfaces on Web 2.0 social networking platforms such as Facebook or Orkut. Users can sign up for specific Web 2.0 application or service, while donors can sign up on a central page rather than signing up directly at the Web page provided by the Desktop Grid. Because the Web2Grid platform can utilize different Desktop Grid middleware, some kind of centralized administration of donor data is required.

The Payment system depicted in Figure 13.5 is an optional component. The developer of each Web 2.0 application decides whether she wants to charge for the service. Currently, the Abaqoos micro-payment system by E-Group is supported. The payment service, where all payment transactions are authorized, is not part of the Web2Grid platform; rather, it is an external service. Any registered user of the platform who wants to use a non-free Web 2.0 service needs to register at his/her first transaction at the Payment system, too.

13.5.4   Grid

The Grid system executes the computation-intensive tasks created by the Web 2.0 services. It has three main components (that are all parts of SZDG): (1) the gUSE workflow enactor; (2) the BOINC Desktop Grid middleware and (3) the 3G Bridge service connecting the previous two components. Each Web 2.0 application is represented by a workflow that consists of several nodes. The gUSE workflow engine stores and executes workflows, where each node represents one or more computation tasks for the (Desktop) Grid. These may be executed on arbitrary supported distributed computing infrastructures (DCIs), in this case on the supported Desktop Grid middleware. An instance of the workflow is started for each order arriving from the Web 2.0 application. Each instance contains the unique id assigned by the Coordinator module to the order representing the workflow. Tasks belonging to these workflow instances are then executed by the Desktop Grid. The 3G Bridge is used to interface the different components and Grid middleware. In the Web2Grid project, the gUSE Submitter is used to send tasks to the 3G Bridge and a BOINC destination plug-in is used to submit those jobs to a volunteer BOINC Grid. 3G Bridge also provides an interface to query the status of jobs and retrieve their outputs once they are finished. This means the 3G Bridge hides the specific features of the underlying computing infrastructure from the Submitter component as described in detail in [FKBG10]. Although 3G Bridge allows connecting arbitrary Desktop Grid middleware, the Web2Grid platform requires three additional services from any connected middleware. These services are used by the Accounting system. First, an interface for administering donors is required. This interface is used to keep the donor account synchronized in the system. The donors never register directly at the middleware but rather at a central interface operated by the Web2Grid platform. The second interface is provided for accounting information. This interface publishes data at a predefined interval (usually hourly) for the tasks executed and finished by the Desktop Grid middleware in the last period. Finally, a service is required for each application hosted by the Desktop Grid that is able to check the correctness or validity of the task computed on volunteer (non-trusted) resources. The next subsection details this concept.

13.5.5 Remote Validation of Tasks

The result checking procedure is referred as validation and it can be done in two different ways. First, only the format of the returned output files is checked, for example, if an image is expected, then a check is run to ensure that indeed an image was returned with the correct size, color depth, etc. The second method is to use redundancy: multiple instances of the same work unit are sent to different resources and the returned results are compared. If a predefined number of results of the same task match, then they are accepted and declared as valid. Matching results is an application-specific task, as depends on the application what differences are acceptable and what are not. The Web2Grid platform uses the second method and every Web 2.0 application needs to have a result validator module developed (as shown in Figure 13.5). BOINC provides a framework for developing result validator modules. For each application deployed at the BOINC server there is a running validator module (Validator daemon). The task of such a daemon is to perform the validation procedure described above. To be able to perform the validation, it accesses the output files belonging to the results from the data server of BOINC and retrieves the required metadata from the BOINC database. The problem with this original BOINC approach is that the validator component is running on the Desktop Grid server, but in reality it should be part of the Web 2.0 application. A component that was developed by a third party (the application developer) is running on the BOINC server can be considered as a security risk because the validator daemon 1) must be validated itself, that it contains no bugs or harmful code; 2) has full access to the data server and is able to read output files of results belonging to other applications, and 3) is running with the same privileges as the other server components. One of the goals of the Web2Grid project is to make the Grid technologies available for as many Web 2.0 applications as possible but this cannot compromise security. The first step to avoid this is to remove or separate all third-party components from the Desktop Grid servers in a way that guarantees security but still provides a transparent solution for the application and validator developers. As a consequence, the monolithic single component architecture of the validator framework was replaced by a server/client architecture. In this way the Desktop Grid server only runs the client part that contains no application-specific elements; all application specific codes reside on the server part. This guarantees that the BOINC server only runs trusted code. Application developers need only to develop a server component by using a framework provided by Web2Grid, and only a couple of application-specific methods need to be implemented. The resulting server component may run on any host; it is always contacted by the client component running on the Desktop Grid server.

13.6 Conclusions

The Web 2.0 term is used as the name for the collection of second-generation services, which are primarily based on communities. This means that members of the communities create the content together or share information. In such systems, the owner of the server only provides the framework; the content is uploaded, created, or shared by the community. Users communicate and create links among each other. Usually, a huge relational network is created by the members that causes a heavy load on the server because of the huge data and relations to be handled. Some of the Web 2.0 applications themselves require relatively big computational capacity, such as video sharing, processing, and conversion. So, in the Web 2.0 system the need for a computational Grid system is a natural requirement due to the numerous tasks related to media conversion and advice systems that are popular nowadays. At the same time, we expect new requirements to be raised, such as digital signature and watermarking. The survey we have run in Hungary has shown many important social aspects and possible impacts of integrating Web 2.0 and DG resource provider communities. It has also revealed the importance of viable business models and the need for financial motivation of DG resource providers. In this chapter we showed a framework for how Facebook applications can utilize an underlying Desktop Grid system with its computational power supported by the individuals in the community. In our system we have integrated a demo Facebook application for watermarking, an accounting system handling micro-payment if needed, and the Grid system. The underlying Desktop Grid system is SZDG, which is represented by a Web service behind which there is a workflow enactor system called gUSE, a bridge, and a volunteer Desktop Grid system based on BOINC. The overall infrastructure is operated by the consortium to demonstrate the solution and to collect experimental information. The Web2Grid project has proved that SZDG is flexible enough to support various Web 2.0 applications by volunteer DG systems. The developed Web2Grid framework is generic enough so that various Web

2.0 user communities can either directly use it or can easily adapt it to their special requirements.

Acknowledgment

The research has been supported by the Hungarian National Office for Research and Technology (NKTH) under grant No TECH_08-A2/2-2008-0097 (WEB2GRID) and has received funding from the European Union Seventh Framework Programme (FP7/2007-2013) under grant agreement no 261556 (EDGI).

Bibliography

[A⁺09]   B.P. Abbott et al. Einstein and Home search for periodic grav-
            itational waves in early S5 LIGO data. *Phys.Rev.*, D80:042003,
            2009.

[ACK⁺02]   D.P. Anderson, J. Cobb, E. Korpela, M. Lebofsky, and
            D. Werthimer. Seti@home: An experiment in public-resource
            computing. *Communications of the ACM*, 45(11):56–61, Novem-
            ber 2002.

[Alm]   AlmeGrid. Almeregrid homepage: http://www.almeregrid.nl/.

[And04]   D. P. Anderson. Boinc: A system for public-resource computing
            and storage. In Rajkumar Buyya, editor, *GRID*, pages 4–10.
            IEEE Computer Society, 2004.

[BEJ⁺]   Adam L. Beberg, Dan L. Ensign, Guha Jayachandran, Siraj
            Khaliq, and Vijay S. Pande. Folding@home: Lessons from eight
            years of volunteer distributed computing. In *8th IEEE Inter-
            national Workshop on High Performance Computational Biol-
            ogy (HiCOMB 2009) in conjunction with the IEEE International
            Parallel and Distributed Processing Symposium (IPDPS 2009)*.

[CBA⁺06]   Walfredo Cirne, Francisco Brasileiro, Nazareno Andrade, Lauro
            Costa, Alisson Andrade, Reynaldo Novaes, and Miranda Mow-
            bray. Labs of the World, Unite!!! *Journal of Grid Computing*,
            4(3):225–246, 2006.

[EDG]   EDGI. The european desktop grid initiative (edgi) project home-
            page: http://edgi-project.eu/.

[EGI]   EGI. The egi organization homepage: http://www.egi.eu/.

[Fac]   Facebook. Facebook homepage: http://facebook.com.

[FGNC01] G. Fedak, C. Germain, V. N'eri, and F. Cappello. Xtremweb: A generic global computing system. In *Proc. of CCGRID2001 Workshop on Global Computing on Personal Devices*. IEEE Press, May 2001.

[FKBG10] Z. Farkas, P. Kacsuk, Z. Balaton, and G. Gombás. Interoperability of BOINC and EGEE. *Future Generation Computer Systems*, May 2010.

[Hag] Hagrid. The hungarian advanced grid: http://hagrid.econet.hu.

[Int] Intel. Intel's progress thru processors: http://www.facebook.com/progressthruprocessors.

[iWi] iWiW. iwiw homepage: http://iwiw.hu.

[KKF+] Peter Kacsuk, Jozsef Kovacs, Zoltan Farkas, Attila Marosi, Gabor Gombas, and Zoltan Balaton. SZTAKI desktop grid (SZDG): A flexible and scalable desktop grid system. *Journal of Grid Computing*.

[KKF+11] P. Kacsuk, J. Kovacs, Z. Farkas, A. Marosi, and Z. Balaton. Towards a powerful european dci based on desktop grids. *Journal of Grid Computing*, 9:219–239, 2011. 10.1007/s10723-011-9186-z.

[KKL11] József Kovács, Peter Kacsuk, and Andre Lomaka. Using a private desktop grid system for accelerating drug discovery. *Future Gener. Comput. Syst.*, 27:657–666, June 2011.

[MBK09] Attila Csaba Marosi, Zoltan Balaton, and Peter Kacsuk. Genwrapper: A generic wrapper for running legacy applications on desktop grids. In *Proceedings of the 2009 IEEE International Symposium on Parallel&Distributed Processing*, pages 1–6, Washington, DC, USA, 2009. IEEE Computer Society.

[O'R05] Tim O'Reilly. What is web 2.0 - design patterns and business models for the next generation of software, September 2005.

[Ork] Orkut. Orkut homepage: http://orkut.com.

[PDS08] R. Kápolnai P. Dóbé and I. Szeberényi. Saleve: toolkit for developing parallel grid applications. *Infocommunications Journal*, LXIII(1):60–64, January 2008.

[PGAB+08] Zeeshan Patoli, Michael Gkion, Abdullah Al-Barakati, Wei Zhang, Paul F. Newbury, and Martin White. How to build an open source render farm based on desktop grid computing. In Dil Muhammad Akbar Hussain, Abdul Qadeer Khan Rajput, Bhawani Shankar Chowdhry, and Quintin Gee, editors, *IMTIC*, volume 20 of *Communications in Computer and Information Science*, pages 268–278. Springer, 2008.

[Roa]   Roadmap.   Desktop   grids for escience - a road   map:
             http://desktopgridfederation.org/downloads/-
             /document_library_display/7tei/view/57919.

[SE10]   B. Schott and A. Emmen. Green methodologies in desktop-grid.
             In *Computer Science and Information Technology (IMCSIT)*,
             Proceedings of the 2010 International Multiconference on, pages
             671 –676, Oct. 2010.

[Sil11]   Mark Silberstein. Building an online computing service over vol-
             unteer grids. In *Proceedings of the 5th Workshop on Desktop
             Grids and Volunteer Computing Systems (PCGrid 2011)*, 2011.

[TTL04]   D. Thain, T. Tannenbaum, and M. Livny. *Distributed Computing
             in Practice: The Condor Experience*, 2004.

[UKF+09]   Etienne Urbah, Peter Kacsuk, Zoltan Farkas, Gilles Fedak, Ga-
             bor Kecskemeti, Oleg Lodygensky, Attila Marosi, Zoltan Bala-
             ton, Gabriel Caillat, Gabor Gombas, Adam Kornafeld, Jozsef Ko-
             vacs, Haiwu He, and Robert Lovas. EDGeS: Bridging EGEE to
             BOINC and XtremWeb. *Journal of Grid Computing*, 7(3):335–
             354, September 2009.

[Wcg]   Wcg. The world community grid homepage:
             http://www.worldcommunitygrid.org/.

[Web]   Web2grid. Web2grid homepage: http://web2grid.econet.hu.

[You]   Youtube. Youtube homepage: http://youtube.com.

Chapter 14

Programming Applications for Desktop Grids

Tamas Kiss

University of Westminster, London, United Kingdom

Gabor Terstyanszky

University of Westminster, London, United Kingdom

The success and importance of a production infrastructure is measured by the interest and engagement of user communities when utilizing it for their own benefit. It is crucial for a new infrastructure and technology to attract users. However, the utilization of complex infrastructures such as Desktop Grids is far from being trivial. Porting applications to this platform requires specific tools and approaches that end users may find challenging to engage. Therefore, it is crucial to provide support for potential targeted user communities, and to port and aid the deployment and running of their applications on the target platform. This chapter gives an overview of the challenges and available tools and solutions when porting applications to Desktop Grid infrastructures, especially when these infrastructures are extended with additional resources from service grid environments. The chapter introduces the EDGeS Applications Development Methodology (EADM) that provides a generic framework for porting applications to Desktop Grids and combined Desktop and service grid platforms. It also analyzes the requirements toward applications to be run on combined desktop and service grid platforms, and describes the most common user scenarios in these infrastructures. Finally, it illustrates the utilization of the methodology, the proposed tools and user scenarios via experiences gained through real-life case studies.

14.1 Generic Methodology for Porting Applications to Desktop and Service Grid Infrastructures

14.1.1 Motivations for Designing a Methodology for Grid Application Porting and Development

Grid application development efforts very often use ad-hoc approaches when porting applications to the Grid. Developers do not follow any suggested methodology and this may result in poorly documented systems that do not fulfill user expectations. These systems are hard to maintain, and their further development or technology-driven upgrade is often impossible. These challenges are even further emphasized when people responsible for a particular application or development leave the organization. Poorly documented and ad-hoc application development efforts are hard to follow up on by the new personnel.

In order to avoid the above trap, and also to support application developers and provide guidelines when porting applications to Desktop Grids and to combined service and Desktop Grid (SG/DG) platforms, the EDGeS Application Development Methodology (EADM) has been specified. The methodology proposes phases of the development process, identifies what aspects the developer team must address in these phases, and suggests tools to be utilized in each phase.

When designing the methodology, several aspects of grid application development and porting have been taken into consideration. The methodology should provide ways of helping identify the problem domain and analyze user requirements. Capturing the needs and requirements of users and providing seamless communication between developers and end users are crucial for the success of the development/porting process. The methodology should also define ways how the application can utilize the infrastructure, and how additional lower- and higher-level tools can be used to port the application to the target grid platform.

Systems development methodologies are widely utilized in software and information systems development to guide systems analysts and developers through the development lifecycle. Some examples for widely used systems analysis and design methodologies include SSADM (Structured Systems Analysis and Design Method), DSDM (Dynamic Systems Development Method), or Checklands Soft Systems Methodology (for more details on systems design methodologies, please see for example G. Curtis and D. Cobham 2005). While these methodologies all differ in the techniques they use and the stages they recommend, they all target generic problems and aim to give a full recipe for systems analysis and design. In the case of the EDGeS Application Development Methodology, the focus is more on specific requirements of porting applications onto DG and SG infrastructures. While the EADM follows the usual stages of the above-mentioned methodologies and recommends widely utilized tools (such as UML diagrams) to aid the process, it concentrates only on specific aspects of the application that may require consideration and modification when porting to the Grid infrastructure.

The EDGeS Application Development Methodology was specifically designed to aid the porting of applications to DG and SG platforms. Following the methodology assures that the necessary questions of application porting are addressed and the design and development process results in an optimum solution for the end user, within the existing technical constraints and limitations, of course. Several applications have already been ported to Desktop Grids, and to combined service and Desktop Grid infrastructures using the EADM. These application porting activities provided valuable feedback and experiences when validating the methodology and defining best practices for its application.

14.1.2 EADM Participants and Roles

EADM distinguishes between five main roles when describing the participants of its stages. Although these roles are different from each other, one particular person or group may play multiple roles in the process (especially in the case of smaller projects and application domains).

- End users: Those who utilize the application in their scientific research or industrial/business conduct. End users may have very different technical skills from grid experts to only basic computational skills. EADM targets this very diverse group with different solutions.

- Developers/system administrators of original application: Either application programmers who developed the current non-grid-enabled version of the application, or in the case of commercial applications, the system administrators who are responsible for the installation and administration of the software.

- Systems analysts: They are responsible for capturing user requirements and for the conceptual design specification to port the application to the target grid platform. The capability to communicate with the previous two groups, and also a full understanding of systems development methodologies and the target desktop and/or service grid platforms are required.

- Application programmers: They are responsible for the implementation of the migration of the application to the target grid platform. In most cases, this group is envisaged as collaboration between developers/experts of the original application and programmers with grid-specific knowledge.

- Grid operators: They are responsible for operating the grid on which the ported applications are running. They should be able to assess the application characteristics, such as security properties and runtime properties for scheduling and monitoring purposes.

14.1.3   EADM Stages and Outcomes

EADM aids application developers throughout the whole lifecycle of application porting, from identification of potential applications to providing support and upgrades for end users. EADM identifies well-defined stages that have a suggested logical order. However, the overall process is in most cases non-linear, allowing revisiting, and revising the results of previous phases at any point. An overview of the EADM stages, participants and outputs is provided in Figure 14.1. A short overview of the methodology is given below. For detailed manual and document templates, please see the EADM Manual [kis08].

14.1.3.1   Analysis of Current Application

The analysis of the current application phase describes the existing application in detail. This phase identifies the target user community, the problem domain, and the typical use cases and functionalities of the system. It also captures technical characteristics, such as the type of computing platform and the way parallelism (if any) is utilized by the current application, data access volume and methods, memory and hard disk usage, programming language, operating system, or security solutions. An Application Description Template has been developed to capture the above information mainly from the operators of the existing application.

Typical questions to be answered at this stage:

- Identification of the target user community:
  - Who are the target user communities?
  - Details of primary contact representing the users.
- Identification of problem domain addressed by the application.
  - What is the application area?
  - What are the limitations of the current application?
- Type of computing platform currently utilized by the application.
  - For example, single computer, local cluster, supercomputer, service grid, Desktop Grid.
- Type of parallelism:
  - What type/level of parallelism is currently supported by the application? (e.g., no parallelism, parameter sweep, master/worker, MPI style inter-process communication).
- Data access:
  - The nature and size of input/output data currently processed by the application (data volume size of input/output files, data sources utilized).
- Functionalities:
  - What are the functionalities currently supported by the system?
  - What are the typical use cases?
- Other factors:
  - Licensing issues (open source, source code available or not).
  - Programming language utilized.
  - Operating system.
  - Memory usage.
  - Security solutions.
  - Ethical or gender issues.

14.1.3.2 Requirements Analysis

The aim of the Requirements Analysis stage is to identify how the target user community will benefit from porting the application to a desktop or service grid platform. The requirements toward the ported application concerning efficiency of execution and data access are analyzed from a user perspective. The target computing platform (either SG or DG) that the user wants to access as an entry point when executing the application, and the desired user interface are also identified in the User Requirement Specification document.

Typical questions to be answered at this stage:

- User requirements:

  - How will the porting of the application enhance the user experience?
  - What does an average user expect from the ported application?
    * Shorter running time
    * Extended range of problems to be solved/analyzed
    * Access more data or in a more convenient way

- Functionalities:

  - Is there any new functionality required from the ported application?
  - Are there any new use cases to be considered?
  - Do any of the current functionalities require modification during the porting process?

- Efficiency of execution and data access:

  - What are the user requirements concerning efficiency of execution?
    * Expected/desired execution time of experiments.
    * Parameter range to be analyzed. (Will and how will the porting of the application extend the parameter range that can be analyzed?)
    * What is the requested speedup compared to the current application?
    * Are there requirements to improve the data access efficiency of the application?
    * Trade-offs between computation and data transfer. How to optimize data transfer for different types of grids?

- Target computing platform:

  - What is the computing platform the user prefers/capable to use as an entry point? Would the user prefer to

      * Submit the application to a DG system that utilizes SG resources at the background
      * Submit the application to SG that utilizes DG resources at the background
      * Submit the application to a platform-independent gateway (portal) that seamlessly distributes the jobs

- Required user interface:

    – What is the preferred user interface to run the application from?
      * Submit the application from a command-line interface
      * Submit the application from a generic portal interface
      * Submit the application from a custom user interface (portal or client-side application)

- Other factors:

    – Is there any need and, if yes, is it possible to change licensing of the application?
    – Will the porting of the application raise any new licensing issues?
    – Is it required to clarify/solve any ethical or gender-related issues before porting the application?
    – Will the porting of the application raise any new ethical or gender issues?
    – Will it be required to run the application on any additional operating systems?
    – What are the security requirements toward the ported application?

14.1.3.3 Systems Design

The Systems Design phase outlines the proposed structure of the system. The target computing platform, the type of user interface, and the parallelization and data access principles are designed, taking both user requirements and technical feasibility into consideration. The outcome of this stage is a Systems Design Specification that identifies at a high-level how the ported application will work regarding the above aspects. The use of structural diagrams, such as a system block diagram and UML diagram techniques, is highly recommended at this stage.

Typical questions to be answered at this stage:

- Target computing platform:

    – Identify target computing platform, taking both user requirements identified and technical feasibility/constraints into consideration. Analyze from the usability and also from the technical feasibility point of view whether the application will be submitted to

* A DG system that utilizes SG at the background
* An SG system that utilizes DG resources at the background
* A platform-independent gateway (portal) that seamlessly distributes the jobs.

- Type of user interface:

  - Identify user interface solutions, taking both user requirements and technical feasibility/constraints into consideration. Analyze from both the usability and the technical feasibility points of view whether the application will be submitted from:
    * A command-line interface
    * A generic portal interface
    * Custom user interface (portal or client-side application)
  - Outline the requirements towards the user interface

- Type of parallelism to be utilized by the new (ported) application:

  - Do we have to change the nature of parallelism utilized by the application in order to run it on the target platform and to fulfill the user requirements identified in Stage 2?
    * Is the change parallelization technically feasible?
    * Is the change parallelization economically feasible (time and budget constraints)?

- What type/level of parallelism would be required to enhance the performance of the application?

  - Is this parallelization technically feasible?
  - Is this parallelization economically feasible?
  - Are the requested enhancements in parallelism supported by the target platform?
  - Outline the suggested parallelization principles for the ported application.

- Data access:

  - Design of what data is needed and how it will be accessed by the ported application.
  - Data transfer: What network speed is required to upload and to download the data? Is it possible to tune this to the available network speed?

- Other factors:

  - Possible design-specific implications of issues identified in this section during previous stages (e.g., licensing, ethical, and gender issues, and operating systems).

14.1.3.4   Detailed Design and Implementation

The Detailed Design stage provides class-level specification of any required modification in the original application when porting it to the target grid platform. This stage results in a Technical Design Specification that forms the basis of the Implementation. EADM recommends specific tools for the detailed design and implementation phases that were specifically created or enhanced to serve desktop and/or service grid migration tasks. These tools include the Distributed Computing (DC-API) [MGBK08], the Generic Wrapper (Gen-Wrapper) [MBK09], and the XtremWeb APIs for application development, an extended version of the WS P-GRADE Grid portal [Kac11] to support the transparent exploitation of SG and DG infrastructures at the workflow level.

Typical questions to be answered at Detailed Design stage:

- Target computing platform:

  - The EADM considers the following main scenarios:
    * The application is registered and used on a Desktop Grid that utilizes service grid resources through the desktop-to-service grid bridge
    * The application is submitted to a service grid and utilizes Desktop Grid resources through the service to Desktop Grid bridge
    * The application is submitted to a platform independent gateway (e.g., the WS P-GRADE grid portal) that seamlessly distributes the jobs to both service and Desktop Grid resources

  - In each scenario the ported application should be capable of utilizing and running on all the target computing platforms.

  - Each scenario can utilize different tools and offers different implementation alternatives.

- Type of user interface:

  - Command line interface: The majority of current SG applications are submitted from a command-line interface directly utilizing the job submission commands and capabilities of the underlying middleware. DG applications are typically developed and registered by native APIs. A large number of users are happy to use these interfaces and would prefer to utilize the ported applications this way. In this case the EADM aims to minimize the changes from the user's point of view when migrating an application. The entry point will be in all cases the platform the user is most familiar with, and the bridging mechanism will be transparent from the users' point of view.

  - Generic portal interface: A growing number of end users, especially those with limited computing and grid-specific knowledge, prefer to run their applications from grid portal interfaces. Users can utilize any of these platforms as entry points to their application and access the infrastructure directly or through the corresponding bridging mechanism.

- Custom user interface: Some applications may come with a custom user interface and end users prefer to utilize this interface when running the ported application on the target platform. Although the EADM does not provide any direct support for these types of applications, it keeps this option open. However, it must be noted that application porting can be significantly more time and resource intensive in this particular case. As a recommendation, the EADM suggests creating a prototype solution considering any of the previous approaches (command line or generic portal) and developing the custom user interface after that.

- Type of parallelism:

  - As the ported application has the potential to run on both SG and DG platforms, the EADM considers two major types of parallelism: master-worker (MW) and parameter sweep (PS) applications. Due to the nature of Desktop Grid systems, applications where intensive inter-process communication is required (e.g., MPI applications) are not considered by the methodology (or at least the MPI parts of the application have to be run on an SG and only the other parts requiring solely MW or PS parallelism can run on the hybrid resources).

- Other factors:

  - Technical details concerning factors listed in this section in previous stages (e.g., licensing, ethical and gender issues, and operating systems).

14.1.3.5 Further Stages of EADM

During Testing, both the functionalities and the performance of the ported application are evaluated and compared to the identified user requirements. The outcome is documented in a test report that is made available for validators and grid system operators.

The aim of the Validation phase is to ensure that the application causes no harm to the computers of Desktop Grid donors, and also that it conforms to the generic aims of the target Desktop Grid system. This stage is inevitable in order to deploy the application on a DG platform where individuals or institutions offer their volunteer resources for the computation. Validation is crucial in the case of Desktop Grid applications as Desktop Grid platforms trust the applications and not the users, when compared to the most widely used security models of service grid systems. Finally, the application is published in an application repository and deployed to be utilized by end users. It is also desirable to give organized support for end users after deployment to support and maintain the application.

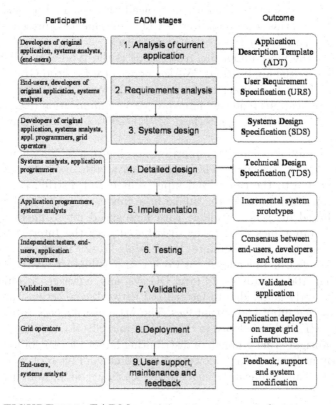

Participants	EADM stages	Outcome
Developers of original application, systems analysts, (end-users)	1. Analysis of current application	Application Description Template (ADT)
End-users, developers of original application, systems analysts	2. Requirements analysis	User Requirement Specification (URS)
Developers of original application, systems analysts, appl. programmers, grid operators	3. Systems design	Systems Design Specification (SDS)
Systems analysts, application programmers	4. Detailed design	Technical Design Specification (TDS)
Application programmers, systems analysts	5. Implementation	Incremental system prototypes
Independent testers, end-users, application programmers	6. Testing	Consensus between end-users, developers and testers
Validation team	7. Validation	Validated application
Grid operators	8.Deployment	Application deployed on target grid infrastructure
End-users, systems analysts	9.User support, maintenance and feedback	Feedback, support and system modification

FIGURE 14.1: EADM participants, stages, and outcomes.

14.2   Requirements toward Applications to Be Run on Desktop Grid Infrastructures

The EDGeS Application Development Methodology, introduced in the previous section, provides a framework for supporting the whole lifecycle of application porting when targeting combined desktop and service grid infrastructures. Combining resources of desktop and service grids significantly increases the number of available resources that applications are able to utilize. Larger numbers of resources could result in higher throughput and better performance. However, using a combined infrastructure also imposes some limitations regarding the applications. In the case of a combined SG/DG infrastructure, the application needs to be executed on both types of grid systems. As service grid infrastructures are capable of running a wider range of applications than Desktop Grids, it is typically the Desktop Grid system that limits the range of target applications. If an application is capable of running on Desktop Grid resources, then it can typically be executed on service grid resources too. In this section an overview of requirements is presented regarding target applications for Desktop Grid platforms.

One of the most important restrictions is the type of parallelization implemented by the applications. Worker nodes of a Desktop Grid are typically isolated, allowing no direct communication between the workers. As a consequence, applications requiring inter-process communication (e.g., applications using the Message Passing Interface MPI [Fos95]) are not suitable for Desktop Grids.

Data handling is another important and limiting factor when running applications on DG resources. The bandwidth available in a public Desktop Grid is limited due to the centralized input source and distribution mechanisms of Desktop Grid middleware. Unlike in a cluster or traditional service grid, where shared disks are available with high-speed interconnects, in a Desktop Grid, input files need to be distributed over the Internet to each and every volunteer node. Therefore, the feasible size of input or output files to be downloaded or uploaded by a worker in a volunteer DG is typically not larger than 100 megabytes. In local DG systems, however, this file size could be significantly larger due to faster local network connections, allowing even gigabyte-sized input and output files. The Attic peer-to-peer data distribution system [AHET11], developed in the EDGeS and EDGI European projects, was built to mitigate this problem of distributing larger files on volunteer Desktop Grids. By utilizing multiple download endpoints and file replicas to distribute data, Attic can significantly increase the feasibility of data-intensive application scenarios.

Confidentiality of data can also be an issue in public DGs. As the data is downloaded to computers of non-trusted volunteers, public DGs may not be able to provide the level of data protection required by many applications. A local Desktop Grid infrastructure could provide a solution for the data confidentiality problem as the data is not passing company or institutional firewalls in this scenario.

Besides the above limiting restrictions, there are also a few generic recommendations that should be applied in order to achieve higher performance. The execution time of individual work units of a DG application need to be well balanced. Also, very short work units, in the range of a few seconds or minutes, could significantly increase the overhead and decrease performance. On the other hand, work units that are running for several hours are likely to be interrupted by the donors. In the case of work units running for a couple of hours or longer, it is essential that the application developer implements internal check-pointing allowing the application to seamlessly resume after user interruption. Finally, as a very large proportion of volunteer computers is running some flavor of the Windows operating system (when compared to the typically Linux-based computing clusters), in most cases it is essential to develop Windows (and/or MacOS)-based client applications.

14.3   Application Scenarios for Combined Desktop and Service Grid Infrastructures

Requirements for porting to and running an application on an SG/DG infrastructure could differ, depending on the primary and secondary target platform. The primary target platform is the one that is directly utilized by the end user, while the secondary platform is used in a user-transparent way utilizing the bridging mechanisms, for example the bridges developed by the European EDGeS and EDGI projects. Based on primary and secondary target computing platforms, three generic user scenarios have been identified and analyzed.

In this section, besides the description of these scenarios, justification for their feasibility and usefulness is given, and illustrated via examples. Differences in the application porting process regarding the different scenarios are also highlighted. This section builds on experiences that were gained through the porting and deployment of over thirty applications to combined desktop and service grid infrastructures within the framework of several European projects (EDGeS, EDGI, and DEGISCO).

14.3.1   Applications Running through the Desktop-to-Service Grid Bridge

In Scenario 1, illustrated in Figure 14.2, Desktop Grid applications can utilize resources from a service grid system, for example from a specific Virtual Organization (VO) of the European Grid Infrastructure (EGI grid), via the desktop-to-service (DG to SG) bridge. The scientific end user in this scenario runs a DG application, and the utilization of SG resources is completely transparent from the users' point of view. The DG-to-SG bridge acts as a powerful Desktop Grid worker, pulling work units to the dedicated service grid virtual organization and sending the results back to the Desktop Grid server.

This scenario is useful when the end user has access to a relatively small local or institutional Desktop Grid system that can significantly be extended with service grid resources. An example of this scenario is the ViSAGE video stream analysis application by Correlation Systems Limited from Israel (ViSAGE 2010). Local users can access the application via a custom user interface and can analyze pairs of frames in a video recording on the local BOINC-based Desktop Grid system. However, the Correlation Systems DG connects approximately twenty local PCs only. Therefore, sending jobs to a specific virtual organization of the European Grid Infrastructure that contains thousands of processors can significantly speed up the computation.

The application porting and deployment process for Scenario 1 is relatively straightforward. If a DG version of the application exists and runs on a public Desktop Grid, then it is very likely that the worker applications are supporting a wide range of Linux flavors including the one applied by the target SG platform. If the application is running on a local Desktop Grid, it may be necessary to develop Linux worker applications that can be submitted to the target SG platform.

Once an executable for the target SG platform has been compiled and tested, the host DG system must be connected to the DG-to-SG bridge. Setting up this connection includes the registration of the local Desktop Grid and the target application with the bridge. The owner of the DG application also requires a valid proxy certificate as the most common mechanism for user authentication in service Grid systems. This certificate is utilized by the bridge when submitting the application to SG resources.

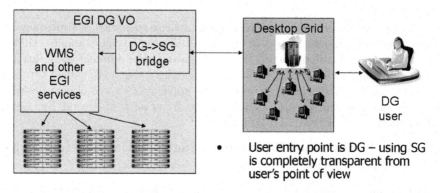

FIGURE 14.2: Applications running from DG to SG.

14.3.2   Applications Running through the SG-to-DG Bridge

Scenario 2 is illustrated in Figure 14.3. The scientific end user in this scenario accesses the SG platform directly and submits the application to service grid resources using the standard job submission mechanism of the target platform. For example, in the case of a gLite [lau06] based grid, the user utilizes a gLite User Interface machine and prepares a suitable job description file for the submission. If the virtual organization where the job has been submitted includes an EDGI-modified computing element, then jobs sent to this computing element will be executed on Desktop Grid resources using the SG-to-DG bridge.

Using Desktop Grid resources could significantly increase the processing power of service grid systems by extending them with potentially millions of worker nodes. As a high percentage of current SG applications are parameter sweeps, these applications can be efficiently redirected to DG resources, freeing up service grid systems for more specific MPI applications, for example.

Several EGI user communities have been supported by the EDGeS and EDGI projects to port and run their applications through the SG-to-DG bridge. An example of these efforts is the VisIVO (Visualization Interface to the Virtual Observatory) application [BCG06]. VisIVO is a suite of software tools for creating customized views of 3D renderings from astrophysical data tables. The application previously run on gLite based service grid systems and has been ported to the SG-to-DG bridge.

Porting an application that utilizes the SG-to-DG bridge starts with the development of a Desktop Grid version of the application. As the application is submitted from the service grid, this side is responsible for the creation and orchestration of parameter sweep jobs or work units. It is only a Desktop Grid client application that needs to be developed. However, as it was analyzed in a previous section, this client is ideally compiled to different target platforms including Windows and MAC OS.

The development of the client is supported by several high-level tools and APIs. These tools include the Distributed Computing API (DC-API) that enables us to execute the ported application on multiple Desktop Grid middleware (e.g., BOINC, XtremWeb or Condor) [MGBK08] without any modification, and the Generic Wrapper (GenWrapper) tool that facilitates the porting of legacy applications onto BOINC-based Desktop Grid platforms [MGBK08].

Once the application is ported to the target DG platform, it must be validated. Validation is required due to the different security models of SG and DG platforms. While SGs trust the user and require certificates, DG systems trust the application. The validation process aims to ensure that the validated application is causing no harm to the Desktop Grid donors.

Once validated, the application is published in the publicly available EDGI Application Repository (EDGI AR). This repository is accessed by three different types of actors. Desktop Grid administrators can browse the repository and download the DG version of the applications that they aim to support. Service grid users can find and download the SG version of the application. Finally, the SG-to-DG bridge uses the reference provided by the user to one of the deployed applications in the repository, and enables the bridging of this application to the target DG systems.

14.3.3 Applications Using Specific Job Submission or Scheduling Systems to Utilize to SG/DG Resources

There are use case scenarios when the target application already utilizes specific high-level user environments or lower-level middleware and job submission frameworks. In these cases, application developers and end users may wish to exploit both desktop and service grid resources without compromising the current user experience. Exploiting the EDGI bridges directly may not be suitable in these scenarios because of some preliminary assumptions regarding the way SG or DG jobs are submitted.

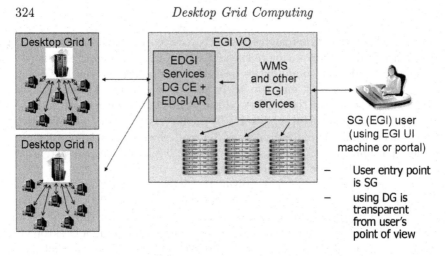

FIGURE 14.3: Applications running from SG to DG.

In addition to the production bridges, EDGI also provides building blocks and components of these bridging solutions that could be integrated into other grid middleware or user environments. This integration ensures that the user community of the target middleware or user environment will utilize the combined SG/DG infrastructure in a transparent way without changing the frameworks they are accustomed to.

An example of this scenario is illustrated in Figure 14.4. The Wisdom production environment [Jac08] has been extended with a DG submitter module. The WISDOM project is an international initiative to enable a virtual screening pipeline on a grid infrastructure. The project has developed its own meta-middleware that utilizes the EGI production infrastructure, and capable to submit and execute very large number of jobs on EGI resources using a pilot job submission mechanism. The newly developed Desktop Grid submitter uses EDGI components, such as the 3G Bridge and its WS Submitter [FKS08] to access Desktop Grid Resources. The DG submitter pulls WISDOM jobs from the WISDOM task manager exactly the same way as in the case of EGI jobs. Therefore, the Desktop Grid submission did not require modification of the original WISDOM architecture, and is completely transparent from the end-user's point of view.

Application porting in this case includes the implementation and validation of a Desktop Grid version of the application, similar to scenario 2. Instead of the EDGI application repository, the middleware may use its own repositories to store and submit applications and work units. In the presented example, the WISDOM Job and Task managers are responsible for this functionality.

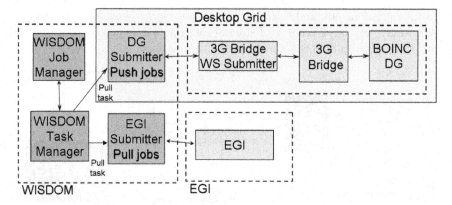

FIGURE 14.4: Extending the WISDOM environment with a DG submitter.

14.4 Molecular Docking Simulations: Application Case Study

14.4.1 Modeling Carbohydrate Recognition

Carbohydrate recognition is a phenomenon critical to a number of biological functions in humans, including highly specific responses of the immune system to pathogens, and the interaction and communication between cells. This type of study could enable bio-scientists to understand how pathogens bind to cell surface proteins and aid in the design of carbohydrate-based vaccines and diagnostic and therapeutic agents. Computer-based simulation programs can successfully aid the study of such complex processes, reducing the cost of wet laboratory experiments and improving time efficiency.

The ProSim (Protein Molecule Simulation on the Grid) [KGH⁺10] project has successfully created a grid-based parameter sweep workflow for the above described scenario using the WS P-GRADE portal [Kac11]. The workflow realizes a complex user scenario, as illustrated in Figure 14.5. First, the receptor protein and the glycan are modeled separately and energy minimized with the help of the Gromacs [LHvdS01] molecular dynamics software (represented by phases 1 and 2, respectively, in the figure). In Phase 3, the putative active binding sites involved in the recognition need to be identified using docking simulation software, such as AutoDock [MGH⁺98] (Phase 3). Finally, molecular dynamics simulations are applied again to dissect the mechanism of recognition (phase 4).

Phase 3 of the above workflow uses the AutoDock molecule docking simulation software to identify the putative active binding sites involved in the recognition based on a genetic search algorithm. The efficiency of this docking phase can be further enhanced by randomly executing a large number of docking experiments, and selecting the lowest energy level solutions out of a potentially very large number of simulation runs. These random dockings are perfect target candidates for a Desktop Grid platform.

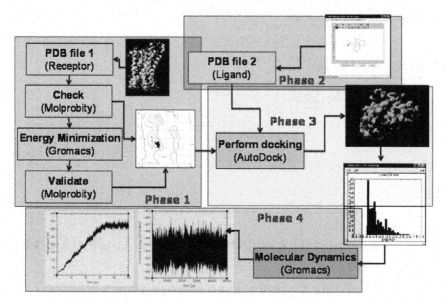

FIGURE 14.5: User scenario for modelling receptor–ligand interaction.

In order to achieve the above objective, AutoDock simulations must be executed multiple times, in the range of thousands or tens of thousands, with the same input files. One simulation run takes minutes or hours on a stand-alone PC, depending on the complexity of the molecules. However, these simulation runs are independent of each other and can potentially utilize a very large number of computing resources where docking simulations are executed on different nodes at the same time. Using a large number of resources significantly reduces the overall execution time, allowing bio-scientists to analyze and compare much larger numbers of scenarios within the same timeframe. A combined desktop and service grid infrastructure is an ideal candidate for providing the very large amount of resources required for the docking phase of the ProSim workflow.

14.4.2   Docking Simulations on Desktop and Service Grid Resources

The original ProSim workflow uses resources from the UK National Grid Service (NGS) to execute its different components. Therefore, in order to run AutoDock simulations on the combined SG/DG platform, the application needed to be ported to a Desktop Grid, and also required to be tested and executed on EGI resources.

Three scenarios are presented in this section. In the first scenario, the application is executed on a BOINC Desktop Grid platform that utilizes EGI resources via the BOINC-to-EGI bridge. In this scenario, a BOINC master application orchestrates the execution of work units on the DG. In the second scenario, a parameter sweep workflow in the WS P-GRADE portal replaces the DG master application when executing the work units on the Desktop Grid. Finally, the AutoDock simulations are submitted to the EGI grid and utilize DG resources via the EGI-to-BOINC bridge.

14.4.2.1 AutoDock as a BOINC Master Worker Application

In this first scenario, the AutoDock application has been ported to the BOINC Desktop grid platform as a stand-alone, traditional BOINC application. The aim of this work was to investigate the suitability of AutoDock to a DG and also to the Desktop Grid-to-EGI bridge.

The AutoDock software suite consists of two main components. AutoDock performs the docking of the ligand to a set of grids describing the target protein, while AutoGrid pre calculates these grids. AutoGrid must be executed only once in an experiment, and finishes rather quickly. However, as described earlier, AutoDock requires very large amounts of computation-intensive runs.

When porting the application to BOINC, a BOINC master has been implemented that deals with the DG-specific problems, such as work unit generation, distribution, validation and assimilation. AutoGrid is also run by the master application on the DG server machine. The worker applications process the individual work units and produce a docking log file that contains the results of the computation. These log files are then bundled into a single archive by the master to ease further processing. The implementation uses the DC-API to create a wrapper application around AutoDock.

In order to send work units through the BOINC-to-EGI bridge, Linux versions (32 bit, 64 bit) of the worker application had to be developed too. These Linux versions use only statically compiled libraries, as they have to run on different machines with different configurations. The application has been deployed on the University of Westminster Local Desktop Grid (WLDG), and has also been registered to the production BOINC-to-EGI bridge. After configuration, the bridge pulls AutoDock work units from the BOINC server and executes them in the EGI Desktop Grid Virtual Organisation in a user-transparent way.

Several performance tests have been carried out to determine the potential speed-up provided by the Desktop Grid-based implementation. The single processor performance of a test PC was compared to a much larger but non-deterministic number of computers (a maximum of 1,600 PCs) used in the Desktop Grid experiment. The test work units used the same set of input and docking parameter files, while the number of work units was increased continuously to see how the application scales. Each experiment was repeated several times, and the medians of the measured values were calculated. The best test results were provided in case of larger numbers of work units. In the case of 3, 000 work units, for example, the experiment achieved 190 times speedup using 469 nodes.

Implementing this first scenario proved that using Desktop Grid resources significantly improves the performance of the application, especially in the case of very large experiments with thousands of work units. On the other hand, the integration of this native BOINC implementation to the ProSim Workflow would have been rather difficult. Therefore, it needed to be modified and executed from the same WS P-GRADE environment where other parts of the workflow are implemented.

14.4.2.2   AutoDock as a BOINC Application in the WS P-GRADE Portal

As the ProSim workflow is implemented in the WS P-GRADE portal, the Desktop Grid-based docking simulations also needed to be ported to this environment. The WS P-GRADE portal, the latest incarnation of the P-GRADE grid portal family, supports the execution of parameter sweep workflows not only in service, but also in BOINC-based Desktop Grids.

When porting the application to the portal, the original DG worker application, based on the DC-API, could be used without any modification. However, the master component of the application is replaced with a Generator and a Collector job of the P-GRADE workflow. The Generator is responsible for the execution of AutoGrid and for the creation of the work units. These work units are submitted to the DG server by the second job in the portal. This job runs as many times as many work units are specified by the portal user. Once the jobs have finished, the portal is responsible for retrieving the results from the DG server. Finally, the Collector compresses the outputs of all simulations and makes this compressed file downloadable for the end user. If the target Desktop Grid, in our case the WLDG, is connected to the BOINC-to-EGI bridge, then work units submitted from the portal can cross the bridge exactly the same way as was described previously. Figure 14.6 illustrates how a WS P-GRADE portal user can execute AutoDock simulations on combined desktop and service grid resources.

This solution is now suitable for integration to the ProSim workflow. This integration is simple from a technical point of view as the AutoDock workflow can easily be embedded into the more complex ProSim workflow. However, complexity arises when analyzing the usability of the solution and the quality of the results from the end user's perspective. The brute-force Monte-Carlo simulations, implemented in the DG version, seem to provide less reliable results than the evolutionary algorithm currently applied in ProSim. Therefore, simply changing the Phase 3 docking job to the Desktop Grid-based version would, in most cases, provide less accurate result,, even for a very large number of iterations. The DG solution could increase the efficiency of the docking phase but only if it was used as a supplementary solution for the evolutionary algorithm.

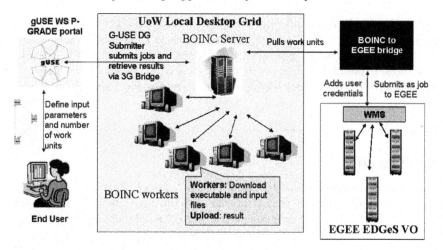

FIGURE 14.6: Running AutoDock on the WLDG and through the BOINC-to-EGI bridge from the WS-P0GRADE portal.

14.4.2.3 Running AutoDock on EGI Resources Supported by Desktop Grids

The third scenario considers the EGI grid as the primary target platform for the application and uses DG resources via the EGI-to-DG bridge. This solution is less relevant from the ProSim workflow's perspective. As WS P-GRADE is capable of submitting to Desktop Grid resources directly, it is more efficient to use BOINC resources as the primary target platform than send the jobs through the bridge. However, AutoDock is widely used by several EGI user communities, for example, in the WISDOM project. Making AutoDock available to be executed in the EGI-to-DG direction has importance for a large number of potential users. To make AutoDock available for EGI users, the application had to be validated and placed into the EDGI application repository.

The AutoDock application can be executed via the EGI-to-DG bridge from an EGI VO if that VO recognizes a specific computing element that includes the bridge client. AutoDock jobs can currently be submitted to the EGI Desktop Grid VO and supported by the WLDG.

14.5 Conclusion

This chapter gave an overview of challenges and possible solutions when porting or migrating applications to Desktop Grids, or to combined desktop and service grid platforms. The EDGeS Application Development Methodology has been introduced as an application porting methodology specifically supporting application porting activities to desktop and service grid platforms. Requirements for typical Desktop Grid applications have been analyzed and scenarios for executing applications on desktop and service grid platforms have been explained. Finally, a case study has been presented, explaining the applicability of tools, methodologies, and scenarios for a specific application domain.

Bibliography

[AHET11] Ian Kelley Abdel Hamid Elwaer, Andrew Harrison and Ian Taylor. Attic: A case study for distributing data in boinc projects. *Parallel and Distributed Processing Workshops and PhD Forum, 2011 IEEE International Symposium on*, 0:1863–1870, 2011.

[BCG06] U. Becciani, M. Comparato, and C. Gheller. Visivo: a vo enabled tool for scientific visualization and data analysis. *Memorie della Societa Astronomica Italiana Supplementi*, 9:427, 2006.

[CC05] G. Curtis and D. Cobham. *Business information systems: analysis, design and practice*. Pearson Publication Company, 2005.

[FKS08] Z. Farkas, P. Kacsuk, and M.R. Solar. Utilizing the egee infrastructure for desktop grids. *Distributed and Parallel Systems*, pages 27–35, 2008.

[Fos95] I. Foster. *Designing and building parallel programs: concepts and tools for parallel software engineering*, chapter 8. Addison-Wesley, 1995.

[Jac08] Salzemann J. Jacq F. Legr Y. Medernach E. Montagnat J. Maa A. Reichstadt M. Schwichtenberg H. Sridhar M. Kasam V. Zimmermann M. Hofmann M. Breton V. Jacq, N. Grid-enabled virtual screening against malaria. *Journal of Grid Computing*, 6(1):29–43, 2008.

[Kac11] P. Kacsuk. P-grade portal family for grid infrastructures. *Concurrency and Computation: Practice and Experience*, 23(3), 2011.

[KGH⁺10] T. Kiss, P. Greenwell, H. Heindl, G. Terstyanszky, and N. Weingarten. Parameter sweep workflows for modelling carbohydrate recognition. *Journal of Grid Computing*, pages 1–15, 2010.

[kis08] *EDGeS Application, Development, Methodology - EADM Manual*, 2008.

[lau06] Programming the grid with glite. *Computational Methods in Science and Technology 12*, (1):33–45, 2006.

[LHvdS01] E. Lindahl, B. Hess, and D. van der Spoel. Gromacs 3.0: a package for molecular simulation and trajectory analysis. *Journal of Molecular Modeling*, 7(8):306–317, 2001.

[MBK09] A.C. Marosi, Z. Balaton, and P. Kacsuk. Genwrapper: A generic wrapper for running legacy applications on desktop grids. 2009.

[MGBK08] A.C. Marosi, G. Gombás, Z. Balaton, and P. Kacsuk. Enabling java applications for boinc with dc-api. *Distributed and Parallel Systems*, pages 3–12, 2008.

[MGH+98] G.M. Morris, D.S. Goodsell, R.S. Halliday, R. Huey, W.E. Hart, R.K. Belew, and A.J. Olson. Automated docking using a lamarckian genetic algorithm and an empirical binding free energy function. *Journal of Computational Chemistry*, 19(14):1639–1662, 1998.

[vis10] *Video Stream Analysis in a Grid Environment*, Visage, 2010.

Chapter 15

Network Awareness in Volunteer Networks

Jon B. Weissman

University of Minnesota, Twin Cities, Minnesota, USA

Jinoh Kim

Lawrence Berkeley National Laboratory, Berkeley, California, USA

15.1 Introduction

In this chapter, we focus on the issue of network performance in volunteer networks. For some applications, communication performance plays a nontrivial role in the overall performance. It is this class of applications that motivates our work. Network performance can become important when peers or volunteers are interacting with each other and/or with other nodes, for example, servers or external nodes. We focus on both techniques to estimate network performance efficiently and how to use such performance in a variety of volunteer computing scenarios. We first consider how time-bounded execution of data-intensive services can be achieved across volunteers using a form of communication makespan estimation relying on application-specific communication measurements. This problem reduces to a server selection problem where the data is replicated across a set of servers. Next, we consider how second-hand communication information can be estimated and propagated to peers through the OPEN framework. Finally, we describe a tool for improving communications in volunteer networks, the network dashboard. This tool can identify superior application-level network routing paths in volunteer networks based on a variety of different metrics. We present results that indicate that superior paths in terms of TCP/UDP bandwidth, delay, and jitter can be found.

15.2 Communication Makespan

A fundamental challenge for the deployment of services such as BLAST (Basic Local Alignment Search Tool) [bla, HFTC09] in large-scale computing infrastructures is the efficient distribution and dissemination of data to the computation nodes; for example, decomposing a BLAST query across a grid typically requires that large databases (with sizes on the order of several gigabytes) be split up and sent to a large number of compute nodes to enable fast parallel execution. Such a requirement makes efficient data download crucial for the success of end-to-end computation.

In this section, we consider the problem of concurrent downloading by a number of compute clients working on the same service request. This challenge is complicated by the extreme time-varying *heterogeneity* of large-scale systems, where data servers have widely different capacity, bandwidth, and latency with respect to a downloading client. Simultaneous downloading from central data servers can lead to bottlenecks, due to capacity and geographic constraints. Because worker nodes can be dispersed worldwide, the download times of some distant and poorly connected nodes might overwhelm the overall execution time of the service request.

Communication makespan is a metric to measure performance of collective data access. Similar to the metric *makespan*, referring to the overall execution elapsed time of a group of tasks, communication makespan refers to the download time of the slowest node in the computation. Minimizing the makespan is a challenge, due to the heterogeneity of the data servers and the possibility of communication load imbalance (if large numbers of concurrent workers happen to pick the same data server). In this setting, simple strategies, such as minimizing round-trip time, do not work well.

We investigate this problem in the context of two distributed computing infrastructures: BOINC (Berkeley Open Infrastructure for Network Computing) [And04], a compute network, and Pastry [RD01], a data network. BOINC is a pull-based system upon which SETI@home was based. In our context, compute nodes pull the distributed work associated with service requests. The compute nodes then retrieve the needed data files from the Pastry network, a peer-to-peer DHT-based storage system. In a data network, it is assumed that the data are highly replicated across a data network and that clients make

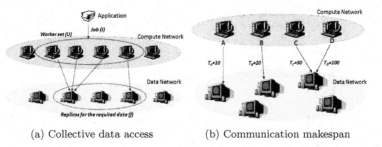

(a) Collective data access   (b) Communication makespan

FIGURE 15.1: Collective data access and communication makespan.

local decisions to select a server for download. In this section, we first define communication makespan and then analyze server selection heuristics that can address the dynamic and heterogeneous nature of the grid environment.

15.2.1   Definition of Communication Makespan

As mentioned earlier, we assume a collective data access environment. All of the data objects required for computation are assumed to be replicated across multiple servers in the data network. As shown in Figure 15.1a, the application submits a job (J) to a set of worker nodes $(U \subseteq W)$, each of which then attempts to download the associated data object (f) from one of its replicas. The submission of the job would be system specific, for example, by using a central scheduler or any distributed manners. To download the data object, each worker node $u_i \in U$ queries the data network for a set of replicated servers $(R \subseteq S)$ holding the associated data, along with their current state. The server state might include attributes such as the server capacity and its round-trip latency from the worker node, among others. In response to the query, the data network returns the replica set to the worker node. The worker node then uses a *server selection heuristic* to select a server from the replica set for the actual download.

Minimizing the *makespan* is key, as the service request will not be complete until all tasks are finished. Because data download is a key component of the job execution time, we define the *communication makespan* to be the maximal download time for job J:

$$makespan(J) = \max_{u_i \in U}(cost(u_i, r_i)) \qquad (15.1)$$

Here, r_i is one of the replicated servers chosen by u_i to download the data (i.e., $r_i \in R$), while *cost* refers to communication cost, such as downloading time.

Figure 15.1b shows an example of the communication makespan. In this example, four worker nodes need to download files for their computation work. Although worker nodes A, B, C complete downloading early within 30 time

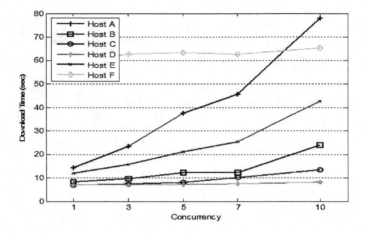

FIGURE 15.2: Heterogeneity of servers.

units, the communication makespan becomes 200 time units due to slow downloading by worker node D. This one late communication could affect overall job completion, particularly for applications relying on a collective performance metric.

We next discuss how to reduce the communication makespan by selecting "good" data servers. A challenge is that the individual compute workers are distributed and isolated from each other. Collecting global state dynamically to improve server selection is neither scalable nor practical. On the other hand, a greedy server selection technique might choose the best server for each node locally without consideration of the other workers. Figure 15.2 shows how such a greedy approach might degrade the download performance of servers by increasing the concurrency of downloads. This experiment uses a set of nodes in PlanetLab [pla]. Another point to be noted from this graph is the heterogeneity of nodes in PlanetLab—each server has a different level of sensitivity with respect to concurrent downloading requests, indicating the difference in their capacities. We introduce an algorithm for server selection that incorporates server heterogeneity to do local server selection while avoiding poor global decisions next.

15.2.2 Server Selection Heuristics

We first investigate different metrics that affect the efficiency of data downloading. To explore metrics that can potentially affect collective performance, we conducted experiments with forty-three nodes in PlanetLab to determine the various parameters that affect download performance. Several measures are explored, and we find strong correlations not only between round-trip time (RTT) and download performance, but also between network bandwidth and download performance. RTT is gathered from our deployed data network,

Pastry [RD01], while Iperf [ipe] statistics are used to determine network bandwidth. We made the following important observations:

- *Observation 1:* In the case of RTT, the vast majority of data download times for each data size are *lower-bounded* by a linear curve, indicating the presence of a near-linear relationship to RTT. However, the variation in the observed download times suggests the impact of other parameters.

- *Observation 2:* In the case of bandwidth, we observed that the *lower bound* on the download times for each data size has an exponential relationship to bandwidth. In other words, servers with fairly large bandwidth (e.g., those over 10Mbps) do not show considerable difference among their download time trends, while low-bandwidth servers (e.g., those under 1Mbps) show a sharp increase in the download time as the bandwidth decreases. However, again, the variation in the observed download times suggests the impact of other parameters.

- *Observation 3:* We also observed that system load and concurrency are correlated to download time (the effect of concurrency is illustrated in Figure 15.2), but we did not find any correlation to other parameters such as CPU power, size of memory, etc. These factors may impact the performance if too many concurrent downloads occur from the same server simultaneously. Such concurrency may happen due to race conditions, where independent workers making independent download decisions might select the same "desirable" server, in turn overloading it. Such overloading should be avoided to minimize the communication makespan.

Based on these observations, we gain the following insights into making server selection:

- Servers with low bandwidth (e.g., under 1Mbps) should be avoided, even if their RTT is small.

- Servers with relatively high-bandwidth (e.g., over 10Mbps) should be preferred, and should use RTT as a discriminator.

- Servers with medium-bandwidth (e.g., between 1–10Mbps) should be discriminated by load or concurrency.

We use these insights to derive a *cost function* that is used by a worker i to quantify the desirability of a server j for data download:

$$cost(i, j) = \alpha_j \cdot rtt(i, j), \qquad (15.2)$$

α_j is a weight used to incorporate other server parameters, defined as follows:

$$\alpha_j = e^{(k_j/bw_j)}, \qquad (15.3)$$

where, bw_j is the bandwidth of the server, and k_j is a (server-dependent) constant that incorporates parameters such as load and concurrency, as discussed next.

This cost function has the following desired properties based on our observations. First, the cost function is proportional to RTT (Observation 1), such that the proportionality constant is the weight α_j, which incorporates the effect of other server parameters. Second, the cost function has an exponential relation to the server bandwidth (Observation 2). Finally, we define the constant k_j to incorporate factors, such as load and concurrency (Observation 3). Note that the values returned by the cost function are not meant to be absolute (i.e., these values are not used for predicting the actual download times), but their relative values can be used for ranking multiple servers in the order of their selection desirability.

We define three heuristics for server selection that use different values for k_j:

- BW-ONLY: Uses $k_j = constant$. We use $k_j = 1$ in our experiments.

- BW-LOAD: Uses $k_j = load_j$, where $load_j$ is the 5-minute average system load on the server.

- BW-CAND: Uses $k_j = num\_response_j$, where $num\_response_j$ is the number of times the servers has responded as a replica server within the past 15 seconds.

The heuristic BW-ONLY uses only the RTT and the bandwidth metrics for selecting a server, while the other heuristics BW-LOAD and BW-CAND also use average system load and concurrency information, respectively. For BW-LOAD, we use a 5-minute system load as the load metric, which is obtained by the Linux `uptime` command. As the load value grows, the weight becomes large, and the predicted download cost goes up. BW-CAND uses the number of times the server has responded as a replica within a predetermined time window. In the experiments, we set the time window to 15 seconds, which is equal to the search time we used in the DHT ring. Using the heuristic BW-CAND, servers that have responded as a replica several times recently are penalized because they are more likely to be selected by multiple workers, and tend to be concurrently serving data in the near future.

15.2.3   Performance Evaluation

To evaluate the server selection heuristics, we conducted live experiments on PlanetLab with a set of randomly selected nodes. For data replication and download, we implemented a data network over FreePastry [fre], a public Pastry implementation developed by Rice University. We conducted each of our experiments as follows: (1) data files are distributed over the data network at the beginning of each experiment; then (2) data queries are generated for downloading these data files; and then (3) for each data query, a set of worker nodes is selected randomly to request the same designated file concurrently.

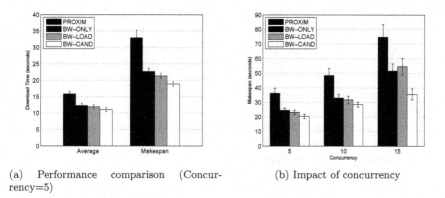

(a) Performance comparison (Concurrency=5)

(b) Impact of concurrency

FIGURE 15.3: Experimental results.

We used 2 MB data, and placed ten replicas for each data item. For fair comparison across the different server selection heuristics, queries are interleaved: for example, each set of worker nodes downloads the files first with the PROXIM (latency-based heuristic) selection, followed by the BW-ONLY selection, etc. Some queries might fail due to reasons such as churn (e.g., nodes going down) or query incompletion (e.g., message routing failure in the DHT ring). If any query fails in the interleaved set of queries, the result is discarded in our analysis.

Figure 15.3a compares the server selection heuristics for concurrency=5. using the aggregated results of all the experiments that used the same concurrency and data size. The figure plots the average download time and communication makespan, respectively, for the various heuristics. The first observation we make from the figure is that the bandwidth-based heuristics perform much better than latency-based server selection in terms of both the average, as well as the makespan. In the figure, we can see that the gaps in performance are greater in the case of the makespan (~30–45%) than in the mean download time (~20–30%).

To see the impact of concurrent downloads for the same files, we used concurrency values of 10 and 15 in addition to the value of 5 used in the above experiment. Because the replication factor for data placement is set to 10, race conditions would be unavoidable in this experiment with clients selecting the same server for download in several cases. Figure 15.3b shows the results in such diverse concurrent downloading environments. In the figure, we can see that the bandwidth-based heuristics consistently outperform latency-based techniques. Moreover, we see that as the concurrency increases, BW-CAND starts outperforming the other heuristics, indicating that avoiding overloads by reducing concurrent data downloads from the same server is important.

In sum, this section focused on the *server selection problem* in collective data access environments: how do individual nodes select a server for downloading data to minimize the *communication makespan*—the maximal download time for a data file? The communication makespan is an important measure because the successful completion of jobs is driven by the efficiency of *collective data download* across compute nodes, and not only the individual download times. Server selection heuristics that incorporate several metrics, such as server bandwidth, load, and download concurrency, improve performance compared to conventional latency-based server selection by at least 30% for communication makespan.

In the next section, we deal with individual communication performance, and introduce a framework that provides scalable network performance estimation based on a passive approach.

15.3   OPEN Framework

Large-scale computing infrastructures such as grids [TTL05, glo] and desktop grids [KCC04, LZZ+04, AF06] are attractive due to their scalability and cost-effectiveness. However, the loosely coupled nature of many of these platforms often makes them unpredictable in their resource availability and performance, particularly in terms of data access. Despite their rich set of computational resources, the unpredictable nature of large-scale computing platforms makes it hard to deploy such data-intensive applications or limits the size of data access, making them inefficient to deploy. Thus, providing predictability in data access is a vital requirement for enabling such data-intensive tasks on large-scale systems.

A key requirement for achieving data access predictability is the ability to estimate network performance for data transfer, so that computation tasks can take advantage of the estimation in their deployment or data source selection. In other words, network performance estimation can provide a helpful guide to run data-intensive tasks in such unpredictable infrastructures having a high degree of variability in terms of data access.

Active probing can be an option for estimation, but is unscalable and expensive in using back-to-back measurement packets. Passive estimation is attractive for its relatively small overhead, and thus could be desirable for many networked applications that do not require an extremely high degree of accuracy such as that needed by network-level applications like network planning. For example, a substantial number of networked applications, such as Web server selection and peer selection for file sharing, rely on *ranking*. According to a peer-to-peer measurement study in [NhCR+03], the second-place peer performance is only 73% of the best peer performance. This significant gap implies that some degree of estimation inaccuracy would be tolerable for such ranking-based applications. A potential problem of passive estimation is that it can suffer from estimation failure due to the unavailability of past measurements. This problem can be mitigated by sharing measurements among nodes; thus, a node can estimate performance even against a server it has never

TABLE 15.1: Degree of Measurement Sharing

Degree	Non-sharing	Pair-level	Domain-level	System-wide
Approach	On-demand measurement	Statistical estimation Time-series forecast	Sharing in a LAN Sharing in a domain	Sharing in a system
System/ Technique	Pathchar [Dow99] Packet pairs [Kes95] bprobe/cprobe [CC97]	NWS [WSH99] HB prediction [HDA05]	SPAND [SSK97] Webmapper [ASS+02]	OPEN

contacted. In previous work, such as [SSK97, ASS+02], however, the sharing was restricted to specific underlying topologies such as a local network, thus limiting scalability.

To realize this goal, there are two important challenges. The first challenge is the *characterization* of a node in terms of its data access capability to enable it to utilize others' measurements for its own estimation. This characterization is key for topology-independent utilization of second-hand measurements. The other important challenge is how to facilitate local measurements to be globally available to other nodes in the system for system-wide sharing. Any server-based techniques for storing global information are limited by well-known problems of scalability and fault tolerance. At the other end of the spectrum is flooding-based dissemination, which while fully distributed, has high network overhead.

In this section, we introduce *OPEN (Overlay Passive Estimation of Network performance)*, a scalable framework for end-to-end network performance estimation. OPEN provides a correlation-based second-hand estimation with empirical node characterization and *proactive dissemination* of measurements with limited overhead by diverse optimizations. We first compare second-hand estimation with existing estimation approaches, and then introduce the OPEN framework, followed by the experimental results for evaluation.

15.3.1 Secondhand Estimation

We classify estimation techniques into several categories, based on the *degree of measurement sharing* for their estimation: *pair-level, domain-level*, and *system-wide*, in addition to the non-sharing model, as summarized in Table 15.1.

Pair-level sharing only utilizes the direct (*firsthand*) measurements made by a *specific pair* of nodes for their network path estimation. Many statistical or time-series forecasting techniques, such as exponential moving average, belong to this class. Previous studies, such as [WSH99], showed the high accuracy of these techniques, but this class requires $O(n^2)$ measurements for estimation between all pairs for an n-node system.

In contrast, some estimation techniques enable nodes to utilize indirect (*secondhand*) measurements provided by other nodes for their own estimation. In domain-level sharing, past measurements in a domain (e.g., a single network) are shared between nodes in the domain. In SPAND [SSK97], nodes in a single network share past measurements for Web server selection. Webmapper [ASS+02] shares passive measurements to select a Web server based on a logical group clustered by IP prefixes. By sharing the measurements in a domain, it is possible to estimate performance if any node in the domain has communicated with the server. Again, however, the sharing is restricted to a single domain. In addition, the underlying assumption of existing techniques belonging to this class is that the nodes in a domain have closely similar characteristics in network access. If this is not the case, sharing measurements without considering node characteristics may cause inaccuracy in estimation.

Unlike the above two classes of sharing, system-wide sharing has no constraints on sharing measurements across the system. In other words, if any measurement against a server is available in the system, any other node can utilize that information for its own estimation to that server. Thus it is possible to perform any-pair estimation with $O(n)$ measurements. Because it does not rely on topological similarities, node characterization is essential to utilize others' experiences. In addition, efficient sharing is also a key for this approach. We next discuss how OPEN realizes those key functions.

15.3.2 OPEN Framework

The OPEN framework provides functionality for passive estimation based on system-wide sharing of measurements. OPEN consists of two core mechanisms, *secondhand estimation* of end-to-end network performance and *proactive dissemination* of observed measurements, described next in detail.

15.3.2.1 Secondhand Estimation Method

OPEN utilizes two types of measurements for estimation for a node: *local measurements* ($\mathcal{L}$) measured directly by the node, and *imported measurements* ($\mathcal{I}$) obtained from other nodes. OPEN makes an estimation by comparing and combining the node capability in data access from its local measurements to the imported measurements of other nodes, as will be discussed. Table 15.2 summarizes the attributes defined in the local and imported measurement records. We use *dot(.)* notation to refer to each attribute in the record, and the cardinality (e.g., $|\mathcal{L}|$ and $|\mathcal{I}|$) represents the number of records stored in the node.

We first characterize a node based on the local measurements at a node. *Download power (DP)* is a quantitative metric empirically formulated to characterize a node (h) based on prior local data access behaviors, and computed as follows:

$$DP_h = \frac{1}{|\mathcal{L}_h|} \sum_{i=1}^{|\mathcal{L}_h|} (\mathcal{L}_h^i.throughput \times \mathcal{L}_h^i.distance) \qquad (15.4)$$

The above equation implies that a greater throughput and distance pro-

TABLE 15.2: Attributes of Measurement

Attribute	Description	Which Record
id	Unique ID	Both $(\mathcal{L}, \mathcal{I})$
client	Measurement node	Imported $(\mathcal{I})$
server	Data server	Both $(\mathcal{L}, \mathcal{I})$
distance	Distance to server	Both $(\mathcal{L}, \mathcal{I})$
throughput	Measurement throughput	Both $(\mathcal{L}, \mathcal{I})$
DP	Download power	Imported $(\mathcal{I})$
timestamp	Time stamp	Both $(\mathcal{L}, \mathcal{I})$

duce a greater characterized metric. Thus the node with a greater DP is considered better in data access capability. Intuitively, throughput has a degree of correlation with distance; for example, we could expect a higher throughput for a closer server if all other conditions are the same. By combining those two factors in Equation (15.4), the download power provides a node capability in terms of data access independent of the distance factor; thus we can compare this metric without considering the distance to server. As seen from Equation (15.4), $DP \propto throughput$ captures how fast a node can download data in general. Further, we also have $DP \propto distance\ to\ the\ server$, which implies that for the same download speed to a server, the download power of a node is considered higher if it is more distant from the server.

With the characterized metric DP, we compute the similarity between two nodes i and j by the following equation:

$$\mathcal{S}(i, j, s) = \frac{DP_i}{DP_j} \cdot \frac{distance(j, s)}{distance(i, s)} \tag{15.5}$$

The scaling factor $\mathcal{S}$ is used to compare the download characteristics of any two unrelated nodes in the system to enable the appropriate scaling of imported measurements for estimation. $\mathcal{S}(i, j, s) = 1$ means that two nodes i and j are exactly the same with respect to data retrieval from server s. If the scale value $= 2$, it means that the node has a factor of two capability in accessing a given server. Hence, $\mathcal{S}(i, j, s) < 1$ indicates that node i is inferior to node j in accessing server s, and vice versa. In the equation, $distance(a, b)$ refers to the distance metric from a to b, and we empirically set $distance(a, b) = \sqrt{RTT(a, b)}$.

Based on the scaling factor, OPEN produces the estimate by utilizing the imported measurements from nodes to the same server. The following equation is used to estimate the expected throughput (T) between a node (h) and a server (s) with an imported measurement m:

$$T_{h,s}(m) = \mathcal{S}(h, m.client, s) \times m.throughput \tag{15.6}$$

It is possible that there exist multiple imported measurements to the same server. To combine multiple estimates in such cases, we investigated a rich set of statistical techniques and observed that the median performs best.

Analytically, OPEN estimation is based on a property of *proportionality of estimates* with different measures to an identical server for any distinct node. For simplicity, we define $f_k(m) = T_{k,m.server}(m)$, as the secondhand estimate is a function of a measure. According to the property, if $f_k(m_1)$ and $f_k(m_2)$ are secondhand estimates made to a server s from a node k using two measures m_1 and m_2, and $f_l(m_1)$ and $f_l(m_2)$ are the corresponding estimates made to s at another node l using these measures, then $f_k(m_1)/f_k(m_2) = f_l(m_1)/f_l(m_2)$. We omit the proof in this book.

To evaluate the accuracy of the OPEN estimation, we performed simulation with actual downloading traces [Rid] collected from PlanetLab. In the simulation, 10,000 estimations were made by Equation (15.6), and the *relative error* (RE) is computed by the following equation:

$$RE = \frac{|estimated\ value - measured\ value|}{\min(estimated\ value,\ measured\ value)}$$

Thus, it implies perfect estimation when the relative error is zero, while the relative error 1 means that the estimated value is a factor of two, either smaller (*underestimation*) or greater (*overestimation*), of the measurement value.

Figure 15.4 illustrates the cumulative distribution of the OPEN estimation results. The x-axis is relative error of estimates. Figure 15.4(a) shows the impact of the number of secondhand measurements, while Figure 15.4(b) shows the impact of the number of local measurements when the number of secondhand measurements is 4. As can be seen in Figure 15.4(a), the estimation with four second-hand measurements approximates the estimation results with the greater number of measurements, yielding roughly 90% of estimations located within a factor of two. The accuracy drops quickly with a single second-hand measurement. In Figure 15.4(b), we can see that the estimation with only one local measurement poorly performs. However, it performs quite well with five local measurements. It continuously improves with ten local measurements, but does not further improve with more than ten. These results indicate that the OPEN framework enables nodes to participate in estimation without a costly learning phase. In addition, it implies that storage requirements can be small.

In Equation 15.6, all the terms except *distance*(h, s) can be obtained from past measurements. Because we compute distance from end-to-end latency, it is possible to employ a lightweight latency prediction technique, such as network coordinate systems [NZ02, DCKM04]. For example, Vivaldi, which is also used in the SWORD resource discovery tool [OAPV05], predicts end-to-end latency based on piggybacking, thus minimizing explicit probing.

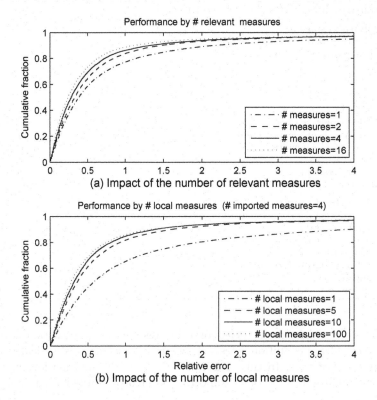

(a) Impact of the number of relevant measures

(b) Impact of the number of local measures

FIGURE 15.4: Relative error of OPEN estimates.

15.3.2.2 Measurement Sharing Method

For secondhand estimation, it is necessary that second-hand measurements are globally visible, so that any other nodes can make their own estimations by referring to the shared measurements. Any server-based techniques for storing global information are limited by well-known problems of scalability and fault tolerance. For this reason, OPEN relies on *probabilistic dissemination*, taking advantage of cost-effectiveness and fault resilience. Many optimizations may be possible for probabilistic dissemination; as in [KMG03, VvS07, DXL+06, HHL06, KCG06], our intention is not to make further optimization for dissemination protocols, but to provide insights for application-oriented optimizations for efficient dissemination, particularly for the OPEN framework.

An optimization technique we introduce in this section is selective dissemination based on *deferral and release conditions*, which define whether new information can be deferred (for its dissemination) or released (to the system). If a "deferral" decision is made for any new measurement, the source node does not emit it into the system until the corresponding "release" con-

---

Algorithm 5 Selective deferral and release

```
1: initiate(message m):
2: if deferral_cond(m) == true then
3:    deferredList.append(m);
4: else
5:    forward(m);
6:    release_test(m);
7: end if

8: receive(message m):
9: if message ∉ historyList then
10:    historyList.append(m);
11:    forward(m);
12:    release_test(m);
13: end if

14: release_test(message m):
15: D ← deferred messages to the same server as m from deferredList;
16: for all d ∈ D do
17:    if release_cond(m) == true then
18:       forward(d);
19:       deferredList.delete(d);
20:    end if
21: end for
```

---

dition is met. Thus, the deferral condition tests if new information is critical, while the release condition retests if deferred information is critical based on the passage of time. In this technique, any deferred information will either be disseminated if it becomes important later or discarded when it becomes stale.

The basic idea of this technique is to distribute a newly collected measurement only if it offers unique information different from past measurements. For example, suppose node A makes an estimation of 100KB/s for end-to-end throughput to node B based on past shared measurements. Now assume node A just downloaded a data object from B with 100KB/s throughput. Then node A may not want to disseminate such redundant information to others (*deferral*). However, this cold information can be changed to hot as more measurements are collected in the system. Continuing with the above example, suppose node A later sees its estimation to B with newly collected information to be significantly different from its own past measurement. For example, for a new measurement of 10KB/s, node A may want to tell other nodes about the deferred experience (*release*).

Algorithm 5 illustrates details of the selective deferral and release technique. A node performs **initiate** when it obtains a new measurement, while non-source nodes perform **receive** when they receive dissemination messages from neighbors. If the measured information is hot to the system (i.e., *deferral_cond(m) == false*), it is immediately disseminated; otherwise, it is put in the deferred list, as seen in **initiate**. Any receiving node stores new information and simply forwards it if it has not seen the information before, as shown in the **receive** function. In both **initiate** and **receive**, a release test follows after new information is forwarded. This checks whether any prior deferred information is now hot and can be distributed, as shown in **release_test**. Although not shown explicitly in the algorithm, deferred messages will be purged, based on their age.

We define a deferral condition and a release condition based on the difference between new measurement and current estimation derived from prior measurements. The ratio between two estimates in a node is equal to the ratio between two estimates in any other node if it estimated with the identical measures. As can be seen below, the deferral and release conditions are established based on a ratio between the median estimate and the new measure. Thus, it would be possible to filter redundant information for estimation by deferral (or to restore essential information by release). However, defining these conditions is application specific.

To define a deferral condition, let us suppose that *observed* is a newly measured throughput to a specific server, and *expected* is the estimated throughput to that server based on past measurements. The deferral condition is then defined as follows:

$$\text{Deferral condition: } \frac{|observed - expected|}{observed} < \tau_1$$

If this condition is true, we defer dissemination for the given information. Thus, $\tau_1 = 0$ means no information will be deferred, whereas any arbitrary large value of τ_1 (e.g., $\tau_1 = 100$) may defer most of the newly collected measurements.

The release condition is similarly defined with the deferral condition by comparing deferred measurement (*deferred*) and current estimation (*expected*), as follows:

$$\text{Release condition: } \frac{|deferred - expected|}{deferred} \geq \tau_2$$

Because *expected* is the estimated throughput with all past relevant measurements, it can be different from the estimated value computed in the deferral phase. By this condition, if the deferred measurement has information distinct from the current estimation, it begins to be disseminated.

Defining τ-values may depend on application or system requirements. In the evaluation subsection, we examine how τ-values impact performance and dissemination overhead.

15.3.3 Evaluation of OPEN-Based Selection

For evaluation, we compare our OPEN-estimation-based selection (OPEN) with a diverse set of selection techniques, based on trace-based simulation with our 10-month data traces [Rid] collected from PlanetLab. The trace data sets include downloading statistics for a diverse size of files from 1 MB to 16 MB. We assume overlay systems with random topology, in which each node has two to eight neighbors by default. We considered two selection applications: *replica selection* that chooses a replica server for downloading data, and *resource selection* that chooses a compute node for assigning jobs. These are common in distributed computing. For replica selection, we use r for replication factor and for resource selection, we use c for the number of candidate resources in question. The selection techniques include *random selection* (RANDOM) that randomly selects a node, and *latency-based selection* (PROXIM) that finds a client-server pair with the smallest RTT. In addition, we consider selection based on several *pairwise estimation* techniques that use only firsthand measurements. These techniques include statistical mean, median, exponential smoothing, and last value; we choose the *best* one of this group and call it PAIRWISE. For example, if selection by median yielded the best result at a round, we take that result (by median) as PAIRWISE for that round.

Unlike RANDOM and PROXIM, PAIRWISE and OPEN can suffer from estimation failures due to a shortage of relevant measurements, as it is usually impossible to make estimates for any passive estimation technique if there exists no past relevant measure. In contrast, PROXIM does not fail because the trace data includes latency information. To avoid meaningless estimation values from impacting the selection algorithm, we use the PAIRWISE and OPEN estimation techniques only if at least half of the measurements required for estimation are available, based on our observation that performance gets degraded if we perform selection with less than half; otherwise we assume that the selection using these techniques falls back on latency-based selection. Thus, *fallback ratio* refers to the fraction of fallback out of the entire selection.

To compare the different selection algorithms, we mainly use the metric *Optimality Ratio* (O.R.) [NhCR⁺03], where optimal is an oracle-based algorithm that chooses the best from a given set of trace data for each selection with a-priori knowledge. Thus, $O.R. = 1$ means the selection technique chooses optimal. Because we used *mixed* data sets with diverse file sizes in simulation as mentioned, relative comparison is also more meaningful than providing absolute download times. We also examine *overhead* of dissemination. For this, we evaluate *number of messages* generated for dissemination of measurements to share in the system. The *normalized* number of messages refers to the average number of messages per round at each node.

Figure 15.5 shows the results for both replica selection ($r = 8$) and resource selections ($c = 8$) for small S; ($n \approx 250$), medium M; ($n \approx 1000$), and large L; ($n \approx 10000$) systems. As expected, PROXIM works better than random choices. However, PAIRWISE does not work much better than PROXIM, except for the small system. This is because there is a high probability that the pairwise techniques fail to see relevant measurements in their estimations, and hence will fall back to PROXIM. In replica selection, the fallback ratio to PROXIM is 15% in the small system, but it increases 95% in the medium system. In the large system, the PAIRWISE fallback ratio reaches almost 100%, indicating that no pairwise estimation was made due to lack of pair-level measurements. In contrast, OPEN falls back to PROXIM 0.5% in the small system, 2% in

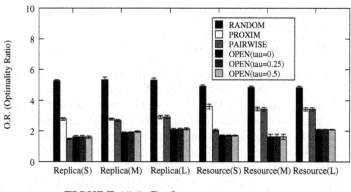

FIGURE 15.5: Performance comparison

the medium system, and 18% in the large system. This result emphasizes again why secondhand estimation is attractive for large-scale systems. Fallback ratios for PAIRWISE are slightly greater in resource selection, while OPEN shows similar fallback ratios in both replica and resource selections. In the large system, OPEN requires more rounds to collect measurements for each server. This slightly affects performance in the large system.

As seen in Figure 15.5, OPEN outperforms all other selection techniques in both replica and resource selections. We set up three configurations for OPEN: and OPEN(τ = 0); which disables selective deferral and release; OPEN(τ = 0.25) and OPEN(τ = 0.5), which enable selective deferral and release. In this experiment, we use $\tau = \tau_1 = \tau_2$ to make deferral and release decisions. Smaller τ would make deferral decision less likely, whereas greater τ tends to aggressively defer dissemination of new measurements. Even with deferring dissemination, we can see little performance loss in the figure. Although not shown in the figure, we observed that a substantial number of measurements were deferred with the selective deferral and release. For example, in replica selection, 28% of measurements were in the deferred list for $\tau = 0.25$ at the end of the simulation, while it was 56% for $\tau = 0.5$.

In this section, we introduced a framework called OPEN, which offers end-to-end accessibility estimation, based on secondhand measurements observed at other nodes in the system. To share secondhand measurements, OPEN proactively distributes newly collected measurements by a probabilistic dissemination technique. The experimental results show that resource and replica selections with OPEN consistently outperform selection techniques based on statistical pairwise estimations, as well as latency-based selection. In addition, OPEN provides a dissemination technique to reduce dissemination overhead in sharing secondhand measurements.

15.4　Network Dashboard

We first examine the opportunity to improve communication between two peers using proxies. Note that the use of a proxy between peers introduces an additional hop at the TCP/IP layer of the networking stack. Despite this, as we will show, proxies can accelerate network communications, sometimes by a substantial factor. To evaluate this opportunity, we have constructed a network dashboard tool[1] for PlanetLab that provides us with the networking statistics for potential proxy nodes in PlanetLab.

Each proxy executes a resource monitor that collects and transmits monitoring data to the entry points and the dashboard control. The resource monitor periodically probes all proxies to measure the following entities, bandwidth (for TCP streams and UDP datagrams), the delay in the arrival of successive datagrams, and the variation of this delay, also known as jitter. Next, we present some interesting results collected by the tool.

We notice that proxies could improve the network characteristics of a number of network endpoints. Approximately $1,600$ pairs of network endpoints were monitored. We calculated the number of alternate paths, formed by routing data through a single proxy, that are superior to the direct path. This analysis was carried out for TCP streams only, as it is the primary protocol used for data and file transfer. The aggregate summary is presented in Figure 15.6. On average, there exist a large number of alternate paths that may benefit a given pair of network endpoints. Furthermore, the benefit of these paths remains constant over a long duration, suggesting that these opportunities, the benefit of alternate paths, are long lived. Looking at these aggregate values, one notes that about 80% of the alternate paths are faster than the direct ones by at least 10%. In the following, we show that specific paths can be accelerated far more.

Next, we drill down and analyze the benefits of introducing a proxy for a pair of network endpoints communicating using a TCP stream. Our observations in Figure 15.7a show that it is possible to accelerate TCP streams using proxies, ranging from 5% to 25%. Numerous proxies also provide a stable and sustained improvement over time. The performance benefit observed if one uses such proxies could be easily computed from historical data with a high degree of confidence.

Next, we examine the benefits of using proxies for the UDP protocol. Many multimedia streaming and remote visualization applications utilize the UDP protocol for data transfer. The use of a proxy can amplify the observed UDP bandwidth between two endpoints. We present an example of one such experiment (of many examples) in Figure 15.7b, where the alternate paths accelerate data transfer by 6% to 50%. Continuing with the benefits seen for

---

[1]Network Dashboard Tool URL: http://netsat.cs.umn.edu.

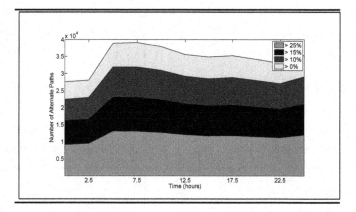

FIGURE 15.6: For a set of 1600 pairs of network endpoints, we plot the number of alternate paths that could improve the TCP bandwidth. The data plotted was collected over a day, and is discretized at intervals of 2.5 hours. For each interval and each pair of network endpoints, approximately forty paths were analyzed, leading to a total of 64k paths analyzed per interval. Each curve indicates the average TCP acceleration as a percentage.

the UDP protocol, one may reduce the network jitter and delay (Figures 15.7c and 15.7d) between the endpoints by using a proxy, to under 10% of the original values. This would benefit human-in-the-loop applications by reducing network delay and jitter for data transmissions. Thus, proxies can be used to reduce the sensitivity of the end user to the vagaries of the network. These observations mirror those seen for the TCP protocol, proxies could be used to improve the network performance and network QoS by a substantial amount that is sustained over time.

The next question is whether the communication performance is predictive. That is, can we exploit prior measurements to select proxies for future optimizations? For this experiment, three pairs of network endpoints are chosen at random. A fixed amount of data (2 MB) is transferred between these pairs using TCP. We measure the bandwidths for both a direct transfer and for data routed through a proxy. These instantaneous values of bandwidth are compared with observed measurements over the last 2.5 hours. We employ a simplistic predictor; the prediction is based on the average of values recorded in the last few hours.

The difference between the predicted and instantaneous values is the error in our predictor. The speedup obtained is the difference between the instantaneous values observed with and without proxies. Because it is notoriously difficult to accurately predict bandwidth due to network volatility, we use the historical values as a hint. A subset of the results, experiments conducted for three pairs of network endpoints, are summarized in Figure 15.8.

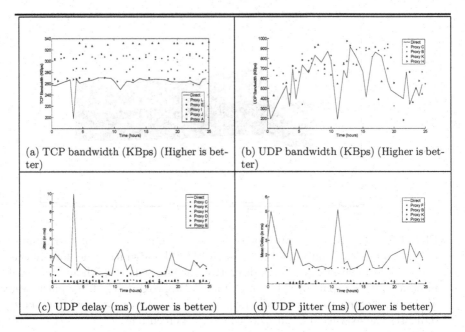

(a) TCP bandwidth (KBps) (Higher is better)

(b) UDP bandwidth (KBps) (Higher is better)

(c) UDP delay (ms) (Lower is better)

(d) UDP jitter (ms) (Lower is better)

FIGURE 15.7: These plots show the networking benefit of using proxies to route data between a particular pair of nodes, for both the TCP and the UDP protocol. The solid line plots values for a direct network path between the endpoints. Other points on each plot indicates values that may be realized by routing data through certain proxies. The plot incorporates data sampled at different times over a day, with more recent values appearing on the right.

The figure shows that the best proxy is able to achieve a 56% improvement, while the prediction error is in the range of 1.6% to 54%. A large error in some predictions can be attributed to the simplistic prediction model employed by our technique. More sophisticated models have been developed [DCKM04, WSH99], and can be used to improve the predictor. Nevertheless, the introduction of proxies leads to an improvement in the bandwidth for most cases.

15.5 Conclusion

In this chapter, we have discussed the important issue of network performance in volunteer networks. For some applications, communication performance plays a nontrivial role in the overall performance; thus compute-only performance is not the dominating factor. For parallel services, network performance can become important when peers or volunteers are interacting with

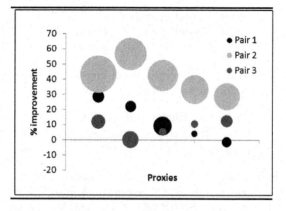

FIGURE 15.8: This plot shows the benefit of routing data through proxies. Six different proxies were chosen for each of the three network endpoints, and are shown on the x-axis. The percentage improvement of using these proxies form the y-axis. The size of each blob on the chart indicates the error in the values predicted.

each other and/or with other nodes, for example, servers or external nodes. We introduced techniques to estimate network performance efficiently and showed how to use such performance in a variety of volunteer computing scenarios. We showed how time-bounded execution of data-intensive services can be achieved across volunteers using a form of communication makespan estimation relying on application-specific communication measurements. We also explored how secondhand communication information can be estimated and propagated to peers through the OPEN framework. Last, we described a network dashboard tool for improving communications in volunteer networks. This tool can identify superior application-level network routing paths in volunteer networks based on a variety of different metrics.

Bibliography

[AF06]   David P. Anderson and Gilles Fedak. The computational and storage potential of volunteer computing. In *Proceedings of the Sixth IEEE International Symposium on Cluster Computing and the Grid (CCGRID '06)*, pages 73–80, 2006.

[And04]   David P. Anderson. BOINC: A system for public-resource computing and storage. In *Proceedings of GRID (GRID '04)*, pages 4–10, 2004.

[ASS+02]   Matthew Andrews, Bruce Shepherd, Aravind Srinivasan, Peter

             Winkler, and Francis Zane. Clustering and server selection using passive monitoring. In *Proceedings of INFOCOM (INFOCOM '02)*, pages 1717–1725, 2002.

[bla]   BLAST: The basic local alignment search tool, http://www.ncbi.nlm.nih.gov/blast.

[CC97]   Robert L. Carter and Mark Crovella. Server selection using dynamic path characterization in wide-area networks. In *Proceedings of INFOCOM (INFOCOM '97)*, pages 1014–1021, 1997.

[DCKM04]   Frank Dabek, Russ Cox, Frans Kaashoek, and Robert Morris. Vivaldi: a decentralized network coordinate system. In *Proceedings of ACM SIGCOMM (SIGCOMM '04)*, pages 15–26, 2004.

[Dow99]   Allen B. Downey. Using pathchar to estimate internet link characteristics. In *Proceedings of ACM SIGCOMM (SIGCOMM '99)*, pages 241–250, 1999.

[DXL+06]   Mayur Deshpande, Bo Xing, Iosif Lazardis, Bijit Hore, Nalini Venkatasubramanian, and Sharad Mehrotra. Crew: A gossip-based flash-dissemination system. In *Proceedings of the 26th IEEE International Conference on Distributed Computing Systems (ICDCS '06)*, page 45, 2006.

[fre]   FreePastry, http://freepastry.org/.

[glo]   The Globus Alliance, http://www.globus.org/.

[HDA05]   Qi He, Constantine Dovrolis, and Mostafa Ammar. On the predictability of large transfer tcp throughput. In *Proceedings of ACM SIGCOMM (SIGCOMM '05)*, pages 145–156, 2005.

[HFTC09]   Haiwu He, Gilles Fedak, Bing Tang, and Franck Cappello. Blast application with data-aware desktop grid middleware. In *Proceedings of the 2009 9th IEEE/ACM International Symposium on Cluster Computing and the Grid (CCGRID '09)*, pages 284–291, 2009.

[HHL06]   Zygmunt J. Haas, Joseph Y. Halpern, and Li Li. Gossip-based ad hoc routing. *IEEE/ACM Transactions on Networking*, 14(3):479–491, 2006.

[ipe]   PlanetLab Iperf, http://www.measurement-lab.org/logs/iperf/.

[KCC04]   Derrick Kondo, Andrew A. Chien, and Henri Casanova. Resource management for rapid application turnaround on enterprise desktop grids. In *Proceedings of the 2004 ACM/IEEE conference on Supercomputing (SC '04)*, 2004.

[KCG06]   Pradeep Kyasanur, Romit Choudhury, and Indranil Gupta. Smart gossip: An adaptive gossip-based broadcasting service for sensor networks. *IEEE International Conference on Mobile Adhoc and Sensor Systems Conference*, 0:91–100, 2006.

[Kes95]   Srinivasan Keshav. Packet-pair flow control. *IEEE/ACM Transactions on Networking*, 1995.

[KMG03]   Anne-Marie Kermarrec, Laurent Massoulié, and Ayalvadi J. Ganesh. Probabilistic reliable dissemination in large-scale systems. *IEEE Transactions on Parallel and Distributed Systems*, 14(3):248–258, 2003.

[LZZ+04]   Virginia Lo, Daniel Zappala, Dayi Zhou, Yuhong Liu, and Shanyu Zhao. Cluster computing on the fly: P2p scheduling of idle cycles in the internet. In *Proceedings of the IEEE Fourth International Conference on Peer-to-Peer Systems*, pages 227–236, 2004.

[NhCR+03] T. S. Eugene Ng, Yang hua Chu, Sanjay G. Rao, Kunwadee Sripanidkulchai, and Hui Zhang. Measurement-based optimization techniques for bandwidth-demanding peer-to-peer systems. In *Proceedings of INFOCOM (INFOCOM '03)*, pages 2199–2209, 2003.

[NZ02]   E. Ng and H. Zhang. Predicting internet network distance with coordiantes-based approaches. In *Proceedings of IEEE INFOCOM (INFOCOM '02)*, pages 170–179, 2002.

[OAPV05]   D. Oppenheimer, J. Albrecht, D. Patterson, and A. Vahdat. Design and implementation tradeoffs for wide-area resource discovery. In *Proceedings of Proceedings of ACM High Performance Distributed Computing (HPDC '05)*, 2005.

[pla]   PlanetLab, http://www.planet-lab.org.

[RD01]   Antony Rowstron and Peter Druschel. Pastry: Scalable, distributed object location and routing for large-scale peer-to-peer systems. In *IFIP/ACM International Conference on Distributed Systems Platforms (Middleware)*, pages 329–350, November 2001.

[Rid]   Ridge planetlab traces available on http://ridge.cs.umn.edu/pltraces.html.

[SSK97]   S. Seshan, M. Stemm, and R. H Katz. SPAND: Shared Passive Network Performance Discovery. In *Proceedings of the USENIX Symposium on Internet Technologies and Systems*, pages 135–146, Monterey, CA, December 1997.

[TTL05]   Douglas Thain, Todd Tannenbaum, and Miron Livny. Distributed
           computing in practice: the condor experience. *Concurrency -
           Practice and Experience*, 17(2-4):323–356, 2005.

[VvS07]   Spyros Voulgaris and Maarten van Steen. Hybrid dissemination:
           adding determinism to probabilistic multicasting in large-scale
           p2p systems. pages 389–409, 2007.

[WSH99]   R. Wolski, N. Spring, and J. Hayes. The Network Weather Ser-
           vice: A Distributed Resource Performance Forecasting Service for
           Metacomputing. *Journal of Future Generation Computing Sys-
           tems*, 15:757–768, 1999.

Index

accounting, 38, 39, 162, 282, 290, 292, 300, 303, 304

accuracy, 174, 188, 198, 202, 203, 340, 341, 344

accurate, 40, 41, 44, 70, 172, 174, 226, 268, 279, 328, 351

active, 30, 31, 65, 201, 231, 298, 325

activities, 66, 208, 213, 240, 246, 253, 311, 330

activity, 38, 65, 66, 288

adaptive, 16, 246, 295

administrator, 38, 75, 80, 81, 83, 84, 87–89, 91, 94, 125, 169, 172, 273, 294, 295, 311, 312, 323

affinity, 253

aggregate, 30, 81, 93, 239, 243, 245, 250, 339, 350

agreement, 96, 171, 229, 230, 305

architecture, 9, 38, 54, 72, 75, 80, 81, 162, 174, 192, 200, 239, 242, 243, 255, 261, 262, 281, 294, 295, 300, 304, 324

asynchronous, 15, 16, 66, 67, 130, 131

attack, 31, 168, 177, 213, 217–220, 224, 229

authenticated, 83

authentication, 82, 83, 94, 220, 293, 322

authorization, 83, 85–90, 93, 158

bandwidth, 4, 10, 43, 73, 81, 239, 242–246, 320, 334, 336–339, 350–352

batch, 15, 25, 70, 247, 300

behavior, 23, 42, 179, 187, 188, 193–195, 197, 198, 212, 213, 215, 224, 225, 227, 229, 230, 278, 280, 342

boinc, 139, 140, 143, 144

bonjour, 138–140, 143, 144, 146, 148

bridge, 71, 75, 92–95, 262, 264, 290, 304, 317, 321–324, 327–329

capacity, 4, 25, 40, 45, 193, 194, 196, 239, 288, 289, 291, 293, 304, 334

centralized, 9, 80, 125, 152, 225, 251, 254, 294, 302, 320

certificate, 83, 91, 93, 94, 299, 322, 323

certification, 82, 220, 230, 231, 251, 254

client, 5–7, 12–14, 16, 17, 19–25, 54, 55, 57, 58, 60, 62, 65, 67, 70–74, 81, 82, 85–88, 90, 94, 130, 131, 140, 142, 144, 146, 148, 157, 170, 172, 212, 216, 240, 264, 294, 295, 297, 304, 315, 316, 321, 323, 329, 334, 339, 348

cloud, 45, 46, 75, 241, 267, 272, 293

cluster, 4, 53, 54, 71, 74, 91–93, 162, 172, 177, 184, 229, 238, 240–243, 248, 252, 253, 255, 265, 280, 288, 313, 320, 321, 342

communication, 6, 10, 12, 18, 19, 22, 23, 29, 47, 55, 62, 64, 82, 83, 90, 92, 94, 130, 133, 155–157, 241, 251–253, 255, 263, 291, 293, 311, 313, 318, 320, 325, 333–337, 339, 340, 350, 351, 353

communities, 5, 80, 262, 266, 267, 288, 289, 292, 294, 296, 297, 300, 304, 305, 310, 313, 323, 329

community, 5, 6, 127, 169, 170, 172,